Performance and QoS of Next Generation Networking

Springer
London
Berlin
Heidelberg
New York
Barcelona
Hong Kong
Milan
Paris
Singapore
Tokyo

Kunio Goto, Toshiharu Hasegawa,
Hideaki Takagi and Yutaka Takahashi (Eds)

Performance and QoS of Next Generation Networking

Proceedings of the International Conference on the Performance and QoS of Next Generation Networking, P&QNet2000, Nagoya, Japan, November 2000

With 175 Figures

Springer

Kunio Goto, Dr of Engineering
Toshiharu Hasegawa, Dr of Engineering
Nanzan University, Department of Information and Telecommunication Engineering, 27 Seirei-cho, Seto, Aichi 489-0863, Japan

Hideaki Takagi, PhD
University of Tsukuba, Institute of Policy and Planning Sciences, Tsukuba-shi, Ibaraki 305, Japan

Yutaka Takahashi, Dr of Engineering
Kyoto University, Department of Systems Science, Graduate School of Informatics, Kyoto 606-8501, Japan

ISBN-13:978-1-4471-1183-2 e-ISBN-13: 978-1-4471-0705-7
DOI: 10.1007/ 978-1-4471-0705-7

British Library Cataloguing in Publication Data
Performance and QoS of next generation networking :
proceedings of the International Conference on the
Performance and QoS of Next Generation Networking,
P&QNet2000, Nagoya, Japan, November 2000
1.Computer networks - Congresses 2.Telecommunication -
Traffic - Congresses 3.Mobile computing - Congresses
I.Goto, Kunio II.Peformance and QoS of Next Generation
Networking. Conference (2000 : Nagoya, Japan)
621.3'821
ISBN-13:978-1-4471-1183-2

Library of Congress Cataloging-in-Publication Data
International Conference on the Performance and QoS of Next Generation Networking (2000 : Nagoya, Japan)
Performance and QoS of next generation networking ; proceedings of the International Conference on the Performance and QoS of Next Generation Networking, P&Qnet2000, Nagoya, Japan, November 2000 / Kunio Goto ... [et al.] eds.
p. cm.
Includes bibliographical references and index.
ISBN-13:978-1-4471-1183-2
1. Computer networks--management--Congresses. 2. Internet--Evaluation--Congresses. 3. Mobile computing--Congresses. 4. TCP/IP (Computer network protocol)--Congresses. 5. Telecommunication--Traffic--Management--Congresses. I. Goto, Kunio, 1957- II. Title.
TK5105.5 .I5725 2000
621.382'1--dc21 00-063773

Softcover reprint of the hardcover 1st edition 2001

Typesetting: Camera ready by contributors

69/3830-543210 Printed on acid-free paper SPIN 10773419

Preface

The advancement of key technologies in communication, such as optical and radio transmission, coding schemes, switching mechanisms etc., has meant that communication networks are quickly growing to a larger-scale and higher speed than was ever anticipated. In terms of usage, Internet and real-time applications are expected to share a significant portion of the bandwidth in the next-generation of communication networks. Therefore, in order to achieve seamless and Quality of Service (QoS)-guaranteed transmission, regardless of source characteristics, extensive research into networking technologies is essential. For the proper design, development and operation of emerging ideas on networking, further studies on the performance modeling and evaluation of networking are also encouraged.

The *International Conference on the Performance and QoS of Next Generation Networking* (P&QNet2000) is being held from November 27 to 29, 2000, in Nagoya, Japan (Seto Campus of Nanzan University). This is the sixth international conference on the performance and other aspects of communication networks. The conference is held once every three years in Japan (1985 in Tokyo; 1988, 1991, and 1994 in Kyoto; 1997 in Tsukuba). The conference is sponsored by the International Federation of Information Processing (IFIP) Working Group (WG) 6.3 Performance of Communication Systems, 6.4 High Performance Networking, and 7.3 Computer System Modelling. Financial supports are given by Commemorative Association for the Japan-World Exposition (1970), Support Center for Advanced Telecommunications Technology Research, and Nanzan University.

The three-day conference consists of the symposium of 2.5 days, and a half-day tutorial. In response to our call for papers for the symposium, we received 33 submissions. Each paper was distributed to three program committee members, each of whom was asked to provide a review report. Based on their recommendations, we accepted 18 papers for presentation at the symposium. In addition, Professor Erol Gelenbe of the University of Central Florida, Dr. Philip Heidelberger of IBM Thomas J. Watson Research Center, and Mr. Hajime Amano of Toyota Motor Corporation presented invited speeches. A tutorial was also given by Professor Kishor S. Trivedi on *Stochastic Petri Nets and Their Applications.*

We are grateful to the members of the international advisory board and program committee, the authors of papers submitted to the symposium, the referees, the speakers at the symposium and tutorial, and all the participants in the conference. Without the dedication of all these people, the conference would not have been a success.

Nagoya Japan
November 2000

Kunio Goto
Toshiharu Hasegawa
Hideaki Takagi
Yutaka Takahashi

Contents

Part I

Invited Paper

Towards Networks with Cognitive Packets

Erol Gelenbe[1], Ricardo Lent[1], and Zhiguang Xu[1]

School of Electrical Engineering and Computer Science
University of Central Florida
Orlando, FL 32816
{erol,rlent,zgxu}@cs.ucf.edu

Abstract. We discuss packet networks in which intelligent capabilities for routing and flow control are concentrated in the packets, rather than in the nodes and protocols. This paper describes a possible test-bed to test and evaluate their capabilities, and presents an analytical model for the worst and best case performance of such systems.

1 Introduction

We propose packet networks in which intelligence is constructed into the packets, rather than at the nodes or in the protocols. Such networks are called "Cognitive Packet Networks (CPN)". Cognitive packets route themselves, they learn to avoid congestion and to avoid being lost or destroyed. They learn from their own observations about the network and from the experience of other packets. They rely minimally on routers. CPNs carry three major types of packets: smart packets, dumb packets and acknowldegments (ACK).

This paper reviews the basic concepts of CPNs, and proposes techniques for packet-based autonomous learning for routing and discuss flow control using adaptive finite-state machines and random neural networks to support these concepts. We describe a possible test-bed to test and evaluate their capabilities, and present analytical models for the worst and best case performance of these networks.

1.1 Networks with Packet-Based Processing Capabilities

Much attention has been devoted recently to networks which offer users the capability of adding network executable code to their packets. Some of these ideas can be used to support a CPN.

A recent survey article [8] is devoted to the *Active Network* concept; discrete (programmable switches) and integrated (capsules) approaches to the realization of active networks are discussed, and a summary of recent research on active networks is given. The potential impact of active network services on applications and how such services can be built and deployed, are discussed in [10]. It is argued that *Active Network Transport System (ANTS)* solves the problem of slow network service evolution by building programmability in the network infrastructure without sacrificing performance and security. Network services provided by ANTS are flexible in that besides providing IP-style routing and forwarding, applications can introduce new protocols. The packets are in the form of capsules as in integrated active networks. Capsule types that share information are grouped together into protocols, while we group Cognitive Packets which share goals and algorithms. Some specific nodes within the network execute the capsules of a protocol and maintain protocol state, similarly to the manner in which CPN nodes execute the code for each CP. The capsule processing routines are automatically and dynamically transferred to the nodes where they are needed. Contrary to CPNs where the code is a field of the CP, in ANTS this is done by a code distribution mechanism.

In [11], Active Congestion Control (ACC), a system which uses Active Networking technology to reduce the control delay that feedback congestion control systems experience, is introduced. Every packet in the system includes the current state of the endpoint's feedback algorithm. When a router experiences congestion, it calculates the new window size and deletes the packets that the endpoint would not have sent and informs the endpoint about its new state. Contrary to CPN where the packets change their behavior according to the state of the network, in ACC the nodes change their behavior. A practical framework which enables the addition of user code to the network element as part of the normal operation of the network is presented in [12], allowing new functionality to be rapidly introduced into the network. Advanced and conventional control architectures can exist together, solving the problem of retaining existing network solutions while at the same time creating innovative control systems for new services.

In [13], the authors describe an object-oriented transport architecture that allows for dynamically binding a variety of protocol stacks on a per-call basis. The architecture, in which the atomic processing entity is based on the consumer/producer model, consists of the model's transport abstraction, called an engine, its control and management abstraction, called front-ends, and a set of controllers implementing network services. In [14], a reference model is defined which separates control intelligence from control mechanisms. The IEEE P1520 standard aims to establish an open architecture in network control, and provide the capability to program the network through the programming interface.

The basic concepts and nomenclature for talking about active networks, and various aspects of the architecture being developed in the Defense Advanced Research Agency (DARPA) program are described in [15]. The functionality of the active network node is divided between the execution environments (EEs) and the node operating system (NodeOS). The EE is responsible for implementing the network API, while the NodeOS manages access to local node resources by EEs. Protocol Booster in [16] is a novel methodology for protocol design aimed at overcoming the slow evolution and inefficiencies associated with general-purpose protocols. It incrementally constructs protocols from elements called *protocol boosters* on an as-needed basis. *Protocol boosters* are transparent to the protocol being boosted. They can reside anywhere in the network or end systems and are designed to improve the performance or features of an existing protocol. Safety and security are major concerns in [17]. The *Secure Active Network Environment (SANE)* architecture, which provides a means of controlling access to the functions provided by any programmable infrastructure, is illustrated in this article./indexSecure Active Network Environment (SANE)

The JAVA programming language and virtual machines combined with the Web as a platform for implementing and deploying protocols offers enticing features including portability, security and object-oriented capabilities [18]. The *Common Object Request Broker Architecture (CORBA)* and the *Distributed Component Object Model (DCOM)* [19] are chosen as two examples of Distributed Object Technology that facilitate open network interfaces. The benefits are abstraction, location independence, modularity, and software reusability. In [9], the impact of mobile agent technology on telecommunication service environments, influenced by the *Intelligent Network* architecture is discussed. The research concentrates on the discrete approach of active networking where service deployment and service processing are separately performed. In contrast to the traditional way of Intelligent Network service implementation in centralized service nodes which control the switching nodes via a dedicated outband telecom signaling network, here the Intelligent Network services are implemented by means of service agents which are software components performing specific tasks, and are divided into parts according to their functionality. Similar to the Cognitive Packets which make their own decision on their routing to choose the best way to reach their destination, agents try to find the best location inside the network to provide the service with minimum usage of the signaling network by moving from one system to the other or cooperating with other agents when they find it necessary.

Much attention has been devoted recently to networks which offer users the capability to add network executable code to their packets. Some of these ideas can be used to support a CPN. A recent survey article [8] is devoted to the *Active Network* concept; discrete (programmable switches) and integrated (capsules) approaches to the realization of active networks are discussed, and a summary of recent research on active networks is given. The potential impact of active network services on applications and how such services can be built and deployed, are discussed in [10]. It is argued that *Active Network Transport System (ANTS)* solves the problem of slow network service evolution by building programmability in the network infrastructure without sacrificing performance and security. Network services provided by ANTS are flexible in that besides providing IP-style routing and forwarding, applications can introduce new protocols. The packets are in the form of capsules as in integrated active networks. Capsule types that share information are grouped together into protocols, as we group Cognitive Packets which share goals and algorithms. Some specific nodes within the network execute the capsules of a protocol and maintain protocol state, similar to the manner in which CPN nodes execute the code for each CP. The capsule processing routines are automatically and dynamically transferred to the nodes where they are needed. Contrary to CPNs where the code is a field of the CP, in ANTS this is done by a code distribution mechanism.

2 Cognitive Packets and CPNs

Learning algorithms and adaptation have been suggested for telecommunication systems in the past [4,7]. However these concepts have not been fully exploited in networks because of the lack of an adequate framework allowing decentralized control of communications.

In Cognitive Packet Networks (CPN) smart packets serve as "explorers" for different source-destination pairs; they rely minimally on routers, so that network nodes only serve as buffers, mailboxes and processors. We use the term Cognitive Packet (CP) an smart packets interchangeably. Smart packets can also be hierarchically structured so that a group of packets share the same goals, and make use of each other's experience. Upon arrival of a smart packet at its destination, the destination node creates an acknowledgement, which will follow the inverse of the route recorded by the smart packet on its way to the destination. The acknowledgement informs the source about the path that should be followed by dumb packets on their way to this specific destination. Dumb packets are given the path to follow to their destination by the source.

CPs store information in their private Cognitive Map (CM) and update the CM and make their routing decisions using the code which is in each packet. This code will include neural networks or other adaptive algorithms which will be decribed below. Figure 1 presents the contents of a Cognitive Packet and the manner in which Cognitive Memory at a Node is updated by the node's processor.

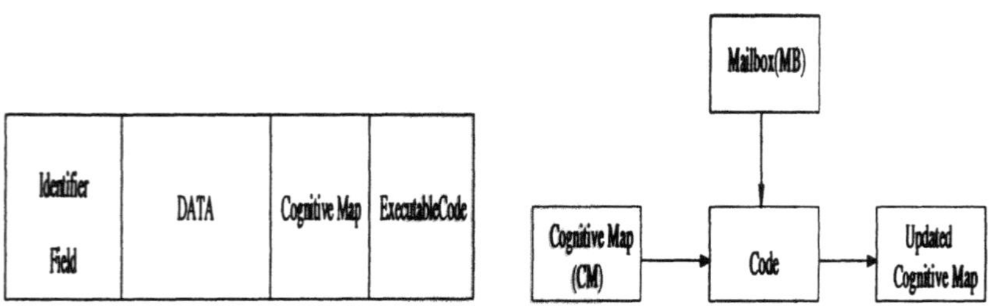

Fig. 1. Representation of a CP (left), Update of a CP by a Node CPN (right)

In a CPN, the packets use nodes as "parking" or resting areas where they make decisions and route themselves. They also use nodes as places where they can read their mailboxes. Mailboxes may be filled by the node, or by other packets which pass through the node. Packets also use nodes as processors which execute their code to update their CM and then execute their routing decisions. As a result of code execution, certain information may be moved from the CP to certain mailboxes. The nodes may execute the code of CPs in some order of priority between classes of CPs , for instance as a function of QoS requirements which are contained in the identification field). A possible routing decision may be simply to remain at the current node until certain conditions in the network have changed. However, routing decisions will generally be result in the CP being placed in some output queue, in some order of priority, determined by the CP code execution. A CPN and a CPN node are schematically represented in Figure 2. CPs are grouped into

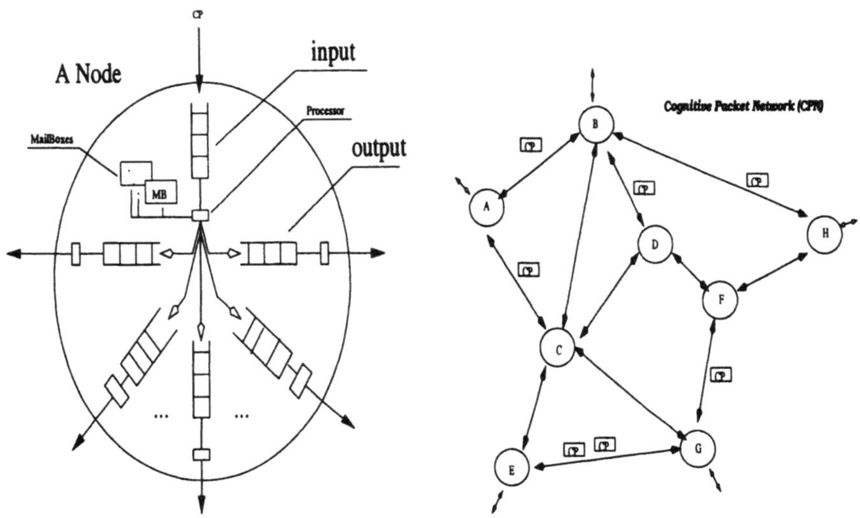

Fig. 2. Schematic Representation of a CPN Node (left), Schematic Representation of a CPN (right)

"CP classes" which share similar characteristics such as quality of service requirements, sets of internal states, control rules, input and output signals, etc.. These "signals" are units of information which CPs use to communicate with each other via mailboxes in the nodes. These signals can also emanate from the environment (nodes, existing end-to-end protocols) toward the CPs. Cognitive packets (CP) contain the following fields:

1. The Identifier Field (IF) which provides a unique identifier for the CP, as well as information about the class of packets it may belong to, such as its quality of service (QoS) requirements.

2. The Data Field containing the ordinary data it is transporting.
3. A Cognitive Map (CM) which contains the usual Source and Destination (S-D) information, as well as a map showing where the packet currently "thinks" it is, the packet's view of the state of the network, and information about where it wants to go next; the S-D information may also be stored in the IF.
4. Executable code that the CP uses to update its CM. This code will contain learning algorithms for updating the CM, and decision algorithms which use the CM.

A node in the CPN acts as a storage area for CPs and for mailboxes which are used to exchange data between CPs , and between CPs and the node. It has an input buffer for CPs arriving from the input links, a set of mailboxes, and a set of output buffers which are associated with output links. Nodes in a CPN carry out the following functions:

1. A node receives packets via a finite set of ports and stores them in an input buffer.
2. It transmits packets to other nodes via a set of output buffers. Once a CP is placed in an output buffer, it is transmitted to another destination node with some priority indicated in the output buffer.
3. A node receives information from CPs which it stores in Mailboxes (MB's). Mailboxes may be reserved for certain classes of CPs , or may be specialized by classes of CPs . For instance, there may be different MB's for packets identified by different Source-Destination (S-D) pairs.
4. A node executes the code for each CP in the input buffer. During the execution of the CPs code, the CP may ask the node to decline its identity, and to provide information about its local connectivity (i.e. "This is Node A, and I am connected to Nodes B, C, D via output buffers) while executing its code. In some cases, the CP may already have this information in its CM as a result of the initial information it received at its source, and as a result of its own memory of the sequence of moves it has made. As a result of this execution:
   - The CM's of the packets in the input buffer are updated,
   - Certain information is moved from CPs to certain MB's,
   - A CP which has made the decision to be moved to an output buffer is transfered there, with the priority it may have requested.

An important issue in the Internet and in future networks is security and dependability, in particular with respect to malicious threats such as viruses, worms, and other types of information warfare threats which may develop in the future. We plan to address these issues in relation to Cognitive Packet Networks in future work. CPs themselves act as autonomous agents and therefore are robust to various forms of network degradation. However our current thinking is that nodes within a CPN could have the power to "clear, or encapsulate, or destroy" packets in a CPN. Each CP could be chacked and cleared if it did not represent a threat. A packet which would seem to represent a threat would be "encapsulated" inside a secure packet and routed to its declared destination or to some specific receiving host. In the extreme case, a packet could be simply destroyed or eliminated by the node.

All CPs have the same packet format, which consist of four sections: a header carrying administrative information for handling the packet, such as quality of service (QoS) requirements, the type of packet and the source and destination addresses; a Cognitive Map (CM), which is used to compute CP routing based on the packets QoS needs; finally executable code, and a payload section. Smart packets make their routing decisions using the executable code which will include neural networks or other adaptive algorithms. The manner in which a CP's Cognitive Memory is updated by the node's processor is shown in Figure 3. Dumb packets follow the paths discovered by the CPs. In a CPN, the packets use nodes as "parking" areas where

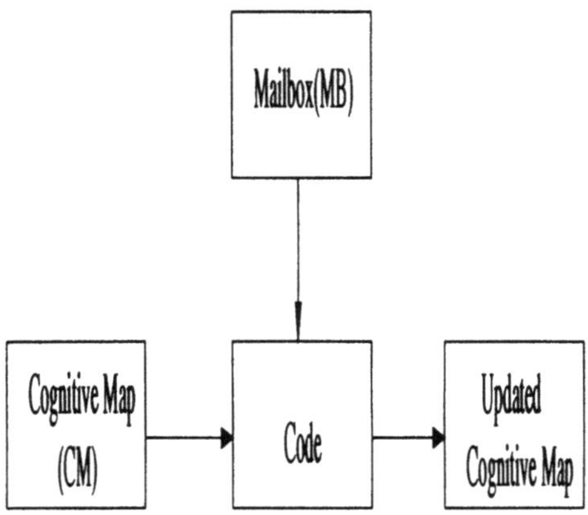

Fig. 3. Update of a CP by a Node

they stop to make decisions and route themselves. They also use nodes as places where they can read their mailboxes. Mailboxes are updated by packets which pass through the node, and in particular by ACKs. Packets use nodes as processors which execute their code to update their CM and then execute their routing decisions. The nodes may execute the code of CPs in some order of priority between classes of CPs, for instance as a function of QoS requirements. Routing decisions will generally result in the CP being placed in some output queue, in some order of priority, determined by the CP's code Each CP entering the network is assigned a Goal before it enters the network, and the CP uses the goal to determine its course of action each time it has to make a decision. We have tested use two learning paradigms for CPs :*Learning feedforward random neural networks (LFRNN)* [6]; these networks update their internal representation (the weights) using gradient based algorithms to improve either their predictive capabilities, or to improve their ability to reach decisions which elicit the maximum reward, and *Random neural networks with reinforcement learning (RNNRL)* [6,22]. In the latter case, a recurrent network is used both for storing the CM and making decisions. The weights of the network are updated so that decisions are reinforced or weakened depending on how they have been observed to contribute to increasing or decreasing the accomplishment of the declared goal.

3 Implementation of a CPN test-bed

We are currently implementating a test-bed to show the effectiveness of the CPN concept in a realistic environment. The test-bed will consist of a network of CPN nodes where we may run current applications and further develop the capabilities of CPN. The CPN code is being implemented in the latest Linux kernel (version 2.2.13). The Linux kernel support for low cost PCs and a growing number of platforms, the freely availability of its source code, and the module support in the kernel, makes Linux an attractive system for the development of a project of this nature. A clear separation between networking protocols in the implementation will make the CPN code independent of the physical layer and the data-link layer, providing flexibility in the interoperation with other existing communication protocols in the kernel. The goal is to produce a simple and compact code that can be easily ported over a wide variety of platforms, ranging from small single-board machines to supercomputers. The networking code in the Linux kernel, and many other operating systems, consists of three layers of software: device drivers, network interface, and protocol layer (see Figure 4). Device drivers perform the I/O operations in physical devices, providing a simplified interface to the protocol layer in the kernel. The network interface is compatible with the popular BSD4.3 socket layer in Linux, and provides a single application program interface (API) for the programmer to access all the protocols in the system. Sockets can be viewed as pipes where information that go into one end, come out at the other end. The socket concept support several types of services under the client-server model. The two principal are either connection-oriented, i.e. allowing the flow of data going orderly in both directions; or connectionless (also known as datagram sockets) allowing the transmission of only one message at a time. The protocol layer consists of several families of protocols: INET for TCP/IP, UNIX for Unix interprocess communications, etc. Data that arrive at this level either from a user application through the socket interface, or from the physical network via a device driver, have an identifier specifying which network protocol they carry. The core of the CPN protocol is being implemented in this layer, and it is able to receive, rewrite, discard and create new packets according to the CPN algorithm. Packets addressed to the local host are passed up to the socket interface. Packets addressed to a remote site are sent to the appropriate device driver for transmission.

The communication between the CPN layer with the socket interface and the device drivers is performed using *skbuff* data structures. A *skbuff* structure contains pointers to a certain buffer area where packets are processed. Using this structure, the headers for each layer can be added or removed as needed while the packet goes up or down in the networking protocol stack. Incoming

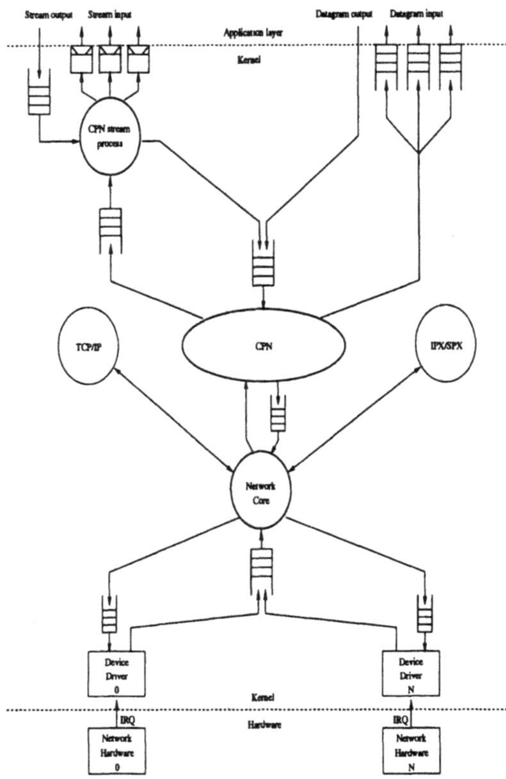

Fig. 4. CPN code in the Linux Kernel

CPN packets are tagged with a unique identifier so that the receiving device driver can use this number to identify the CPN packets. Arriving CPN packets are delivered to the CPN receiving function where the routing decision is performed based on the information stored in the MBs and the CM. When a CPN node receives a new packet it checks the type of packet. If the packet is dumb and the destination local, then an upper layer protocol process the packet. If the destination if not local then the packet is delivered using the information stored in its CM. When a Smart packet is received with a destination address different to the local node, the code stored in the data section is executed. The Smart packet stores information about the current node in its private CM and then takes a path to another node. On the other hand if the current node is the final destination for the Smart packet, then an Ack packet is generated and sent back to the origin with all the information collected by the Smart packet. Each time a CPN node receives an Ack packet the internal MB is updated with the new information.

4 Modeling and Simulation of CPN behavior

The purpose of our modeling and simulation work is to compare and evaluate a variety of CP learning algorithms. Both very simple decision algorithms (such as the Bang-Bang algorithm described below), and more sophisticated algorithms using learning, were tested. CPs used three different paradigms for adaptation, under identical traffic conditions. A single network simulation program representing a rectangular 100 node network was simulated, and three different learning algorithms were used by the CPs. Throughout the simulations, we vary the arrival rates of packets to each input node between 0.1 and 1. All simulations compare the CPs with controls of different kinds, against a static routing policy (marked "No Control" on the figures) where the packet is routed along a static shortest path to the output layer of nodes, and then horizontally to its destination. In our simulations, lost packets are not retransmitted by the source nodes. Loss rates at most of the the network nodes are set to a value of 10%, while the rate is 50% at two specific areas which are unkown to the CPs. These "high loss" areas are in four continiguous nodes given in (x,y) coordinates as (2,0) to (2,3), and also in four other nodes (7,7) to (7,10). These values are very high in practical terms, but are selected at these high values simply to be able to illustrate to be able to observe the effect of the control algorithms. Figures 5 and 6 compares the RNN with Reinforcement Learning (RL) and Bang-Bang when the Goal includes both Loss and Delay. We see that RL and the Bang-Bang algorithm provide essentially equivalent performance.

4.1 Worst-Case and Best-Case Performance

The worst case performance is obtained by considering "smart" packets which simply try to find the route to their destination by moving at random, with two constraints related to the topology of the network which will be discussed below. The network topology that we use to evaluate the worst case is very similar to the one which we have used in our simulations. It has a cylindrical toplogy with "R" circles or rows, each containing "a" nodes, making up the cylinder. The only difference with the network in the simulations is that because we take a cylindrical topology, there are no "edge" nodes in cylinder we use to construct the analytical model. Each node on the two (top and bottom) "end" circles serves both as a source and as a destination for packets. The two routing constraints for packets, which we mentioned above in relation to the topology of the network, are:

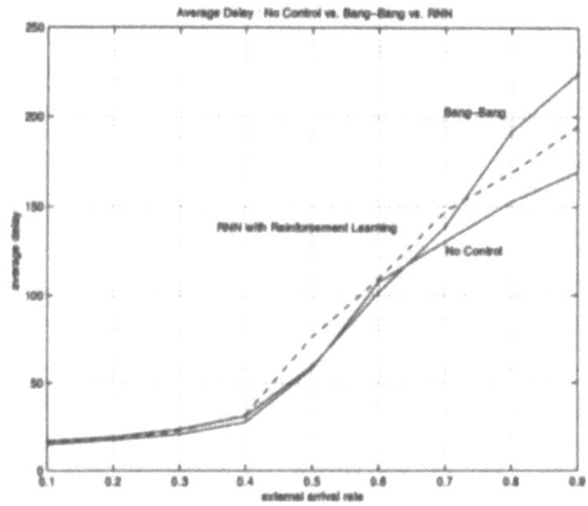

Fig. 5. RL Based Control using Delay and Loss as the Goal: Comparison of Average Delay through the CPN with High Loss

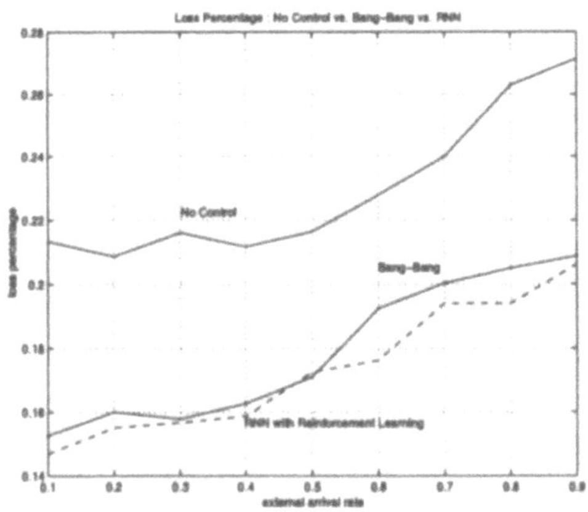

Fig. 6. RL Based Control using Delay and Loss as the Goal: Comparison of Average Loss through the CPN with High Loss

- A smart (or dumb) packet originating at the top row never heads upwards, and if it originates at the bottom row it never heads downwards.
- A smart packet which is one link away from its destination will directly go to its destination without any further random search. This is because in a real network, the outgoing link of a node will carry information concerning the identity of the node which is at the other end of the link.

Our analysis provides the following results under the assumption that packets at any source may head to any destination, and that smart packets do not know the direction they should head in, except that they need to go from the top row to the bottom row (or vice versa).

- Average Number of Nodes Visited in Row i by a Smart Packet When There Are No Losses

$$A(i) = \begin{cases} \frac{1}{f_i} & i \le i \le R-2 \\ \frac{a}{(1+(a-1)f_i)} & i = R-1 \\ \frac{a}{2} & i = R \end{cases}$$

  where f_i is the probability that a smart packet leaves a node in row i to move to the next row.

- Average Number of Nodes Visited in Row i by a Smart Packet with Losses at each Node

$$A_l(i) = \begin{cases} \frac{1}{l_i+f_i(1-l_i)} & i \le i \le R-2 \\ \frac{1}{1-(1-f_i)(1-l_i)(1-\frac{1}{a})} & i = R-1 \\ \frac{a}{2+(a-2)l_i} & i = R \end{cases}$$

  where l_i is the probability that a loss can occur at some node in row i of the network.

- The Probability that a Smart Pcket Is Lost as It Traverses Row i

$$\pi(i) = 1 - \frac{A_l(i)}{A(i)}, \ 1 \le i \le R$$

- The Probability that a smart packet eventually enters row i

$$P_e(i) = \begin{cases} 1 & i = 1 \\ (1-\pi(i-1)) \cdot P_e(i-1) & 2 \le i \le R \end{cases}$$

- The Average Number of Times that a Randomly Selected Smart Packet Visits a Node In the row i

$$e_l(i) = P_e(i) \cdot \frac{A_l(i)}{a}, \quad 1 \le i \le R$$

  under the assumption that the traffic in the network is homogeneous.

- The Probability that a Smart Packet Is Lost as It Traverses the Network

$$P_l^s = 1 - \Pi_{i=1}^{R}(1-\pi(i))$$

- The Effective Traffice from S to D

$$\lambda(S,D) = \frac{\lambda^0(S,D)}{1-P_l^s}$$

  where $\lambda^0(S,D)$ is the offered S to D traffic.

- The Average Traffic of Smart Packets Entering a Node in Row i

  - For unilateral traffic going just from the top to the bottom of the network
    $\lambda_s(i) = \sum_S \sum_D e_l(i) \cdot \lambda(S, D)$
  - For symmetric bilateral traffic going both from top to bottom and vice versa
    $\lambda_s(i) = \sum_S \sum_D (e_l(i) + e_l(R - i + 1)) \cdot \lambda(S, D)$
- The Average Number of Smart Packets at a Node in Row i

  $$R(i) = \frac{\lambda_s(i)}{\gamma - \lambda_s(i)}$$

  where γ is the average service rate at each Node.
- The Overall Average Delay of Smart Packets
  - For unilateral traffic going just from the top to the bottom of the network
    $$r = \frac{\sum_i a \cdot R(i)}{\sum_S \sum_D \lambda^o(S, D)}$$
  - For symmetric bilateral traffic going both from top to bottom and vice versa
    $$r = \frac{\sum_i a \cdot R(i)}{\sum_S \sum_D \lambda^o(S, D) \cdot 2}$$

To illustrate these results, we have varied f_i, the probability a smart packet leaves a node in row i to move to the next row. The numerical results of the leftmost graph in Figure 7 show that the larger the value of f_i, the smarter the packets are, and consequently, the smaller the average response time R.

The *best case* performance can be achieved if the following three conditions are satisfied:

- The traffic load is evenly distributed among all the nodes in the network;
- Each packet takes the shortest path from its source to its destination under the assumption that the number of nodes visited by a packet is the dominant factor among those that determine its delay;
- There is no packet loss.

Take an arbitrary packet, let "s" and "d" represents its source and destination respectively. The length of the shortest path that it can possibly take is $M = |d - s| + R$ with expected value $\overline{M} = \frac{a}{2} + R$. Since all nodes are equally loaded, the packet arrival rate to each node is

$$\lambda_n = \frac{2a\lambda^0(S, D)\overline{M}}{aR} = (\frac{a}{R} + 2)\lambda^0(S, D)$$

where $\lambda^0(S, D)$ is the offered S to D traffic. Similar to the previous worst case analysis, now we can adopt the queueing theory to obtain the best case average packet delay:

$$r = \frac{aR\frac{\lambda_n}{\gamma - \lambda_n}}{2a\lambda^0(S, D)} = \frac{\frac{a}{2} + R}{\gamma - (\frac{a}{R} + 2)\lambda^0(S, D)}$$

The following figures conclude our smart packet analysis. The one on the left shows the best case average packet delay and the one on the right illustrates the performance comparison among the best case, the worst case and the simulation (RNN with reinforcement learning) in terms of the average packet delay.

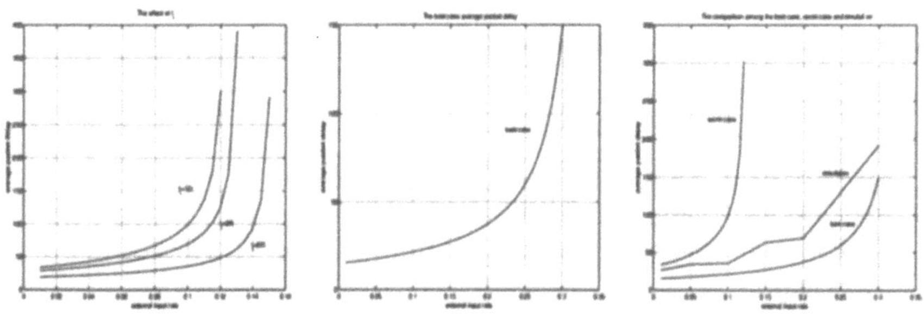

Fig. 7. Comparison between the Worst Case (left), Best Case (center), and the Analytical'Worst and Best Cases Compared to RNN-RL Learning Based Simulation (right)

References

1. E. Gelenbe "Probabilistic automata with structural restrictions", SWAT 1969 (IEEE Symp. on Switching and Automata Theory), also appeared as *On languages defined by linear probabilistic automata*, Information and Control, Vol. 18, February 1971.
2. E. Gelenbe *A realizable model for stochastic sequential machines*, IEEE Trans. Computers, Vol. C-20, No. 2, pp. 199-204, February 1971.
3. R. Viswanathan and K.S. Narendra *Comparison of expedient and optimal reinforcement schemes for learning systems*, J. Cybernetics, Vol. 2, pp 21-37, 1972.
4. K.S. Narendra and P. Mars, *The use of learning algorithms in telephone traffic routing - a methodology*, Automatica, Vol. 19, pp. 495-502, 1983.
5. R.S. Sutton "Learning to predict the methods of temporal difference", *Machine Learning*, Vol. 3, pp. 9-44, 1988.
6. E. Gelenbe (1993) "Learning in the recurrent random neural network", *Neural Computation*, Vol. 5, No. 1, pp. 154-164, 1993.
7. P. Mars, J.R. Chen, and R. Nambiar, *Learning Algorithms: Theory and Applications in Signal Processing, Control and Communications* , CRC Press, Boca Raton, 1996.
8. D. L. Tennenhouse, J. M. Smith, D. W. Sincoskie, D. J. Wetherall, G. J. Minden, *A survey of Active Network research* , IEEE Comm. Magn., Vol. 35, no. 1, pp. 80-86, January 1997.

9. M. Bregust, T. Magedanz, *Mobile agents-enabling technology for Active Intelligent Network implementation* IEEE Network Magn., Vol . 12, no. 3, pp. 53-60, May/June 1998.
10. D. Wetherall, U. Legedza, J. Guttag, *Introducing new Internet services: Why and How* IEEE Network Magn., Vol . 12, no. 3, pp. 12-19, May/June 1998.
11. T. Faber, *ACC: Using Active Networking to enhance feedback congestion control mechanisms*, IEEE Network Magn., Vol . 12, no. 3, pp. 61-65, May/June 1998.
12. S. Rooney, Jacobus E. van der Merwe, S. A. Crosby, I. M. Leslie, *The Tempest: a framework for safe, resource-assured, programmable networks*, IEEE Communications, Vol. 36, No. 10, Oct. 1998.
13. J.-F. Huard, A. A. Lazar, *A programmable transport architecture with QoS guarantee*, IEEE Communications, Vol. 36, No. 10, pp. 54-63, Oct. 1998.
14. J. Biswas, A. A. Lazar, S. Mahjoub, L.-F. Pau, M. Suzuki, S. Torstensson, W. Wang, S. Weinstein, *The IEEE P1520 standards initiative for programmable network interface*, IEEE Communications, Vol. 36, No. 10, pp. 64-72, Oct. 1998.
15. K. L. Calvert, S. Bhattacharjee, E. Zegura, J. Sterbenz, *Directions in Active Networks*, IEEE Communications, Vol. 36, pp. 64-72, No. 10, Oct. 1998.
16. W. Marcus, Ilija Hadzic, Anthony J. McAuley, J. M. Smith, *Protocol boosters: applying programmability to network infrastructures*, IEEE Communications, Vol. 36, No. 10, pp. 79-83, Oct. 1998.
17. D. S. Alexander, W. A. Arbaugh, A. D. Keromytis, J. M. Smith, *Safety and security of programmable networks infrastructures*, IEEE Communications, Vol. 36, No. 10, pp. 84-92, Oct. 1998.
18. B. Krupczak, K. L. Calvert, M. H. Ammar, *Implementing communication protocols in Java*, IEEE Communications, Vol. 36, No. 10, pp. 93-99, Oct. 1998.
19. J.-P. Redlich, M. Suzuki, S. Weinstein, *Distributed object technology for networking*, IEEE Communications, Vol. 36, No. 10, pp. 100-111, Oct. 1998.
20. M. Falousos, A. Banerjea, R. Pankaj, *QoSMIC : Quality of Service sensitive multicast Internet protocol*, Computer Communications Review, SIGCOMM'98, Section 4, Quality of Service, pp. 281-190, 1998.
21. E. Gelenbe, Zhi-Hong Mao, Y. Da-Li (1999) "Function approximation with spiked random networks" *IEEE Trans. on Neural Networks*, Vol. 10, No. 1, pp. 3–9, 1999.
22. U. Halici, *Reinforcement learning with internal expectation for the random neural network* European Journal of Operations Research (in press).

Part II

Internet I

Users' WWW Access Statistics Measured at Proxy Servers: Case Study for Cache Hit Ratio and Response Time

Kunio Goto[1] and Hirona Amano[2]

[1] Department of Information and Telecommunication Systems
Nanzan University, Seto 489-0863, Japan
goto@{it,iq}.nanzan-u.ac.jp
[2] Graduate School of Business Administration
Nanzan University, Nagoya 466-8673, Japan

Abstract. We analyzed the access logs of our WWW proxies used by most of the 500 terminals in our campus. An empirical access frequency distribution to different URLs was derived to compare cache hit ratios with different cache strategies. The distribution was found to be a good approximation also for access log at other sites. Also we investigated the distribution of length and numbers of subsequent accesses by a user.

1 Introduction

WWW has been vital to people not only for work but also for daily life and hobby as means of information sharing and retrieval. HTTP protocol traffic load caused by WWW browsing has been rapidly growing with increase of access and trunk line speed and volume of multimedia data. HTTP traffic load has become a major part of total IP traffic load. Then we should know WWW traffic pattern to reduce unnecessary load.

We need WWW access behavior of an individual user to estimate and reduce the traffic load because of the following two reasons. The first reason is that total WWW traffic load is a collection of HTTP responses from the WWW servers to the requests by individual users. A user's thinking(reading) time and data transfer time affects the time period between successive clicks on a WWW browser. The second is that cache efficiency on a WWW proxy server depends on how many users want to get the same URL. It is, however, not easy to build a theoretical model for user's WWW access behavior since there are many human factors. Fortunately, many organizations have huge amount of data stored as proxy access logs, and can get useful information if the logs are analyzed properly.

A several research papers on WWW traffic load have been published so far. Squid proxy [13] has been the most popular and high performance caching proxy server software and used also in Nanzan University. Workload at WWW server side was investigated in /citeARLI96. Performance and limitation of caching proxy was discussed in [1]. Also workload of caching proxy characterization has been examined by the SPA project [3]. However all of the analyses in these works are rather macroscopic and cannot be used for our microscopic modeling. There are two other related works to be cited. General network load models are discussed in

[12]. [6] defined a detailed model for WWW access but it is not compared with actual data yet, and purely theoretical.

An interesting cache hit ratio analysis based on the distribution of the frequency of users' requests to different URLs was introduced in [11]. The idea in their work is the analogy between frequency of requests to a URL in WWW space and use frequency of a word in a text. The authors tried only one of word frequency empirical formulas in mathematical linguistics and the match does not seem to be good enough.

In this paper, we apply three empirical word frequency formulas in mathematical linguistics [7–9] to the local proxy access log, and find the best matching formula through comparisons. Also access logs of the five other sites are examined to see if they fit the best matching formula. The estimated parameters of the formula for our site are then applied to different cache strategies to compare the efficiency in cases of different cache storage space and number of total accesses. Also we build a simple model for a user's WWW browsing behavior and measured the distribution of a retrieval time, thinking time, numbers of subsequent accesses by a user, and so on. Note that we use almost all available data in our site and expect the proposed model and estimates may apply to other cases after further investigation with the data in other sites.

This paper is organized as the following. In section 2, statistics of IP traffic incoming to Nanzan University campus and WWW proxy configuration are briefly described to show how much bandwidth WWW traffic occupies. This part of the work is based on [5]. Also WWW proxy configuration for HTTP layer multi-homing is illustrated. In section 3, matches for three formulas for the distribution of the frequency of requested URLs are calculated. Also, cache efficiencies at a WWW caching proxy server with different strategies are compared. This part of the work is based on [4]. Section 4 is devoted to the user behavior model and measurement based on [10]. Concluding remarks are provided in the last section.

2 IP traffic incoming to the campus

In this section, we briefly describe IP traffic statistics to show how Nanzan University's campus network is used by 6,000 students and 500 faculty and staff members. This statistics is the motivation of this work.

We have two connections to the global Internet as shown in Fig.4. Our LAN is multi-homed in HTTP layer but not in IP layer. (i.e. It is not a real multi-homed LAN.) Only the statistics for the traffic incoming from the academic ISP (SINET) are included here but not for the commercial ISP(OCN) because they show similar tendency and this paper does not focus on the amount of traffic.

Fig. 1 and Fig. 2 illustrate the incoming IP traffic average over 5 minutes and 30 minutes, respectively, from May 1998 through December 1998. Fig. 3 shows monthly traffic from May 1998 through May 1999 with average over 2 hours. Traffic is measured at the remote router's dedicated line interface to the academic ISP and generated by MRTG using SNMP. Time axes in the figures are from right to left. Note the dedicated line capacity was upgraded from 512kbps to 1.5Mbps at the end of March 1999 and the peak had been bounded to 64kbytes/sec until March 1999.

As the figures show, the peaks are in working hours and weekdays. Traffic tends to increase month by month. Then next question is how much HTTP contributes

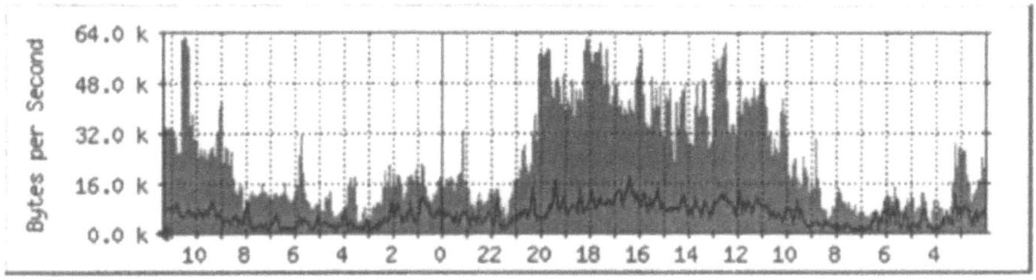

Fig. 1. Daily incoming IP traffic (5 minute average)

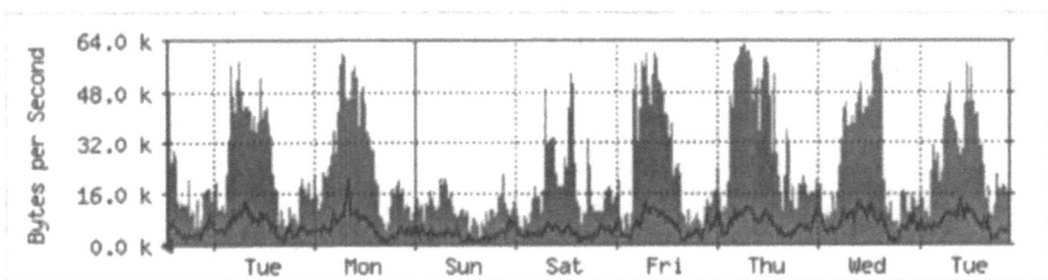

Fig. 2. Weekly incoming IP traffic (30 minute average)

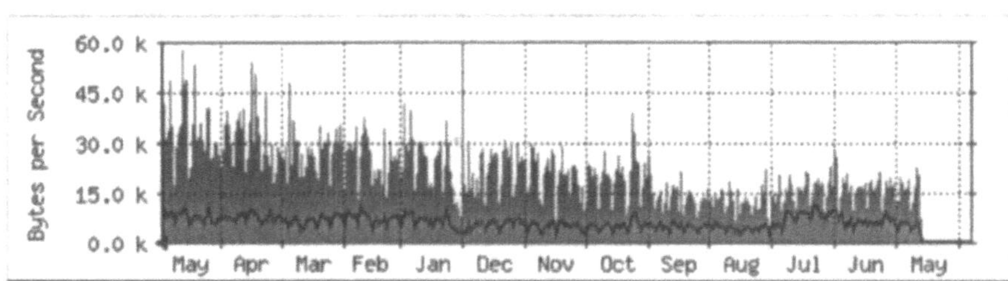

Fig. 3. Yearly incoming IP traffic (2 hour average) 1998 May-1999 May

to the total IP traffic. 500 PCs and EWSs are located in the classrooms and the students may use those terminals anytime from 8am through 8pm.

We set up most of the WWW browsers to use Squid caching proxies to reduce the number of HTTP requests to the WWW servers outside the campus. 4 shows the multilayer configuration used in our campus. Each of WWW browsers uses one of sibling Squid proxies in turn with round robin DNS setup. Note the SINET is the academic ISP operated by the Ministry of Education and the OCN is one of commercial ISPs. An HTTP request originated from a WWW browser is routed to one of the two Internet connections according to the domain name in the URL; a request to non JP or SINET domain (mostly AC.JP) is directed to the SINET, and a request to other JP domains (CO.JP etc.) is directed to the OCN. We will analyze the access logs of these proxies later in the following sections.

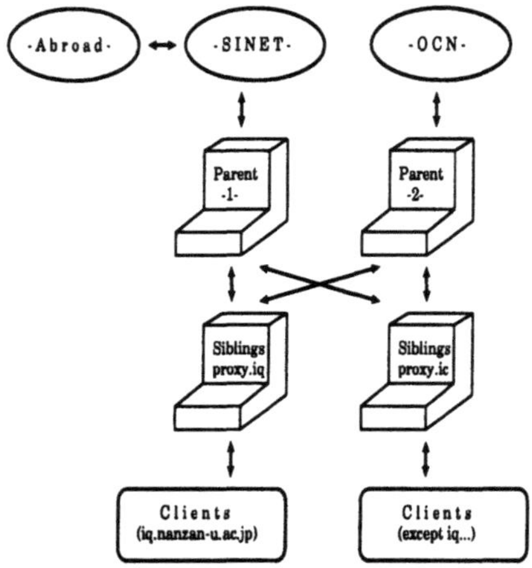

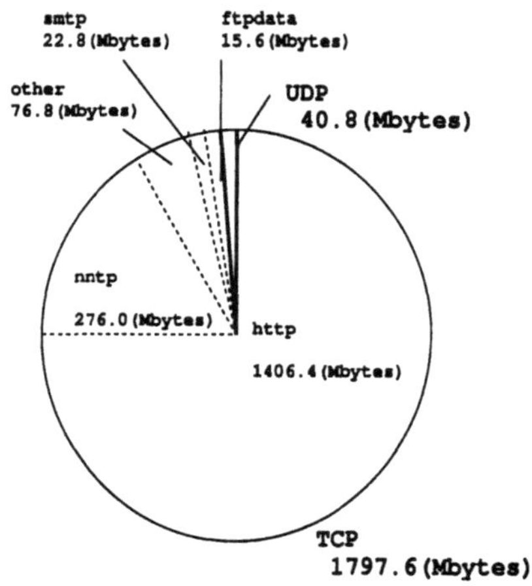

Fig. 4. WWW Proxy configuration(left)

Fig. 5. Total incoming traffic per port in a weekday(right)

We have stored and analyzed the packet headers for 24 hours in a weekday (December 15, 1998) with tcpdump on FreeBSD at the gateway subnet to examine the contribution of HTTP in the total IP traffic.

5 shows that HTTP traffic in a typical 24 hours is about 1.4Gbytes and 75% of total traffic by the amount of data. Therefore, we can conclude that HTTP is the network killer application as it is said to be. This is the motivation of the examination of WWW access statistics to estimate and reduce WWW traffic with a better cache policy. The details of WWW access statistics will be described in the following sections.

3 Access frequency distribution by URL

The cache hit ratio on a caching proxy server heavily depends on user behavior. Assumed that all users want to get only a limited number of very popular pages, the most effective caching strategy might be to get those pages before the users access. It is so called 'pre-fetching'. In contrast, assumed that users' interests are uniformly distributed among all of the web pages in the world, we cannot expect high cache hit ratio at all. The reality is somewhere between these two extremes. And we will see how users interests are concentrated on popular WWW pages.

3.1 Analogy to word frequency in a text

As mentioned in Section 1, the analogy between frequency of requests to a URL in WWW space and frequency of appearance of a word in a text is introduced in [11]. We apply not only one but three empirical word frequency formulas in mathematical linguistics [7–9] to our local proxy access log for a month, and find the best matching formula through comparison.

Zipf law Zipf law in Eq.1 is the formula used in [11]. The result is described here for comparison only since it does not match very well. This formula represents that f_r negative-exponentially decreases rapidly as rank r increases.

$$f_r r^\lambda = C \tag{1}$$

or

$$\log f_r + \lambda \log r = logC \tag{2}$$

where r is the rank of popularity for the URL, f_r is the number of requests (word frequency) for a URL of rank k, and C and λ are constant parameters to be estimated from the sample data in a time period. Fig. 6 shows the relation of f_r and r in linear scale (left) and logarithmic scale with the least square residue line (right) (Eq. 2), respectively.

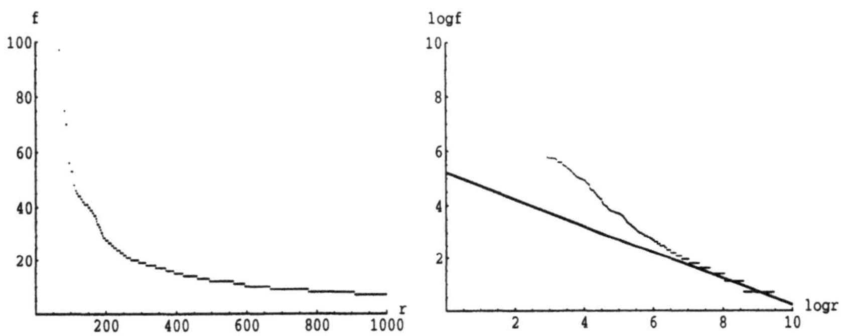

Fig. 6. Zipf: Frequency f_r as a function of rank r

As shown in the right graph in Fig. 6, the frequency of our local users' requests does not match the formula well. It does not seem to match well either in the reference [11]. The frequency for requests to the most popular URLs are underestimated.

Zipf second law Zipf second law is defined as Eqs. 3 and 4 below.

$$k_f f^2 = \delta \tag{3}$$

or

$$\log k_f + 2\log f = \log \delta \tag{4}$$

where f is the frequency of the request to a URL, k_f is the number of URLs with the same frequency f, and δ is a constant parameter. Fig. 7 shows the relation between f and k_f.

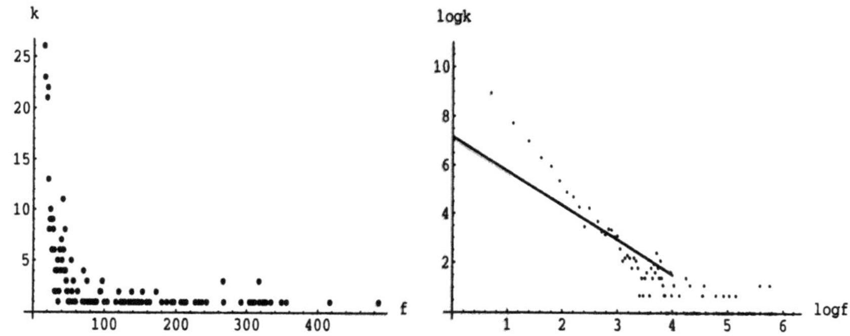

Fig. 7. Zipf 2nd low: k_f as a function of frequency f

This equation represents that the fewer number of URLs with the same frequency exists for more popular URL. Unfortunately the right graph in Fig. 7 shows that the effect of non-popular URLs is underestimated.

Projection function This third formula is the modification of Zipf second law formula and called projection function. The original formula is defined as Eq. 5 below.

$$F(p) = \frac{p}{\alpha p + \beta} \tag{5}$$

where p is the utilization of a URL (or a word in a text) and α and β are constant parameters. $F(p)$ is the ratio of URLs of which utilization is less than or equal to p.

Further we denote the total requests as N, the number of different URLs as L, the number of URLs of which utilization is less than or equal to p as $K(p)$, and again f as the number of requests. Note that $p = \frac{f}{N}$ and $F(p) = \frac{K(p)}{L}$. Then Eq. 5 becomes,

$$\frac{f}{K(p)} = \frac{\alpha}{L} f + \frac{N\beta}{L}. \tag{6}$$

This Eq. 6 is easier to evaluate since it is linear and values of the variables are directly derived by counting the proxy access log.

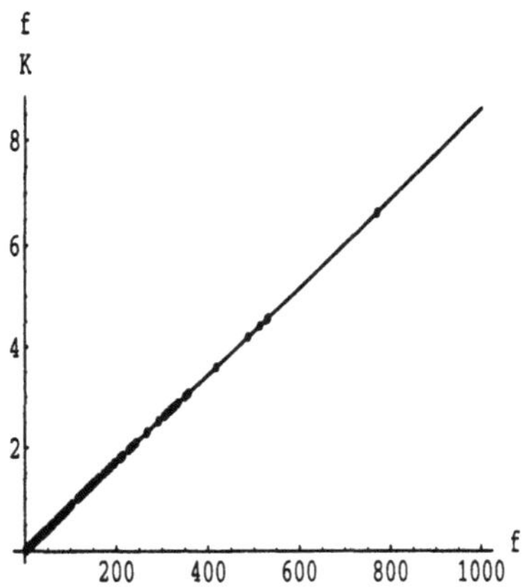

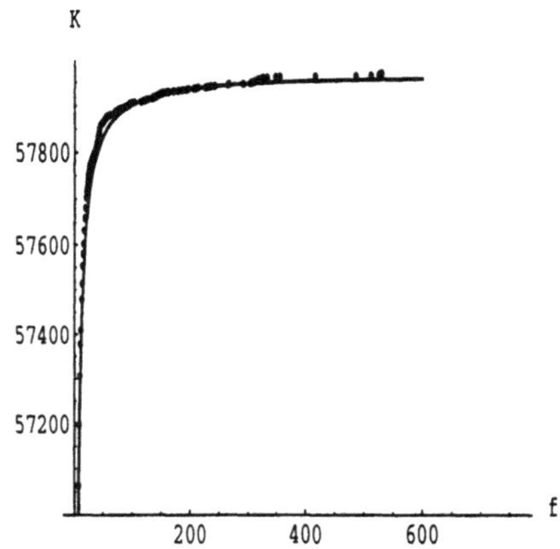

Fig. 8. Projection: $\frac{f}{K(p)}$ as a function of f (left)

Fig. 9. Projection: $K(p)$ as a function of $f = pN$ (right)

Fig. 8 shows excellent match with the least square residue line. Fig. 9 is the accumulative distribution function with the estimated function in the linear scale and shows far better match than Fig. 7. Actually the correlation coefficient is 0.998 and the determination factor is 0.996.

Estimation for α and β are 0.981, respectively and finally we obtain,

$$F(p) = \frac{p}{0.981p + 1.08 \times 10^{-6}} \tag{7}$$

for our sample data for a month.

Fig.11, Fig.13, Fig.14 illustrate that the projection formula is also a good approximation at the five other sites.

3.2 Theoretical cache hit ratio

Three popular cache strategies, random, LRU (Least Recently Used), and frequency are considered. Hit ratio for each of these strategies assumed that WWW access follows Zipf law are derived in [11]. We recalculated the results in the paper with the projection function defined in the previous subsection.

Random Hit ratio for random strategy is simply just the ratio of cache storage capacity, S, to the number of all URLs, L. Note that S denotes cache storage capacity in number of files of 6Kbytes on the average.

$$H_{Random}(S) = \frac{S}{L} \tag{8}$$

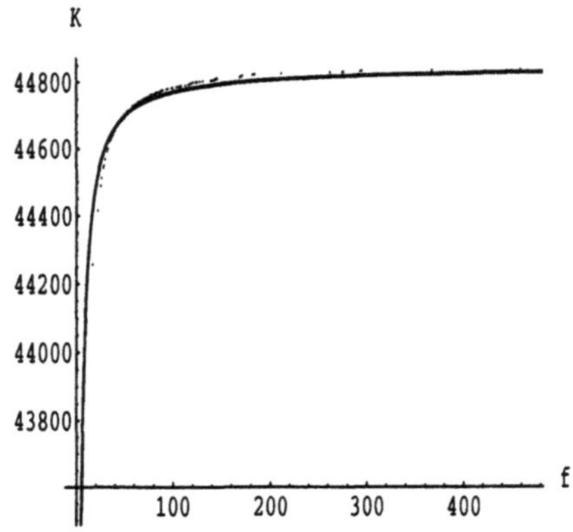

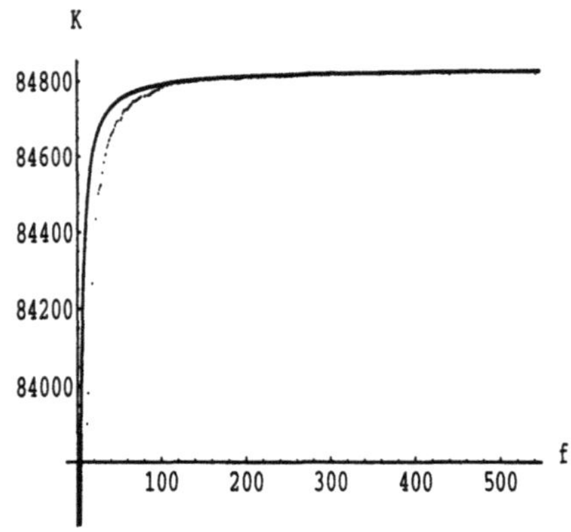

Fig. 10. Projection: $K(p)$ as a function of $f = pN$ at site A (left)

Fig. 11. Projection: $K(p)$ as a function of $f = pN$ at site B (right)

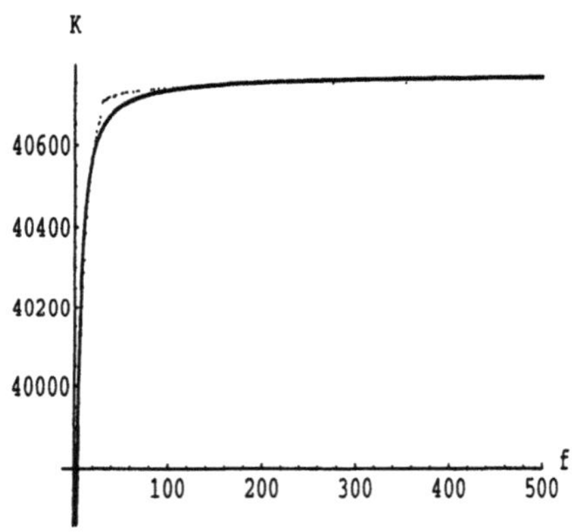

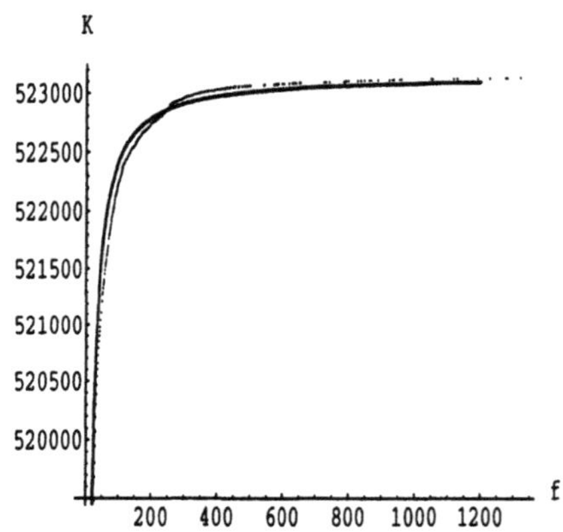

Fig. 12. Projection: $K(p)$ as a function of $f = pN$ at site C (left)

Fig. 13. Projection: $K(p)$ as a function of $f = pN$ at site D (right)

LRU Hit ratio for LRU strategy is expressed in term of the probability that the file is requested, p, which is the utilization defined in the previous subsection. Note that the probability that the cache is not occupied by other data is $1-(1-p)^S$ and the file is cached by one request to the file at probability p. K is again the number of URLs of which utilization is below p, but p is written as a function of K in the following equation because we taking expectation over K. Then we obtain,

$$H_{LRU}(S) = \int_0^L p(K)(1-(1-p(K))^S)dK. \tag{9}$$

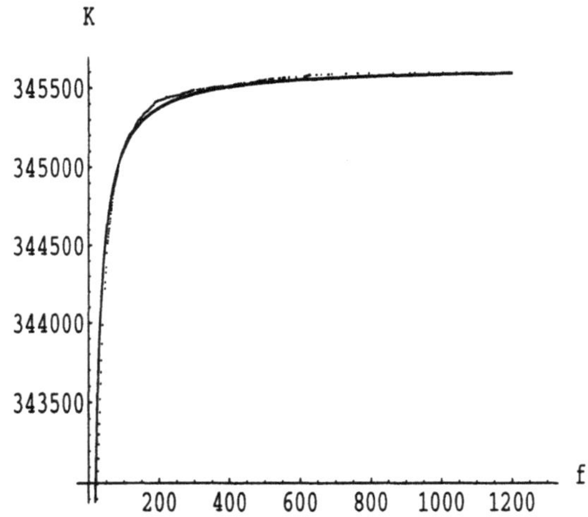

Fig. 14. Projection: $K(p)$ as a function of $f = pN$ at site E (left)

where $p(K)$ is given by $\frac{\beta K}{L-\alpha K}$ from Eq. 5.

Frequency Recall $F(p)$ is the fraction of the URLs which utilization is less than or equal to p. Since URL counts are accumulated in $F(p)$ from that of the highest utilization to lower as p increases, a file is not stored in the cache if $F(p) < \frac{L-S}{L}$. Taking integration over F to count on the number of accesses per URL. Then we get a rough estimation of the hit ratio for cache strategy by frequency as in Eq. 10 below.

$$H_{Freq}(S) = \frac{\int_{\frac{L-S}{L}} p(F)dF}{\int_0^1 p(F)dF} \tag{10}$$

where p is $\frac{\beta F}{1-\alpha F}$ from the definition in Eq. 5.

3.3 Numerical results

The access log for a year on a proxy server has applied to Eqs. 5, 8, 9, 10. Also the corresponding cache hit ratios have been derived by direct analysis of the access log. Table 1 compares the actual value and theoretical hit ratio. Cache storage is about $30,800 \times 6Kbytes = 185MB$. Theoretical values for random strategy and frequency strategy can be conjectured as good approximations. And those for LRU strategy tend to overestimates because aging of LRU cache is not taken into account in this analysis. In the real cache, older data are erased when the prescribed expire time has come. Effect of access locality is not exactly treated but included in access frequency. As locality becomes higher, so does frequency.

It is interesting to the readers to show some numerical results from the theoretical formulas. Fig. 15 shows effect of the amount of cache storage. The lines

Table 1. Comparison of actual and theoretical hit ratio (S=30,800)

	Random	LRU	Frequency
Actual	5.5%	16%	69%
Theoretical	5.7%	20%	73%

correspond to frequency, LRU, and random from the top. Figs. 16 and 17 illustrate number of accesses to LRU, and that to frequency, respectively. As a matter of course, if there are more accesses, higher the hit ratio becomes. However, the number of accesses affects less the hit ratio for frequency strategy than that for LRU. Old Squid implementation was based on LRU. As Fig17 shows, LRU may be as a good choice if a proxy server is heavily used even without a huge amount of cache storage space.

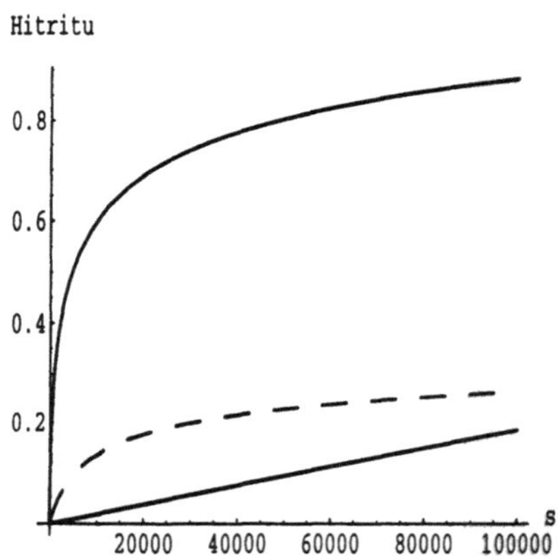

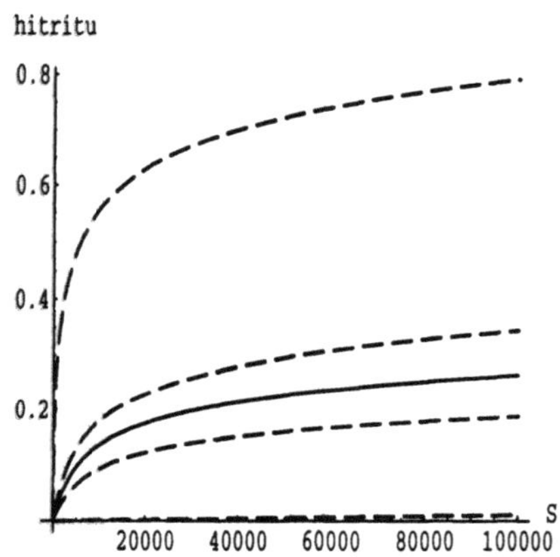

Fig. 15. Cache storage vs. hit ratio(left)

Fig. 16. Number of access vs. hit ratio with LRU cache(right)

4 A user's WWW access behavior

In this section, we focus on the behavior of individual users. The information available from a Squid proxy log line is listed below.

> Timestamp Elapsed_Time Client_Address Log-Tag/HTTP-Code Size
> Request_Method URL Ident Hierarchy-Data/Hostname Content-Type

Our interest are to find how many pages a user requests and reads and how long it takes for a user to get and read a page. Actual HTTP transmissions for

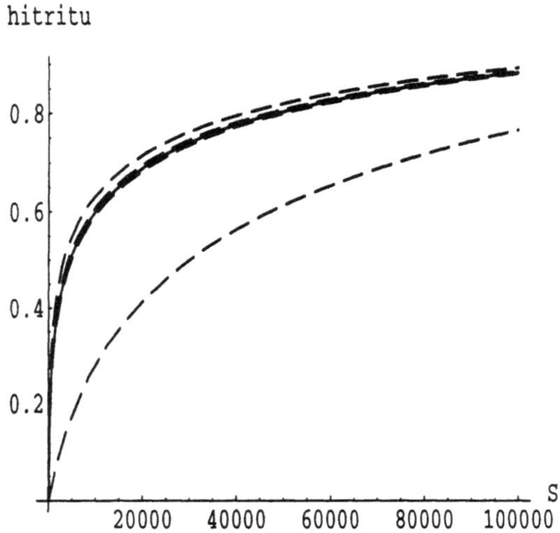

Fig. 17. Number of access vs. hit ratio with Frequency cache

a browser are simultaneous especially for image icons. User name is not recorded unless 'identd' is not running on the browser host which is typically a kind of Unix. We will simplifies the observation and build the access model. Also we compare the behavior of two types of students: students in Dept. of Information Systems and Quantitative Sciences and others since there are significant difference in the total WWW requests per terminal. In contrary to the computer operation skill level, information science oriented students' WWW accesses are roughly half of the others.

4.1 User behavior model

Fig. 18 shows a typical user's WWW access scenario. A user starts a WWW browser and requests a WWW server for transfer one page, reads the received data, and then requests another WWW page or quit using WWW. Actually it is not easy to detect when a user quits browsing, since all the terminals are shared by many students and user name is not recorded in the proxy access log. We have chosen 10 minutes idle to mean the user have quit using the browser by looking at some examples.

In the Fig. 18, **Set** denotes the time period of successive requests and its length is denoted as **User_Think_Time**. And **Term** denotes a time period during a user uses a WWW browser, which consists of a series of **Sets**. Note that it is possible to request another page before all files of the current page have not been received yet. The case is not considered here because it is observed that only few users act so.

There are usually multiple file requests in a **Set** since typical Web page consists of a HTML text file and several image icons and sometimes Java applets. **User_Think_Time** begins when a request to an HTML file is transmitted since a part of HTML text is shown on a browser as soon as it arrives.

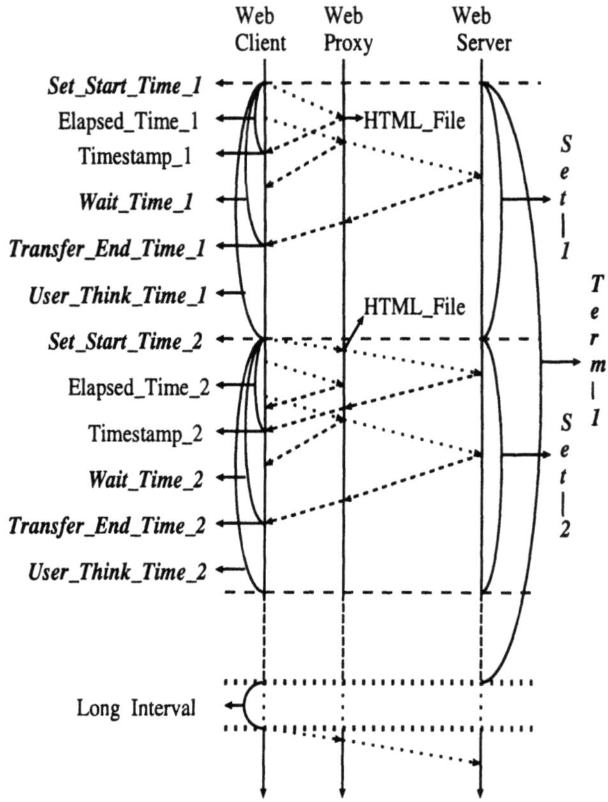

Fig. 18. User's WWW access scenario

4.2 Access statistics based on the model

We will show the result in the order of macro to microscopic viewpoints. Note that dotted lines for exponential distribution with the same mean of the samples are in most of the figures.

Statistics of Term Most interesting measurements for **Term** are length in time, number of requested pages.

Figs. 19 and 20 illustrate the distributions of length and number of requested pages, respectively.

Term length roughly obeys exponential distribution with averages of 647 seconds and 555 seconds for information science students and others, respectively. Distribution of the number of requested pages is close to exponential with average of 7.47 and 8.88 for the two types of students, respectively. Total transfer bytes in a **Term** also obeys exponential distribution with averages of 323KB and 395KB for the two types of students, respectively.

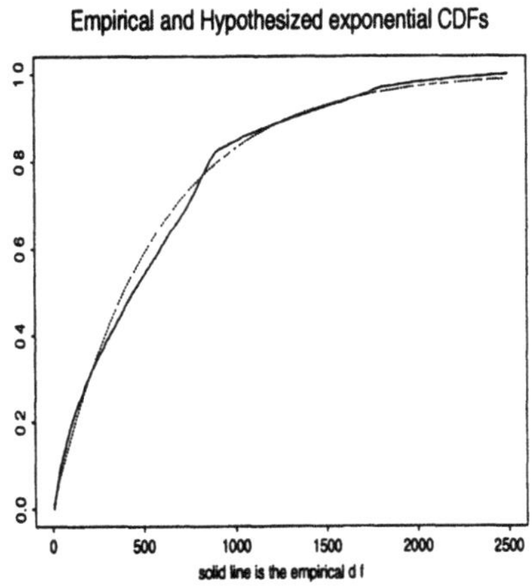

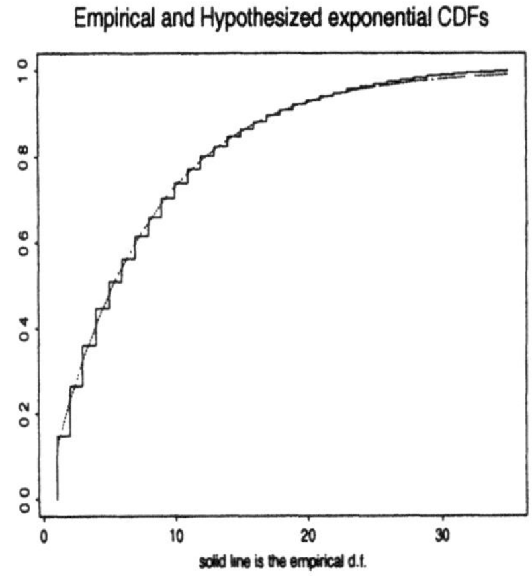

Fig. 19. **Term** length distribution (left)

Fig. 20. Distribution of th number of requested pages (**Sets**) in a **Term**(right)

Statistics for Set A **Set** includes multiple file transfers caused by one click on the browser by a user. Interesting statistics are length of **User_Think_Time** (i.e. length of a **Set**), number and total bytes of requested files, and fraction of **User_Wait_Time**.

Mean number and total bytes of transferred HTML files in a **Set** are 1.57 and 12.0 KB and the standard deviations are 1.06 and 11.8, respectively. Those of image files are mean 5.18 and 28.4KB with standard deviations of 5.19 and 34.4, respectively. Therefore a WWW page consists of one HTML file and 3.3 image files on the average.

Fig. 21 illustrates **Set** length distribution. The averages are 70.0 and 72.8 seconds for the two types of students. Fig. 22 shows the distribution of **User_Wait_Time**, which is the time from the time a user requests a page to the time all the files for the pages are received, within a **User_Think_Time**. Neither of these distribution is not obey exponential distribution. Therefore those distributions are illustrated as probability density functions.

The figures show that approximately 60% of **User_Think_Time** is **User_Wait_Time**.

Individual file transfer statistics Finally, we examine statistics for individual file transfer.

Figs. 23 and 24 show the distributions of the size HTML file and image files in bytes, respectively. HTML file size seems to be exponentially distributed with mean of 6.74KB. Image file size distribution is rather concentrated in shorter sizes with mean 4.7KB. It does not obey exponential distribution.

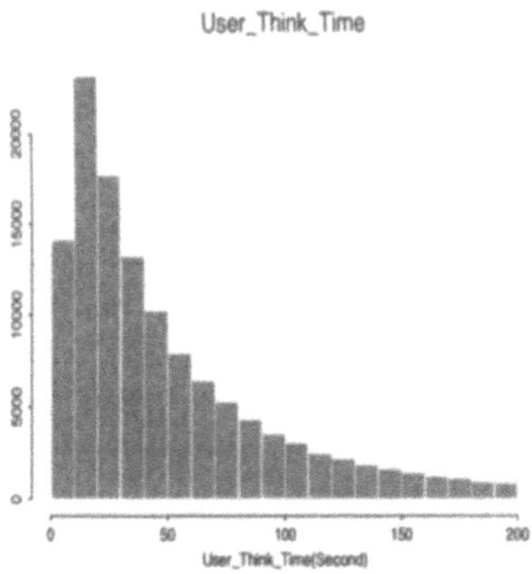

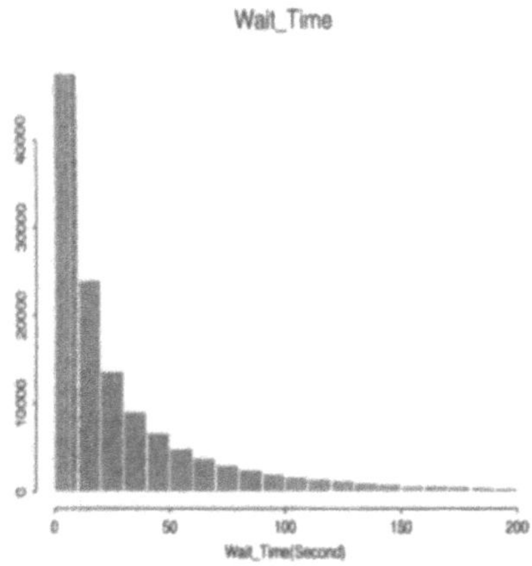

Fig. 21. User_Think_Time distribution(left)

Fig. 22. User_Wait_Time distribution(right)

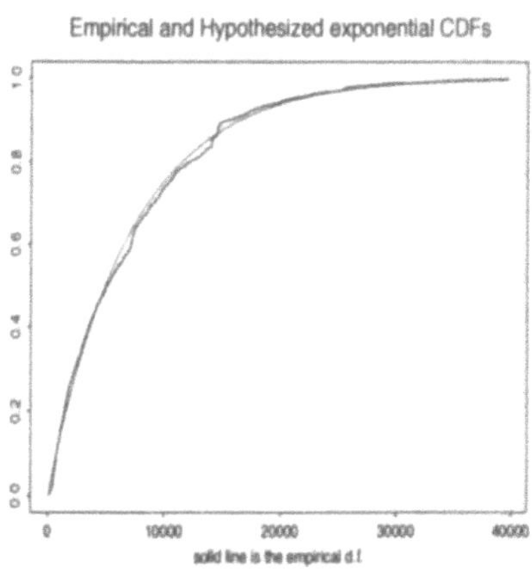

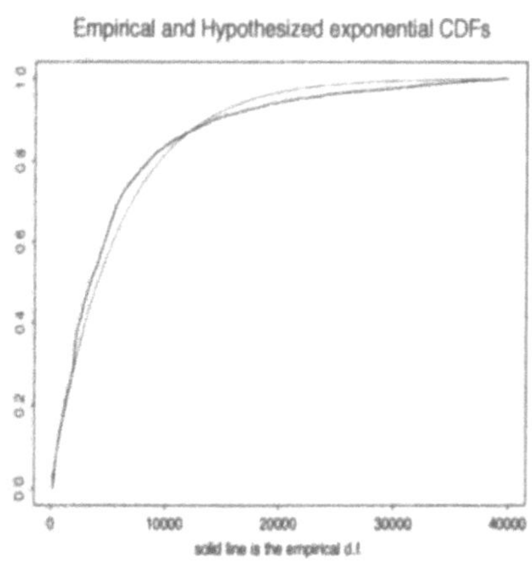

Fig. 23. HTML file size distribution(left)

Fig. 24. Image file size distribution(right)

5 Conclusion

We analyzed the access logs of our WWW proxies used by most of the 500 terminals in our campus. An empirical access frequency distribution to different URLs was derived to compare cache hit ratios with different strategies.

We found the projection formula gives the best approximation for the frequency distribution of requested URLs. Also the formula is shown to be a good approximation at the five other different sites which gave us their access frequency data.

Based on the estimated parameters, we conjectured that the best cache strategy is 'frequency' in general and LRU strategy is very good for a heavily used proxy server even without large cache storage space.

Also we investigated user's access behavior from proxy server log. We found that a typical user reads 8 WWW pages and quit browsing after 600 seconds on the average. Total transfer bytes for the duration is 350KB. A WWW page consists of 1 HTML file of 12KB and 5 image icons of 28.4KB, and it takes 70 seconds for a user to read the page.

We believe the results and the analysis method do not lack generality. However, the parameters are different by site by site and those results are based on access logs only at six sites. Therefore, to be sure about the results or to find the difference from the statistics with other types of users, extensive analyses with sample data at other sites are required in the future study. We will find probability distribution of WWW traffic as a collection of individual users behavior analyzed in this paper. Also the approximations for cache hit ratios might be improved.

References

1. Abrams, M., Standridge, C. R., Abdulla, G., Williams, S. & Fox, E. A. (1995). Caching proxies: limitations and potentials. *Proceedings of 4th Inter World-Wide Web Conference, Boston,* pp.119-133.
2. Arlitt, M. F. & Williamson, C. L.(1996) Web server workload characterization : The Search for Invariants. *Proceedings of ACM SIGMETRICS'96,* pp.126-137.
3. Duska, B. M., Marwood, D. & Feeley, M. J. (1997). The Measured Access Characteristics of World-Wide-Web client proxy caches. *Proceedings of USENIX Symposium on Internet Technologies and Systems,* http://www.cs.ubc.ca/spider/marwood/Projects/SPA/.
4. Fujita, Y., Ito, A. & Matsuo, S. (1998). Analysis of local WWW access traffic and evaluation of caching schemes on a proxy server (in Japanese), *Nanzan University Graduate Thesis.*
5. Ideno, S. & Hosokawa, R. (1998). Internet traffic analysis at Nanzan University (in Japanese), *Nanzan University Graduate Thesis.*
6. Kotsis, G., Krithivasan, K. & Raghavan, S. V. (1997). A workload characterization methodology for WWW applications. *Proceedings of International Conference on The Performance and Management of Complex Communication Networks (PMCCN'97),* pp.145-159.
7. Mizutani, S. (1974). *Five new discovery and re-discovery in Japanese Language (in Japanese).* Sobun-sha.
8. Mizutani, S. (1982). *Mathematical Linguistics (in Japanese).* Baifu-kan.
9. Mizutani, S. (1983). *Vocabulary (in Japanese).* Asakura-shoten.
10. Muramatsu, T. (1998). WWW traffic analysis and users' access model (in Japanese), *Graduate School of Nanzan University Masters dissertation.*
11. Nishikawa, N., Hosokawa, T., Tsuji, H., Mori, Y., and Yoshida K.(1996). WWW traffic analysis and distributed cache (in Japanese). *Proceeding of IPSJ Distributed System Research Meeting.*
12. Raghavan, S. V., Vasukiammaiyar, D., and Haring, G. (1994). Generative networkload models for a single server Environment. *Proceedings of ACM Sigmetrics Conference on Measurement & Modeling of Computer Systems (SIGMETRICS'94),* pp.118-127.

13. Wessels, D., and Claffy, K. (1998). ICP and squid web cache. *IEEE Journal on Selected Areas in Communications, Protocol Architectures for 21st Century Applications,* Vol.16, No 3, pp.345-359. (Software available at http://squid.nlanr.net/Squid/)

Characterisation of End-to-End Performance for Web Based File Server Repositories

Manoel Eduardo Mascarenhas da V.Alves, Sergey Nesterov, and Reginald P Coutts

Centre for Telecommunications Information Networking, The Univerisity of Adelaide,
Level 5, Capita Building, Adelaide, South Australia 5000, Australia
2) 2849.5968 – MS Word + PDF (27 pages)
(use 2849.5948/2849.5948.pdf)
(page 1)

Abstract. The paper suggests an experimental methodology for analysing the performance of Web File Servers in terms of throughput and Client-Server path instability. The experimental results preview the network performance for a Client user in Australia and the influence of interconnecting networks and file server traffic demand for download applications.

1. Introduction

In terms of end user perception, information can be classified as either time-based or non-time based. The former carries within it intrinsic time properties. For example, *video* has information about frame rate display and audio/image synchronisation. Conversely, non-time based information has no essential built-in time property for an adequate display (eg, a written document, a jpeg picture, etc). For the purpose of Quality of Service (QoS) requirements then, applications can be classified as real-time streaming, real-time block transfer and non-real time applications [1].

Real-time streaming applications deliver time-based information in real time over the network. For adequate user perception, the network has to deliver time-based information without changing its built-in time properties. Therefore, certain QoS requirements such as delay, jitter and error rates must be taken into account for adequate provision of these applications. The most well known Internet applications in this group are Web television, Internet telephony and Internet radio.

Real-time block transfer applications deliver either time-based or non-time based information. This group deliver one or more blocks of information within a deadline though, unlike real-time streaming applications, consecutive blocks do not have a time correlation. Internet applications in this group are: Web browsing, client-to-client application sharing, online games, chat and file transfer.

Non-real time applications deliver both time-based and non-time based information without demanding a time delivery deadline. In terms of QoS, the main requirement is delivering error-free information. The most well known application in this group is electronic mail (e-mail).

Given the potential bottlenecks in the access network and taking into account common Internet applications, file downloading (via HTTP or FTP) is likely the most "unpleasant" application for dial-up customers today and we assume that it might be the most important real-time block transfer application for a broadband access platform in the future. An experiment with file transfer will then be a good test for analysing current Internet performance in terms of its ability to deliver different application types. Moreover, to some degree we can consider file transfer application performance as an approximation for data streaming (e.g., video) in a non-congested environment because a Transmission Control Protocol (TCP) session with a reasonable window size has a consistent throughput over time after the slow start phase [2].

The importance of file downloading has increased together with the development of new compression formats because multimedia content such as compressed music, compressed video and general data can be widely found on the Web. Furthermore, NAPSTER[1] and other collaborative download environments increase the popularity and demand for downloading files among a community of Internet users [3,4,5]. Further, the traffic impact of NAPSTER has been such that academic and commercial organisations are prohibiting its use [6,7].

At the same time, the FTP program is the most significant application for transferring scientific data over the Internet. The final report from the "Advanced Networking Infrastructure Needs in the Atmospheric and Related Sciences" (ANINARS) workshop in July 1999 states:

> "... the workhorse networking application in the atmospheric community is still FTP. Aside from [...], FTP is practically the only networking tool used to construct applications in this scientific discipline. There was also a universal cry for FTP to actually deliver the available network bandwidth to the end-user. The lament was that the bandwidth actually obtained is much lower than the apparently available bandwidth.
>
> Most participants thought that the need for bulk data transfer would never go away, even if sophisticated data extraction methods could be developed to extract subset portions of datasets. Such mechanisms would simply supplement the FTP function but not replace it."

The same report continues:

---

[1] See http://www.napster.com

> "FTP (or FTP-like) bulk data-transfer is the most important networking function used to construct applications in this scientific discipline, yet failure to achieve effective bandwidths equal to apparently available bandwidths is most evident with bulk data-transfer applications. A variety of host-software problems contributes to this failure, and programs should be developed to help solve these problems" [8].

Since TCP packets are responsible for 90 to 95 percent of all Internet traffic [9], a TCP file download measurement tool could provide a suitable analysis of bulk transfers over the Internet.

This paper introduces an experimental methodology for analysing the performance of download applications in a broadband access environment from a client-based perspective. Some studies for analysing the performance of networks prioritise network layer measurements and do not focus on application layer analysis [10]. Other studies have monitored traffic without a client perspective at all [11]. In this paper we also consider an end user perspective.

2. Experiment Objectives

This paper describes a new method for analysing the performance of Web File Servers. We introduce a hybrid throughput-path analysis, which allows better understanding of how backbone routes influence throughput performance to the end user.

The first parameter is a throughput variable that depends on traffic conditions within the interconnecting networks and remote server traffic demand (See Sections 4.1 & 6.1).

The second parameter (ε) looks at the instability of a path for any given Client-Server connection[1]. This is achieved by repeated use of the route analysis tool Traceroute[2] [12] and subsequent comparisons of collected traces to look for changes in that path connection (See Sections 4.2 & 6.2).

Once we have established this method to look at the quality of the path from a instability point of view, we then go on to compare paths to different file servers on a regional basis (See Fig. 2). Some studies in network performance measurement have focused on analysing the stability of a certain path of the Internet though did not introduce a quantitative variable for comparing different paths through the Internet [13]. We suggest a quantitative parameter through the definition of the parameter (ε).

---

[1] Client-Server connections are also called *virtual paths*.

[2] For ease reading, we call *Traceroute* as TRACERT

3. Client-Server Internet Connectivity Model

For ease of analysis, the Internet environment can be summarised as a collection of Autonomous Systems[1] (AS), which vary in size, geographical coverage and function. A Client-Server model based on AS connections provides an overview of virtual paths through the Internet and this model can be used for better understanding the influence of different AS on overall path performance. A Client-Server Internet connectivity model is shown in Fig. 1 and its main elements are described below:

A. Client Machine. This is a computer running a standard Operating System and having dial-up or dedicated line access to the Internet. For our study, clients ran NT 4.0 Operating System, had similar hardware configurations (Pentium class) and were connected to the University of Adelaide Internet gateway via a 100 BaseT Local Area Network (LAN).

B. Local Network. This is the AS that provides Internet access to the client machine. For a dial-up user such a network is a traditional ISP, for a dedicated line user a broadband ISP, research/commercial organisation or backbone provides the connectivity. For our study, the local access is via the University of Adelaide Internet gateway comprising the later.

C. Local/Remote Regional Network. This is the AS that provides connectivity to local networks. Depending on the geographic location of the regional network (serving the Client or the Application Server), it can be classified either as local or remote.

D. Backbone. A backbone is a transit AS, providing *regional, national* or *international* coverage. For our study, we define three types of backbone interconnections: the Australian backbone, the US backbone and the remote server backbone.

E. Remote LAN. This network provides Internet connectivity to an Application Server.

F. Application Server. The Application Server is used for storing digital content such as Web pages and program files. File application servers can have mirror sites closer to users, resulting in better response time and higher throughput [14].

This model will be used as the basis of the experiment to assess Internet performance of those terms discussed in Section 2.

---

[1] Autonomous Systems are a group of subnetworks administered by a Single Administrative Authority (SAA) with a set of Interior Gateway Protocols (IGPs). The SAA is normally a Network Service Provider or large organisational network (e.g., campuses and corporate networks).

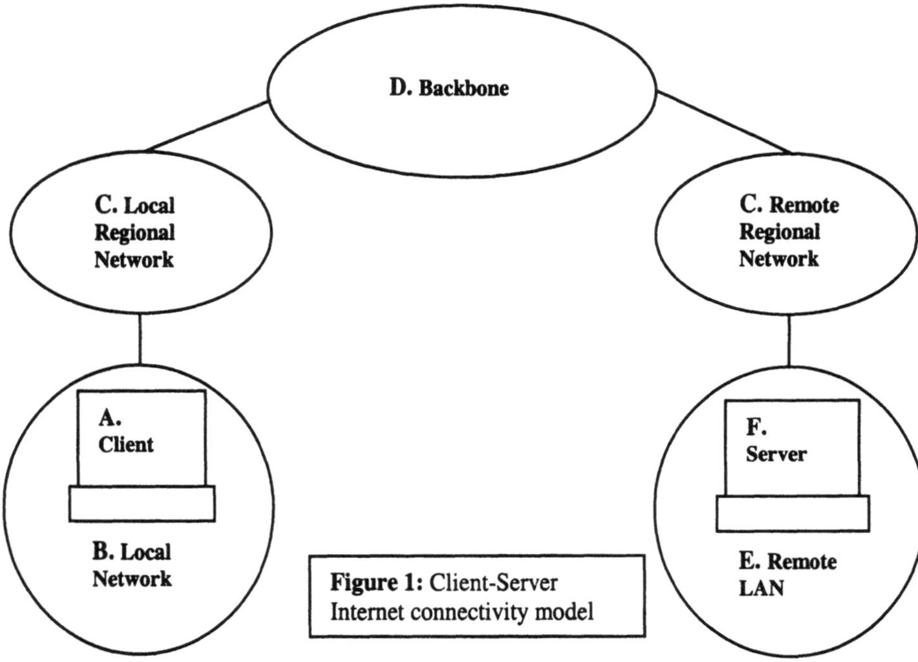

Figure 1: Client-Server Internet connectivity model

4. Path Instability and Throughput Models

4.1. Throughput Model

The Throughput rate for our model follows a general equation as:

$$Throughput \propto \alpha(\beta(\phi(\chi(\pi(\delta(t)))))) \tag{1}$$

Each element of the Equation (1) describes the network capacity of the defined areas of our connectivity model (See Section 3). We postulate the throughput in each network element is inversely proportional to the traffic demand. The named elements of the equation then reflect our model and are:

α : Local Network
β : Local Regional Network
ϕ : Backbone
χ : Remote Regional Network
π : Remote LAN
δ : Application Server

Equation parameters will vary over time (t) and are related to the QoS settings in each network.

Backbone traffic conditions will vary with the influence of national and international backbone characteristics, both at the Client and Application Server sides. In Section 7, our results will also show how backbone traffic conditions are highly influenced by those particular to the USA backbone.

For our study, we assume bottlenecks will result from regional network, backbone or Application Server limitations. Therefore, Equation (1) reduces to:

$$Throughput \propto \beta(\phi(\chi(\delta(t)))) \tag{2}$$

For the case where Application Servers are close to the Client and there are neither backbone nor regional network bottlenecks, Equation (2) reduces to:

$$Throughput \propto \delta(t) \tag{3}$$

4.2. Path Instability Model

The analysis of IP and DNS characteristics over time for all nodes in a virtual path provides a good overview of path instability. This section suggests a new variable for estimating the instability of a virtual path. If N is a number of *Traceroute*[1] (TRACERT) samples collected and each sample has a variable number H of nodes per sample, then:

$$\vartheta = \frac{\sum_{1}^{i=N} H(i)}{N} \tag{4}$$

where ϑ is the average number of nodes

Further, if there are W changes in path configuration during a period of time T, the path instability variable results in:

$$\varepsilon = \frac{W}{\vartheta T} \tag{5}$$

Where ε is the number of changes of node configuration for a certain period of time T. The dimension of ε is $[T]^{-1}$. This variable is calculated by the software *Utility 5*, which is described in Section 6.2. In our study, we do not consider T for the calculation because the experimental analysis period was common for all paths. Path instability (ε) is useful for comparing the instability of different virtual paths.

[1] For ease reading, we call *Traceroute* as TRACERT

5. Data Collection Procedures & Assumptions

For the experiment we collected data to analyse the effects of network interconnections, Application Server traffic demand and time zone differences over file transfer applications. A number of assumptions were made, which are discussed below:

5.1. The file transfer Application Servers are common file repositories used by Internet users (See Fig. 1: F). They have a standard World Wide Web (WWW) interface and are located in several geographic locations worldwide. We selected the popular TUCOWS[1] Web site, which has a number of mirror servers on all continents. The file transfer connections used Hypertext Transfer Protocol (HTTP), which is the standard WWW transfer method. The selected mirror site regions are showned in Table 1:

Table 1: Location for Tucows mirror sites

Region A	Region B	Region C
S. Australia State	USA W. Coast	Argentina
Victoria State	USA E. Coast	Brazil
Hong Kong	Germany	S. Africa
Israel	England	Zimbabwe

Regional classification was based on expected throughput performance for different geographic regions visible from Adelaide, Australia. Time zones were also an influence on the locations selected, as these impact performance because servers and interconnecting network loads depend on local traffic conditions in each region [14].

Three computers were used as clients to run the throughput experiment as shown in Fig.2. In addition, Client 1 ran a software utility collecting paths to each Application Server shown in Table 1 using the TRACERT program. This utility ran in parallel to all sites and sampled respective paths every five minutes. The throughput data collection was carried out from Nov 3 until Nov 22 1999 for the first two clients while for the third client the period was extended until Nov 30 1999. This decision of extending the experimental period for the third client was due to the small number of throughput samples collected during the first experimental period. The path collection ran from Nov 3 until Nov 30 1999.

---

[1] Tucows servers are configured based on a suggested hardware/OS configuration. During this study the suggested characteristics were: Pentium 133 or greater, UNIX OS, 32Mb of RAM or greater, 8Gig hard disk space, T-1 or greater bandwidth.

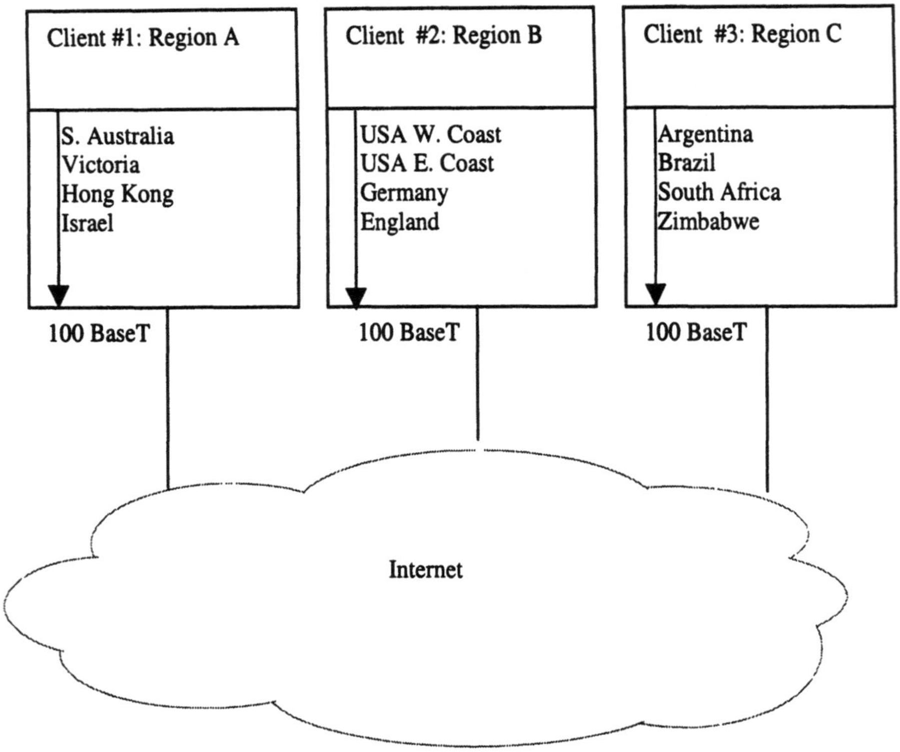

Figure 2: Client servers and their target Application Servers

Client 1 was responsible for servers in three countries, Australia, Hong Kong and Israel, each country having a well-developed national network infrastructure and represented three time zones.

Client 2 was responsible for servers in two European countries, Germany and England, as well as the USA. The servers are located in three different time zones.

Client 3 was responsible for servers in countries with less developed national network infrastructure and with two time zone groups (Brazil and Argentina; South Africa and Zimbabwe).

5.2. TRACERT provides performance information (RTT, packet loss and IP/DNS addresses) for every node within a virtual path. Each node within a path is probed with three echo-request datagrams. For each echo-reply received, TRACERT calculates the estimated RTT and registers it in a log file. A TRACERT sample is shown below:

#3/11/99 0:04:58

*Tracing route to **www.ozbytes.net.au [203.152.224.5]** over a maximum of 30 hops:*

1	*<10 ms*	*<10 ms*	*<10 ms*	*vlan0180.atm2-0.pancho.net.adelaide.edu.au*	*[129.127.180.253]*
2	*<10 ms*	*<10 ms*	*<10 ms*	*lis255.atm1-0.central.saard.net*	*[203.21.37.2]*
3	*<10 ms*	*10 ms*	*<10 ms*	*fa0-0-108.boomerang.internode.on.net*	*[203.16.212.13]*
4	*<10 ms*	*11 ms*	*<10 ms*		*198.32.240.100*
5	*10 ms*	*20 ms*	*	*ser-1-0-bigpipe-grote-adl.bna.com.au*	*[203.34.35.82]*
6	*10 ms*	*<10 ms*	*10 ms*	*nostromo.senet.com.au*	*[203.56.239.98]*
7	*<10 ms*	*10 ms*	*10 ms*	*www.ozbytes.net.au*	*[203.152.224.5]*

The sample above provides information about the local time when the TRACERT is activated and the destination IP/DNS address that TRACERT is probing. For each node within a path, it is possible to obtain estimated RTTs and IP/DNS address details. In the case where echo-request datagrams are lost for a particular node, a star (*) is registered to indicate packet loss. For our experiment, we are interested in IP/DNS details and node number.

5.3. We assume that DNS information obtained via TRACERT sample inspection (See Section 5.2) provides a good estimate of geographic location[1] for each node.

5.4. A Standard Benchmark File (SBF) was used to measure throughput performance from TUCOWS sites and a large size file was selected in order to avoid the initial low transmission throughput due to the TCP slow start mechanism. Netscape Communicator distribution setup file[2] was the selected SBF.

5.5. The file download session was scheduled as described in section 5.1. For each client computer, the SBF was sequentially downloaded from TUCOWS servers as shown in Fig. 2. For all download sessions, the *Start time* (StartT) and *End Time* (EndT) were registered in a log file. These values are used for calculating the average connection throughput, which we calculated as follows:

$$Throughput(Kbps) = \frac{SBF\ \text{size (bits)}}{(StartT - EndT)} \times \frac{1}{1024} \tag{6}$$

In our case we do not consider any bottleneck that would affect the transfer rate in either the client server or the local network (See Fig 1: A & B), as there is a 100 BaseT connection to the University backbone from the client and this gateway does not have any congestion problems. In addition, recent data suggests file servers are responsible for 33% of congestion while 42% is a result of core network problems [3]. The impacts of these assumptions on our results are discussed in Section 7.

---

[1] We have learned from TRACERT sample analysis that normally DNS addresses have information about airport codes *or* city initials.

[2] Version Number: 4.7, byte size: 18.1 MB – (18,968,232 bytes)

network problems [3]. The impacts of these assumptions on our results are discussed in Section 7.

5.6. Further to our discussion in Section 5.5 on the position of any potential bottleneck in our analysis approach, we make assumptions about the influence of interconnecting networks and Application Servers with respect to throughput for any particular Internet virtual path.

We assume that due to the popularity of the TUCOWS file server among Internet users, the influence of traffic demand within the Application Server on throughput will always be high. Further we postulate that if the Client is geographically close to the Application Server, the influence of interconnecting networks is negligible (See Fig 1: C & D). Conversely, if the Client is geographically distant from the Application Server we assert that influence of interconnecting networks will exist.

During the course of the experiment a wide variation of transfer rates was observed. Since a dial-up session under the best circumstance (64 Kbps) would require 40 minutes to download the SBF, we decided to set this as the time limit for the download sessions. To maintain simulating a broadband access environment any transfers taking longer than this were discarded. Given this broadband access definition, we later discuss its statistical significance, because it may have an impact on the median value of throughput transfer rate.

6. Software Supporting the Experiment

Software utilities for analysing the Application Servers' throughput and path were developed to run the experiment as follows:

6.1. Throughput Analysis Utility (See Fig 3)

Utility 1 - Scheduling the downloading sessions. Activates the web browser and directs it sequentially to the web sites described in Fig. 2.

Utility 2 - Timing the downloading sessions. Registers the variables *Start time* and *End time* in a log file, which are used for calculating the duration of the downloading session. *Utility 3* described below does this calculation. For clarity in later discussion, we refer to utilities 1&2 collectively as THROUGHP.

Utility 3 - Calculating the throughput. Calculates the throughput based on the output from *Utility 2* and follows the Equation (6).

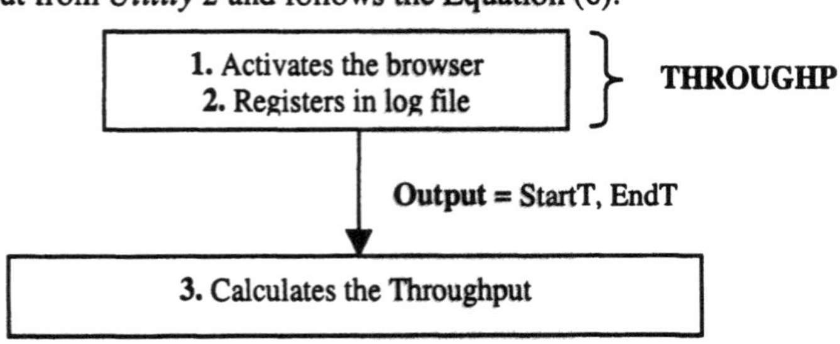

Figure 3: Throughput Analysis Utility

6.2. Path Analysis Utility (See Fig 4)

Utility 4 - Tracing the route to all TUCOWS Web sites every five minutes in parallel. Activates the TRACERT network tool every five minutes, which generates a log file for the path analysis. While a sampling interval of 5 minutes might loose short-lived network phenomena [13], we found this a better granularity because we are interested in longer routing changes for path stability analysis. The log file parameters that we used to analyse the paths are:

- Node number;
- IP address for a particular node;
- Domain Name Server (DNS) for a particular node.

Our path analysis utility output does not guarantee downloaded packets from the TUCOWS servers (downstream) will follow the inverse path described by this tracing software (upstream). However, this utility does provide an extra facility for estimating packet flow behaviour between two Internet destinations. Furthermore, TRACERT might occasionally provide wrong conclusions due to low priority of ICMP packets in relation to other packet types. For example, Matthews & Cottrell reported deterioration in performance from North American sites to Scandinavian sites due to the installation of *Smurf filters* in the connection link [15]. These filters give low priority to ICMP echo requests returning a *Destination Unreachable* message and affecting path analysis as a result.

The aim of developing this client software utility was to verify path instability to a particular Application Server destination. While this does not guarantee that downloaded packets from the TUCOWS servers (downstream) will follow the inverse path described by this tracing software (upstream), the utility gives a good instability preview in terms of Application Server-Client datagram flow. This software will be called TRACER (Fig. 4).

TRACER also allows a better understanding of the Throughput experiment because it enables the analysis of interconnecting paths and therefore a better appreciation of how interconnecting networks can influence transfer rate performance to a Client.

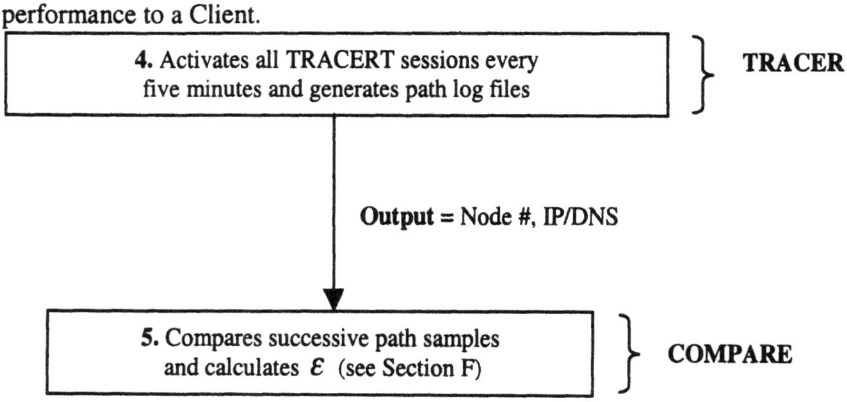

Figure 4: Path Analysis Utility

Utility 5 - Comparing the sampled paths for a particular TUCOWS Web site. Compares sequential samples from a particular TUCOWS log file. We call this utility COMPARE.

Below we describe the manner in which COMPARE measures the difference in path giving successive TRACERT samples to a particular Application Server:

Node changes: If a change in an IP address node occurs in sequential samples, COMPARE registers the sample number and the new node configuration.

Pseudo-node changes for IP addresses in Class C format: IP addresses are classified as described in Table 2. The last 8-bit number field of the IP address in Class C format describes the number of hosts for a certain network address ID. We do not consider a change in path for cases where two IP Class C nodes in sequential samples have common network IDs. In most cases this means a balancing policy has been implemented for routers in the same network and are possibly located physically in the same premises.

Therefore, if a certain node in two sequential samples has an IP address in the Class C range, COMPARE verifies a match in the first three fields of the IP address. If a match occurs, COMPARE does not analyse the fourth field. A change in host ID does not significantly change the Quality of Service (QoS) settings for this network. For Class A and B, COMPARE looks at all fields of the IP address.

Table 2: IP class types

IP Class types	Range of Host addresses	Maximum # of Networks	Maximum # of Hosts per Network
A	1.0.0.0 to 127.255.255.255	128	16777216
B	128.0.0.0 to 191.255.255.255	16384	65536
C	192.0.0.0 to 223.255.255.255	2097152	256

In some cases TRACER may generate "abnormal" messages, for example, timed out or destination unreachable messages, which are discussed here.

Timed Out: If a "Timed out" situation happens in a certain node from a sample, COMPARE verifies the subsequent node matches the IP address of the same node from the previous sample. If this situation happens, the path has not changed between the sequential samples and COMPARE verifies the successive nodes in the samples until the destination host is reached. However, if the IP address does not match the previous sample node IP address, COMPARE considers this a change in the path.

Destination Unreachable: If the message "host or network – unreachable" happens in a certain node, COMPARE compares this path with the previous sample path until the "unreachable" node is processed.

7. Results & Discussion

7.1. Throughput

To assist readers, this Section is discussed on a server-by-server basis, following the order: South Australia, Victoria, USA West Coast, USA East Coast, Hong Kong, Israel, Germany, England, Argentina, Brazil, South Africa & Zimbabwe. We have decided on such a server order that departs from our initial Client organisation (See Figure 2) because we understand the USA backbone has a high influence on overall performance of all Application Servers outside Australia. Therefore, a better understanding of the performance of Application Servers within the US helps us discuss the other Application servers.

All data was measured with the time reference based on the client time zone. In order to have a better understanding on how time zone differences might affect throughput for a certain Client-Server connection, we decided to shift the data time stamp from the local client time to the remote Application Server time.

To compare statistics between different virtual paths, we have classified measured throughput in terms of expected time-of-day patterns. This decision was also supported by other studies that showed a high correlation of data traffic characteristics with time-of-day patterns [12, 16, 17].

Each day was partitioned into three zones that correspond to known profiles of Internet usage throughout the day and this is illustrated in Table 3. Table 3 also shows the expected throughput for a download session within a virtual path where the Client and the Application Server are located in the same geographical region and there are no network bottlenecks between them (See Equation 3).

Table 3: Time-of-Day Patterns

In-day zone	Usage profile description	Expected throughput
00:00-09:00	Off-peak time	High
09:00-18:00	Business usage	Low
18:00-24:00	Family usage	Medium

Based on our time-of-day traffic pattern assumptions and for ease of comparison between different virtual paths, we have plotted graphs of 'median value of transfer rate' against 'day of the week'. These graphs represent a median throughput during a typical week for each Application Server.

For calculating the median throughput curves, we assume backbone characteristics will remain approximately constant for the whole experimental period. While this assumption is adequate for most of the servers, this was not a valid case for the US West Coast server. We observed a major path change for this Application Server on Nov 11 1999 and discuss this Section 7.1.3.

Deviations from the expected time-of-day patterns (See Table 3) are found in virtual paths where influence of regional networks, backbones and server traffic demand is substantial. For ease of understanding, in some cases (USA West Coast & Germany) we have plotted graphs of median value throughput against hours for a typical day of the week. This approach allows a better analysis of throughput behaviour for a typical day of the week. Moreover, by matching throughput information with path analysis, it is possible to infer the causes of a non-standard time-of-day pattern (ie, a pattern that does not follow Table 3).

7.1.1. South Australia

Fig. 5 shows the throughput from the South Australian TUCOWS server follows Table 3. This behaviour suggests a high influence of the file server traffic demand and low influence of the backbone and regional networks with respect to throughput (See Equation 3).

We observe larger variations in throughput rates across divisions of the day on weekdays compared to the weekend. The maximum throughput for the file server is around 1Mbps.

As expected on the weekend, the Business and Family times exhibit similar behaviour in terms of throughput because there is no business traffic demand on the server.

Figure 5 - Median Throughput in South Australia

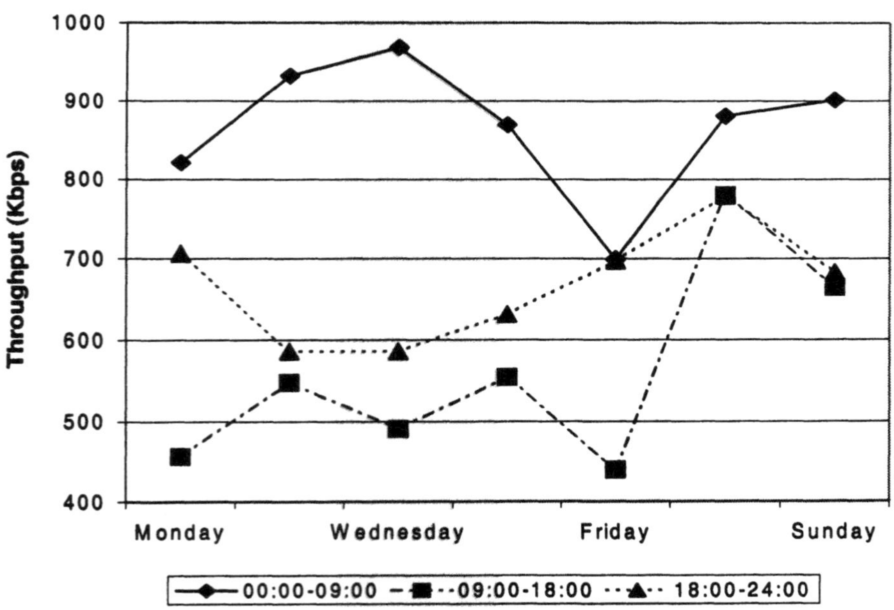

7.1.2. Victoria, Australia

In terms of Internet usage profile we find this site also conforms to our assumptions stated in Table 3, that is, there is a high influence of the Application Server with respect to throughput (See Equation 3).

Figure 6 - Median Throughput in Victoria
Period 03/11/99 - 07/11/99

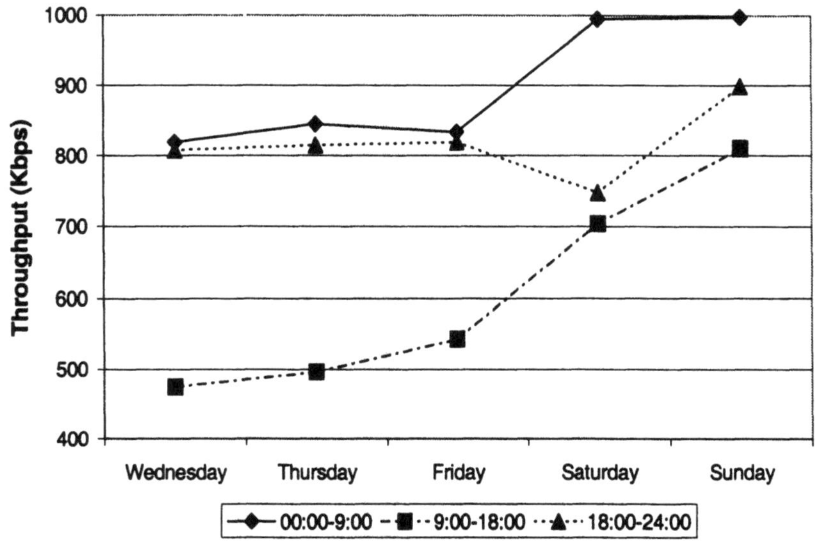

Unlike most other sites, two median values of throughput calculation were appropriate, each equating to separate periods of the experiment (Fig. 6 & 7). We experienced a drop in throughput performance between these periods and suspected this could be result of bottlenecks in the backbone, regional networks or Application Server (See Fig. 1: C, D & F). In fact the path analysis did not show any changes to the Application Server at the time the throughput dropped and suggested a bottleneck in the Application Server. We found this to be the case. Surprisingly the bottleneck was economically imposed. The ISP hosting the Application Server experienced a change of charging structure for bandwidth from the backbone service provider[1]. In an attempt to make its Application Server less attractive in terms of speed for Internet users outside its network, the ISP administrator limited the maximum speed for a download session to 256Kbps.

[1] The ISP network administrator informed us the major backbone provider in Australia, *Telstra,* was charging AU$ 80.00 for every 1GB uploaded from its Application Server which the ISP was not willing to pay.

Figure 7 - Median Throughput in Victoria
(Server after Re-Configuration)
Period 08/11/99 - 22/11/99

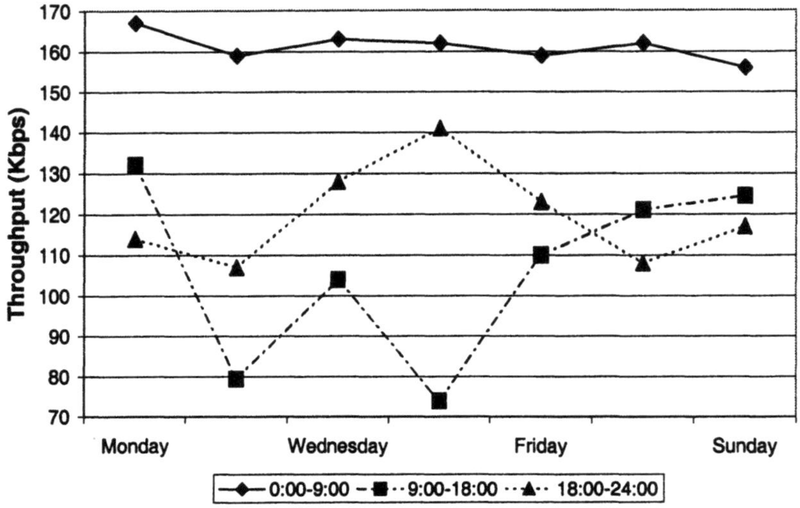

7.1.3. USA West Coast

The median download throughput from this Application Server remained approximately constant throughout the period of the experiment. The throughput does not follow Table 3, suggesting throughput is also influenced by traffic demand within the backbone and regional networks (See Equation 2).

We observed three distinct path configurations persisting during different periods of time for this Client-Server connection. Based on this observation, we decided to trace a typical weekday median throughput curve for each path configuration period.

Figure 8 - Median Throughput for a Typical Weekday
USA WEST COAST Period: 03/11 - 10/11

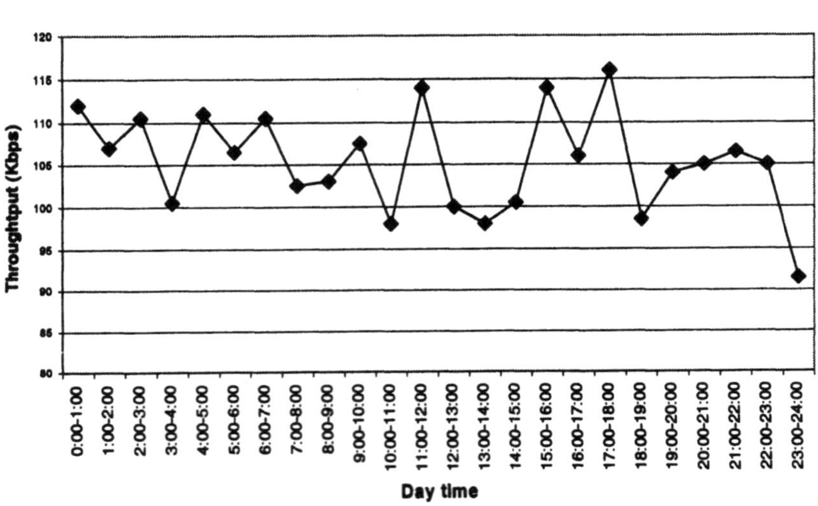

Paths 1 & 3 displayed time dependent route characteristics across the North American continent to reach a given Application Server. From Nov 3 to Nov 11 1999 Path 1 looped between California and New York. By contrast, Path 3 did not loop and was direct to the Application Server. The second path only persisted for two days thus there were not enough throughput samples for plotting a curve. As a result, we have plotted typical weekday throughput curves for the first and last typical paths (Figures 8 & 9).

By observing the throughput behaviour for these typical paths, we find while median throughput mainly varies between 100 Kbps and 115 Kbps for Path 1, throughput variation for Path 3 fluctuates from 110 Kbps and 120 Kbps.

Figure 9 - Median Throughput for a Typical Weekday
USA WEST COAST - Period: 15/11 - 22/11

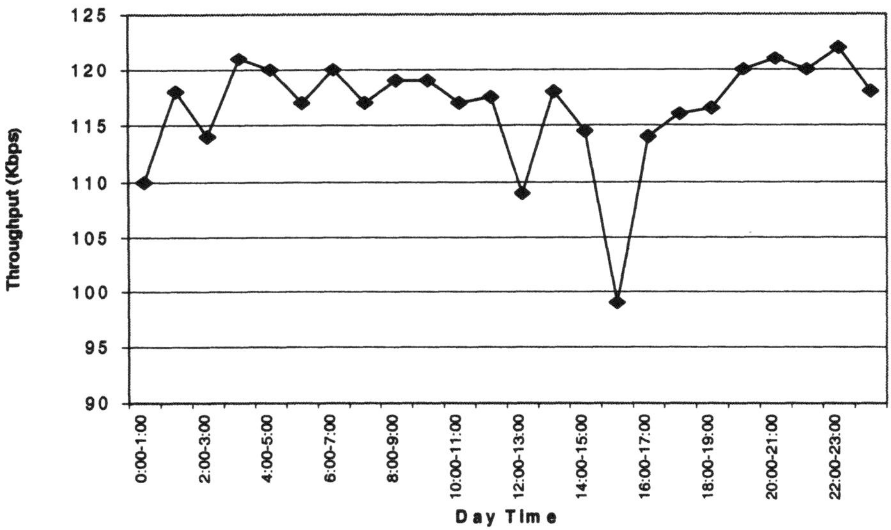

If we assume there are no significant path asymmetries for this server and that traffic demand on Regional Networks, Australian backbone and Application Server has not changed considerably during this experimental period, we can argue this variation is due to packets looping back within the US for Path 1. The drop in throughput performance due to this looping topology might be explained by at least one of the following reasons:

- Increase in RTT what results in longer delays for the receiver to acknowledge data received;
- Higher probability of packets crossing a congested or low bandwidth network.

7.1.4. USA East Coast

This site exhibited a performance characteristic very similar to the US West Coast typical path. Fig 10 does not follow Table 3's scheme completely. In spite of the off-peak time having a median throughput value higher than that during the Family Time, the Business Time throughput sometimes approaches that of Family

Time. This situation leads us to conclude there are contributions from the Application Server, Backbone and Regional Networks to the client throughput.

The median value of throughput is estimated at 100Kbps, which is similar to measurements from the US West Coast where packets were looping back (See Figure 8). We did not observe significant changes in throughput from this value in spite of the Application Server having two different persistent path configurations.

Figure 10 - Median Throughput in the USA East Coast

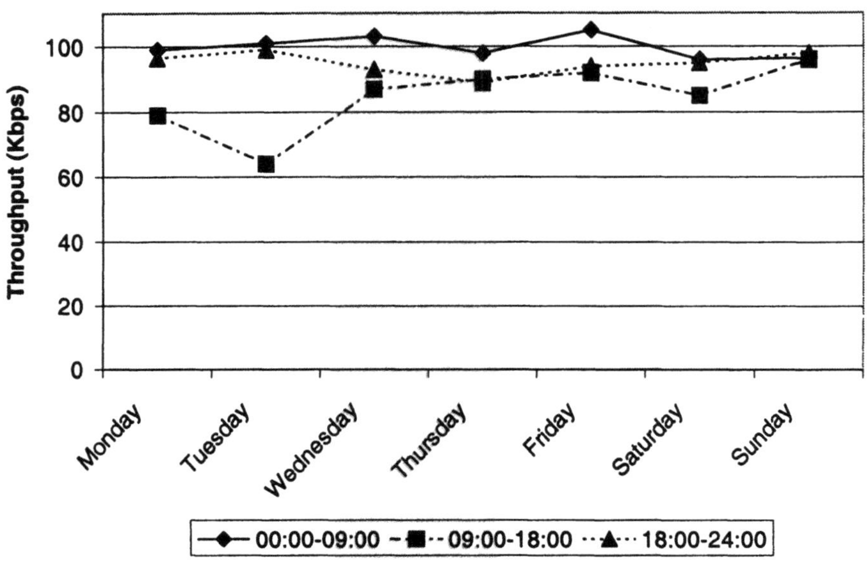

7.1.5. Hong Kong

As with other sites within Client 1 we expected to collect throughput data from the experiment for a period of 20 days (See Figure 2). However, we experienced a drop in throughput performance after the third day, which caused us to limit the data collection to 18 days. To attempt to explain the drop in performance we followed the same investigation for bottlenecks as we did for the Victorian file server (See Section 7.1.2).

While it was difficult to determine a cause for this throughput decrease, a discussion with the backbone service provider highlighted the importance of peering arrangements that needs to be considered during any discussion of such throughput performance. From this we can say the upstream path is likely to be different from the downstream path for this Application Server unlike other cases considered.

Due to service level requirements between AARNET[1] and C&W Optus, traffic to Asia-Pacific destinations is routed via direct links to Asian peers *but* downloaded traffic to Australia is via U.S. peers. Under this situation we cannot

[1] Australian Academic Research NETwork

say to what extent the backbone influences download throughput. However, Figure 11 does in fact exhibit a profile highly influenced by the Application Server and conforms to our assumptions in Table 3.

Figure 11 - Median Throughput in Hong Kong

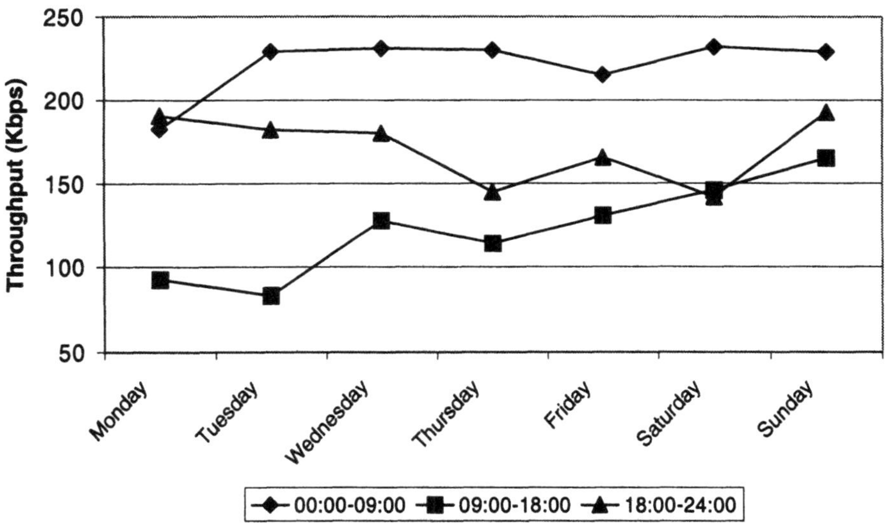

7.1.6. Israel

The Israeli site breaks with our assumptions on Internet usage profiles given in Table 3. This is possibly due to the influence of the US backbone that we have verified by path analysis. Specifically we found the family usage period in the Table 3 gives a transfer rate from the Israeli site lower than that of the business period (See Fig. 12). Normally we would expect the reverse. To emphasise the different usage periods from those expected we reclassify the day into two periods as seen in Fig. 13.

Figure 12 - Median Throughput in Israel
3-Day Periods

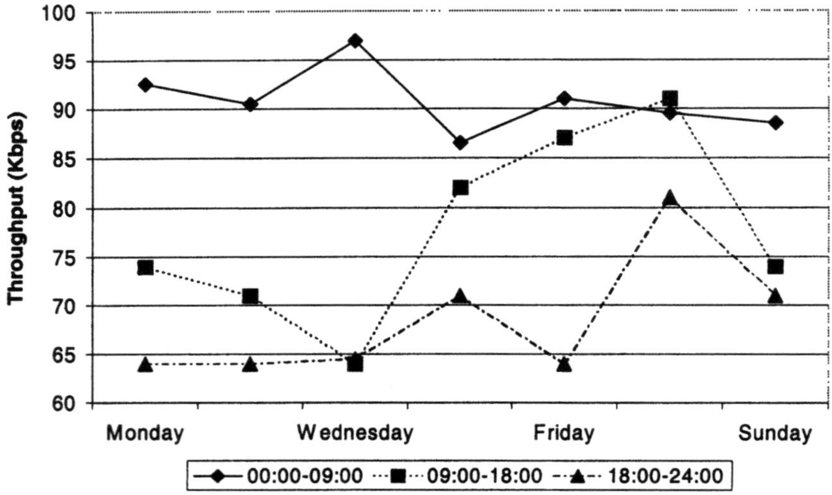

Figure 13 - Median Throughput in Israel
2-Day Periods

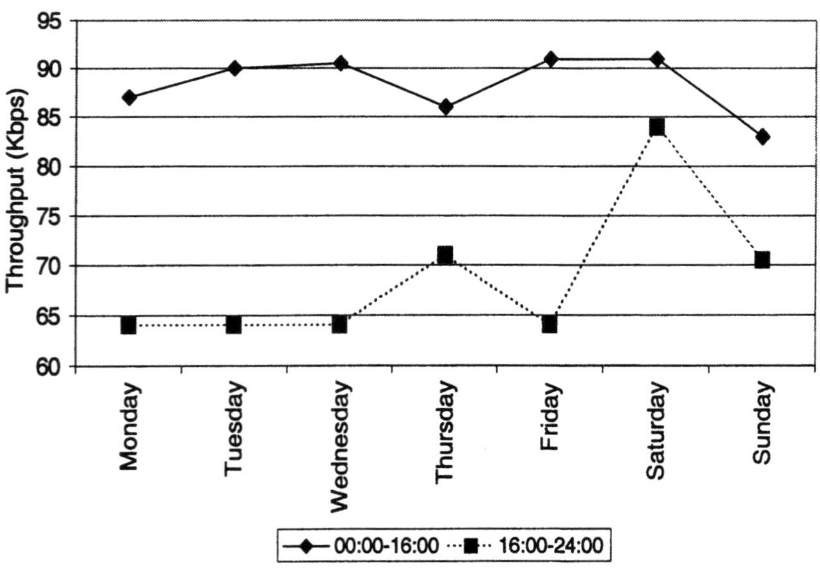

Why did we have a marked departure from our model? By assuming the upstream and downstream paths are the same, we can say local traffic within the

US backbone that requires local access in its own right might be affecting the throughput from the Israeli Application Server to the Client.

We argue that at times when the dominant nodes in the New York region (e.g., NewYork.Teleglobe.net) are at their highest demand locally, this coincides with the Application Server usage family profile in Israel (See Table 3). Therefore during the off-peak time (00:00-16:00), the throughput has a maximum rate, which is the result of low traffic demand in Israel and local demand on the US backbone (specific to the East Coast).

Interestingly, we did not find a higher level of packet loss for the New York regional nodes compared to other regional nodes. A higher packet loss could suggest a higher traffic demand in those nodes. In order to be more definitive in our conclusions, we would require further analysis.

7.1.7. Germany

Similar to the Israeli TUCOWS server, the German server did not follow Table 3. In order to emphasise this, we have chosen to represent a median value of throughput, which corresponds to a typical weekday value (See Fig 14). This graph suggests a high influence of Backbone & Regional Networks on the results of this Application Server.

Figure 14 - Typical Median Throughput weekday in Germany

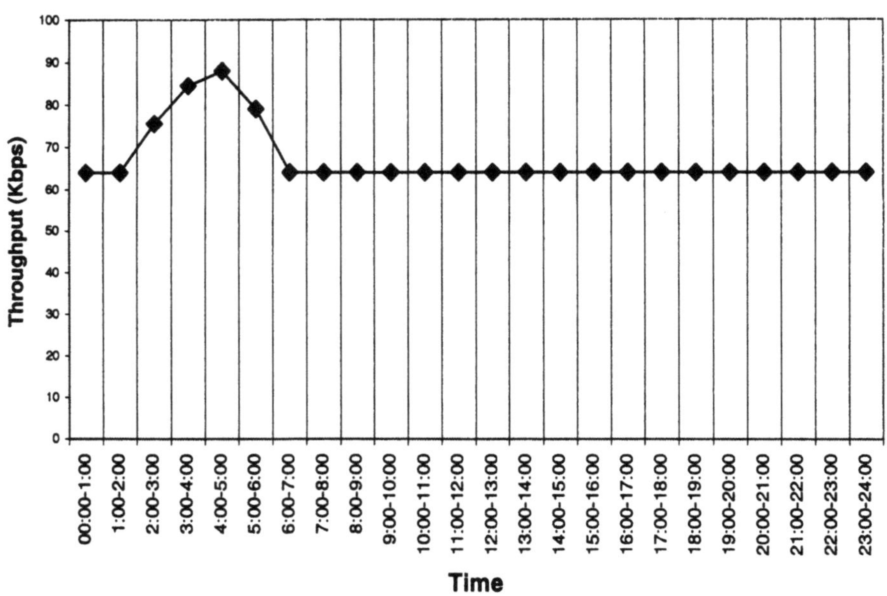

We found 76 percent of all samples for this site equated to a transfer rate of 64Kbps or less, further, these samples lay in the period of Family and Business time. Again these results show the high influence of the US backbone to the extent that the only appreciable rise in throughput rate was observed during the off-peak period as localised to the German Application Server. In order to be more definitive in our conclusions, this site would require further analysis.

7.1.8. England

Fig. 15 suggests that backbone and a regional network influence this Application Server throughput and does not follow Table 3. While it is hard to characterise the throughput behaviour for this server, we would require more samples for a better analysis.

Figure 15 - Median Throughput in England

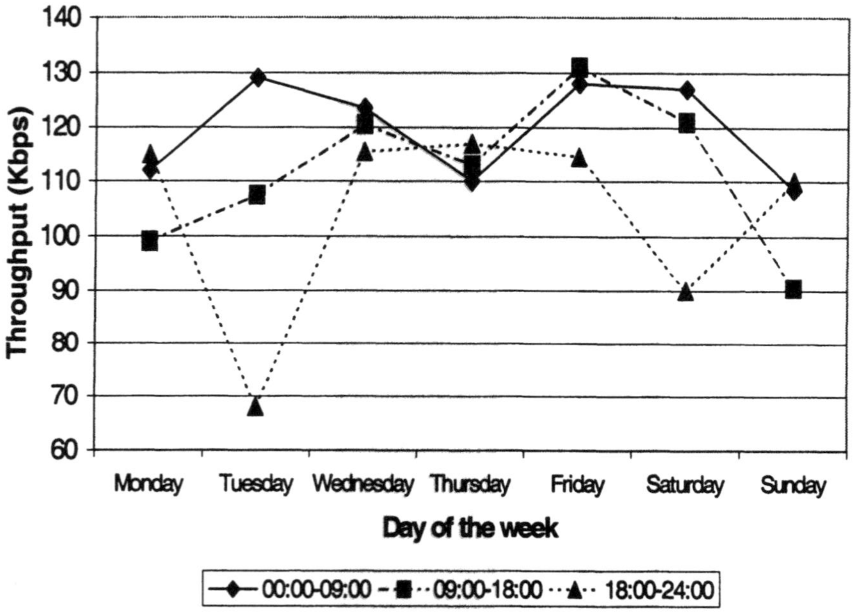

7.1.9. Argentina, Brazil, South Africa & Zimbabwe

We have observed a great many throughput samples for these Application Servers having a performance below 64 Kbps. These servers also had a higher number of "Timed Out" & "No server response" messages. However, it is important to note that this low performance behaviour is different from the Germany server case. While the German server has a daily period with higher throughput (See Fig. 14), these servers have low performance sessions distributed equally over all day periods.

Because paths to most of the servers[1] have similar routes crossing the Australian and the American backbones, we argue that poor performance for Argentina, Brazil, South Africa and Zimbabwe are due to remote regional network/backbone limitations. Moreover, as it is suggested in new research being undertaken, typical paths for these Application Servers have more satellite links than that of other servers' paths. Because some TCP congestion control mechanisms have poor performance over satellite links [18], we argue the low performance from this Application Servers is the result of additional satellite-links.

In the case of Germany, we believe there might be a remote regional network/backbone limitation, however during off-peak periods in Germany the throughput increases significantly (See Fig. 14).

7.2. Path Instability

As described in Section 6.2, we have analysed all paths in terms of our path instability model (See Section 4.2). While the software for path analysis proves to be a useful tool for finding changes in node configuration within a virtual path, we find that in some cases a higher value of node changes has been mistakenly associated with an Application Server path. In some cases, sequential samples contain different IP addresses for a particular node number, however these are *pseudo-node* changes. We describe such pseudo-node changes as follows:

- *Variations in node IP number having similar DNS information*: We observe a number of samples where the IP address of a certain node changes in sequential samples *but* the DNS properties show the same network characteristics/geographical locations. The COMPARE utility does not consider changes where two IP Class C nodes in sequential samples have common network IDs but different hosts IDs. Since COMPARE does not analyse DNS information, this results in *pseudo-node* changes for other IP address cases (See Section 6.2, *Utility 5*).
- *Cyclical Changes of IP addresses without DNS information:* We find a number of samples where the IP address of a certain node changes in a cyclical fashion (i.e., Load Balancing). Variations in Class C format IP addresses with a common network ID are not considered by COMPARE utility as a change in path (See *Utility 5*). However, our software does treat Class A & B IP address cases as path changes.

Path re-configuration complexity in the manner described above forced us to "filter" the results obtained from the COMPARE utility to achieve a fairest path instability (ε) calculation.

In addition, for the filtering process we have not considered changes in path where the *varying* node reports loss of reachability. Nodes returning "Host Unreachable" or Net Unreachable" messages are thus not adequate for analysing changes in path. Administrators of such nodes might be blocking ICMP echo-requests due to traffic policies or valid network problems might be occurring.

The *filtering* process is suitable for analysing data where there are a small number of pseudo-node changes to a Server. In addition, subjective assessment

[1] Except South Australia, Hong Kong & Victoria (Australia)

needs to be made in terms of whether a node change is a pseudo-node change or a valid change. In the next Section we provide an analysis of the "filtered" results covering the period between Nov 3 and Nov 22, 1999. For ease reading, we introduce a general discussion of the main findings and we summarise the results in Table 4.

7.2.1. Discussion

The first important finding in our study is the high stability of all Client-Application Server connections as shown in Table 4. For all Application server paths, path changes affect less than 1% percent of total samples measured. This result confirms research carried out by Paxson [13], Labovitz et al [19] & Chinoy [20].

In addition we find the number of changes in a path is not relevant for previewing throughput performance. An Application Server might have lower throughput but higher number of path changes (or higher path instability) than other servers (eg, See Table 4: Application Server in Brazil).

The South Australian server path has the lowest Path Instability (ε). While we find the COMPARE utility results in a reasonably high number of changes for this site, the path re-configurations do not seem to be result of congestion *but* of a higher level of capillarity within the Regional Backbone. We observe that 20 out of 24 changes for the South Australian path was a result of packets being forwarded to a stand-by node. We did not observe DNS similarities between this IP address and the default node IP address. However, we have found from the Throughput analysis there is a low influence of the Regional backbone/network on our results (See Section 7.1.1). Therefore, we do not consider these changes for the Path Instability (ε) calculation.

For the USA East Coast server, we find a great many changes (260 occurrences) happen in node 7. While the COMPARE utility classifies these as changes, the analysis of the path configuration for this Application Server shows these are pseudo-node changes due to a load balance.

The Server in Israel had 88 changes. The analysis shows that a number of changes are due to nodes responding to "Net/Host Unreachable" messages (26). The other changes are again pseudo-node changes due to variations in node IP number having similar DNS information.

The Server in England had a high number of changes. While analysing the output, we observed this Application Server path had a high number of Net/Host unreachable occurrences within varying nodes. We find that 38 out a total of 73 path changes happened in nodes reporting loss of reachability.

The Server in Argentina had 240 changes. Similarly with the USA East Coast Server, we find a great many changes (151) are due to a load balance implementation in node 15 between Nov 3 and Nov 15 1999. Moreover, some changes are due to variations in node IP number having similar DNS information (41) and others are not clear due to lack of DNS information (13). The rest of the path changes are due to nodes responding to "Net/Host Unreachable" messages and therefore are not considered for the Path Instability (ε) calculation.

The South African Server had 6 pseudo-changes as result of a load balance implementation and 3 changes as a result of variations in node IP number having similar DNS information. The rest of the path changes were not considered for the path instability (ε) calculation have varying nodes responding to "Net/Host Unreachable" messages.

The Servers in the US West Coast, Germany, Brazil, and Zimbabwe had all pseudo-node changes due to variations in node IP number having similar DNS information.

Table 4: Summary of the Path Instability Analysis

Server	Aver. # Nodes	COMPARE Output	Changes after Filtering	$\left(\frac{\text{Filt. Changes}}{\text{Samples}}\right)\%$	Path Instab. (ε)
S. Austral.	7.1	24	4	0.07	0.56
Victoria	6.0	8	8	0.14	1.33
USA WC	18.1	30	24	0.42	1.32
USA EC	16.1	292	27	0.47	1.68
H. Kong	8.0	9	9	0.15	1.12
Israel	17.2	88	48	0.84	2.79
Germany	20.1	41	38	0.66	1.89
England	15.1	73	25	0.44	1.65
Argentina	16.2	240	16	0.28	0.99
Brazil	17.1	35	12	0.21	0.70
South Afr.	19.1	34	19	0.33	0.99
Zimbabwe	17.5	61	40	0.70	2.28

8. Conclusions

Our paper describes a new methodology to assess end-to-end performance analysis of broadband access networks and Internet backbones allowing the user to generate a picture of traffic profiles to chosen Application Servers throughout the world.

While we feel the method holds for the given analysis of path and throughput, more definitive conclusions in terms of these parameters could be further obtained if more samples are collected. For improvement in throughput analysis the SBF could be smaller and/or the Client could run in a dedicated mode for a particular Application Server. In the case of path analysis, the sample collection interval could be decreased.

In the general case, we assumed that upstream and downstream paths are likely the same. Nevertheless, in particular cases such as Hong Kong, the paths are different because of peering agreements between ISPs and their backbone providers. In this case throughput can also be affected by different time zones (ie, USA West Coast) causing inaccurate conclusions. To solve this problem and by way of further work, the path analyser software could be run at the target Application Server instead of running at the Client Machine. This would provide a better estimate of the throughput path in the downstream direction.

The throughput depends not only on technical but also on economic drivers (See the Victorian file server example). The inter-network pricing regime and the way this impacts on available bandwidth are important aspects that were observed in this experiment but are not analysed here.

For other factors that impact on total throughput, we recommend reading [21]. For example, TCP protocol configurations running on different Operating Systems have significant influence on TCP total throughput [22]. We did not investigate this topic in our research.

In terms of path instability analysis, we introduce a parameter (ε) for measuring path instability within a virtual path. Since this parameter can be used for comparing instability of distinct virtual paths, it can be considered a "QoS" analysis parameter. In this study, we were not interested in finding an optimum value for (ε) but in comparing the instability behaviour of the Client-Application Server paths. Further work should be undertaken with a larger number of WWW sites during a longer experimental period for obtaining the optimum path instability value.

The analysis of the throughput and path parameters result in an "Internet weather forecast". These ideas could possibly be used for monitoring or previewing the performance of an internetworking environment.

The experiment suggests current Internet networks and heavily loaded Web File Servers provide a range of transfer rates up to 1 Mbps. Thus, new broadband access technologies such as Asymmetric Digital Subscriber Line (ADSL), Cable Modems and Local Multipoint Distribution Service (LMDS) cannot provide downstream throughput to their full capability in the current Internet environment.

Finally, with the exception of download sessions from the Australian Application Servers, all download sessions were US centric and thus exhibit a high dependence on this backbone.

Acknowledgments

The authors are very grateful to Mr. David Klemitz for his contributions in discussions and in the development of the paper.

References

[1] T. C. Kowk, "Residential Broadband Internet Services and Applications Requirements", IEEE Communications magazine, Vol. 35, No. 6, June 1997.

[2] V. Jacobson, "Congestion Avoidance and Control," in Proceedings ACM SIG-COMM'88, ACM, August 1988, pages 314-329.

[3] B. S. Arnaud, "CANARIE," presented at "The AARNET Advanced Internet workshop", Adelaide, Australia, 10th & 11th March 2000.

[4] J. W. Gurley, "Can Napster be stopped? No! ", April 17, 2000, http://www.news.com/Perspectives/Column/0,176,419,00.html

[5] K. Reichard, "Napster on Linux: From a Whisper to a Scream", March 20, 2000, http://www.linuxplanet.com/linuxplanet/reviews/1617/1/

[6] "Napster Changes its Tune", Wall Street Journal, March 23, 2000 http://interactive.wsj.com/articles/SB953765627187819503.htm

[7] C. Oakes, "Napster not at Home with Cable", April 7, 2000, http://www.wired.com/news/print/0%2C1294%2C35523%2C00.html
[8] "Advanced Networking Infrastructure Needs in the Atmospheric and Related Sciences (ANINARS) report", final draft, July 21, 1999, http://www.scd.ucar.edu/nets/projects/NETSprojectplans/1999.complete.projects/nlanr/final.report.doc
[9] D. Newman & R. Mandeville, "Corporate-Class Internet? Don't Count On it!," http://www.data.com/issue/981107/isp.html
[10] S. Kalidindi & M. Zekauskas, "Surveyor: An infrastructure for Internet performance measurements", presented at INET'99, San Jose, June 1999.
[11] K. Thompson, G. J. Miller & R. Wilder, "Wide-area Internet traffic patterns and characteristics" , IEEE Network magazine, November/December 1997.
[12] W. Richard Stevens, "TCP/IP Illustrated", vol. 1, Addison Wesley, edition 12th, September, 1998.
[13] V. Paxson, "End-to-End Routing behaviour in the Internet", IEEE/ACM transactions on Networking, Vol. 5, No. 5, pp. 601-615, Oct. 1997.
[14] "Australia's international bandwidth", Telecommunication Journal of Australia, vol. 49, no. 1, 1999.
[15] W. Mattews & L. Cottrell, "The Pinger Project: Active Internet Performance Monitoring for the HENP Community", IEEE Communications magazine, Vol. 38, No. 3, May 2000.
[16] V. Paxson, J. Mahdavi, A. Adams & M. Mathis, "An Architecture for Large-Scale Internet Measurement", IEEE Communications magazine, Vol. 36, No. 8, Aug. 1998.
[17] B. Huffaker, M. Fomenkov, D. Moore & E. Nemeth, "Measurements of the Internet topology in the Asia-Pacific Region," in Proceedings Inet 2000, http://www.caida.org/outreach/papers/asia_paper/
[18] Allman, M., Glover, D. and L. Sanchez, "Enhancing TCP Over Satellite Channels using Standard Mechanisms", BCP 28, RFC 2488, January 1999.
[19] C. Labovitz, G.R. Malan & F. Jahanian , "Internet Routing Instability," in Proceedings of ACM SIGCOMM' 97
[20] B. Chinoy, "Dynamics of Routing Information," in Proceedings SIGCOMM' 93, pp. 45-52, September 1993.
[21] "Network analysis times", Vol. 1, Jan 2000, http://moat.nlanr.net/NATimes
[22] "Enabling high performance data transfers on hosts: notes for user and system administrators", http://www.psc.edu/networking/perf_tune.html

Part III

Internet II

Analysis of World Wide Web Traffic by Nonparametric Estimation Techniques *

Udo R. Krieger[1], Natalia M. Markovitch[2], and Norbert Vicari[3]

[1] T-Nova Deutsche Telekom, Technologiezentrum, Am Kavalleriesand 3, D-64295 Darmstadt, and Computer Science Department, J. W. Goethe-University, D-60054 Frankfurt, Germany, E-mail: udo.krieger@ieee.org

[2] Institute of Control Sciences, Russian Academy of Sciences, Profsoyuznay 65, 117806 Moscow, Russia, E-mail: markovic@ipu.rssi.ru

[3] Lehrstuhl Informatik III, University of Würzburg, Am Hubland, D-97074 Würzburg, Germany, E-mail: vicari@informatik.uni-wuerzburg.de

Abstract. The study of measurements of world wide web traffic has shown that different characteristics are governed by long-tail distributed random variables. We discuss the nonparametric estimation of their corresponding probability density functions. Two nonparametric estimates, a Parzen-Rosenblatt kernel estimate and a histogram with variable bin width called polygram, are considered. The proposed estimates are applied to analyze data of real web sessions. The latter are characterized by the sizes and durations of sub-sessions as well as the sizes of the responses and inter-response time intervals. By these means the effectiveness of the nonparametric procedures in comparison to parametric models of web-traffic characteristics is demonstrated.

Keywords: World Wide Web traffic; Nonparametric density estimation; Parzen-Rosenblatt estimate; Polygram.

1 Introduction

Traffic measurements and reliable off-line as well as on-line estimation techniques are required to determine accurately the traffic load and resource usage in current high-speed packet-switched ATM and IP networks with their different service classes. For this purpose, new operational data collection procedures have to be developed and implemented in the network elements that use both measurements scheduled at a regular time basis and special high-resolution measurements triggered by network management actions (cf. [1,8,13,14]). Normally, measurement facilities count events of interest, e.g. incoming or outgoing calls, sessions, frames,

* The corresponding research efforts were performed within the framework of the EU activity COST 257 "Impacts of new services on the architecture and performance of broadband networks" and supported by the European Commission. See http://nero.informatik.uni-wuerzburg.de/cost/Final.

packets or cells, as well as durations of relevant periods, e.g. interarrival times between events, and sizes of corresponding objects, e.g. file lengths, at a specific network element in consecutive time intervals of fixed length.
If we consider particularly the rapid growth of the Internet traffic due to an enormous increase of world wide web applications over the last few years, the effective design of the underlying IP-based transport infrastructure has become an important task of teletraffic engineering. To perform it, the characterization of web traffic by its basic ingredients such as session lengths, page sizes and response times of requests is required. It is based on efficient data gathering and a careful analysis of the underlying random processes and the corresponding random variables (r.v.) associated with these characteristics. For planning and control purposes, it is very important to reconstruct the resulting traffic load accurately and by statistically thorough techniques.

The analysis of existing measurements of web traffic has shown that its characteristic r.v.s are often governd by long-tailed distributions or even follow mixtures of long-tailed distributions due to the heterogeneous sources of the information transfer (see [1,4,12–14,18]). However, from a statistical perspective, mainly parametric modeling techniques including the parameter estimation of distributions or the parametric modeling of the tails of distributions by maximum likelihood and Bayesian techniques have been applied. The resources of nonparametric modeling and estimation techniques have not been exploited up to now. They include effective density estimation techniques based on generalized regression schemes as well as kernel and series estimation procedures (cf. [6,16] - see Fig. 1). In this paper, we point out the difficulties of parametric modeling in the extremely dynamic environment of web applications on the Internet and we present a purely nonparametric approach for traffic characterization. For this purpose, we use a simplified hierarchical model of web sessions and compare our nonparametric estimation techniques with a parametric approach based on traces of web traffic gathered at the University of Würzburg.

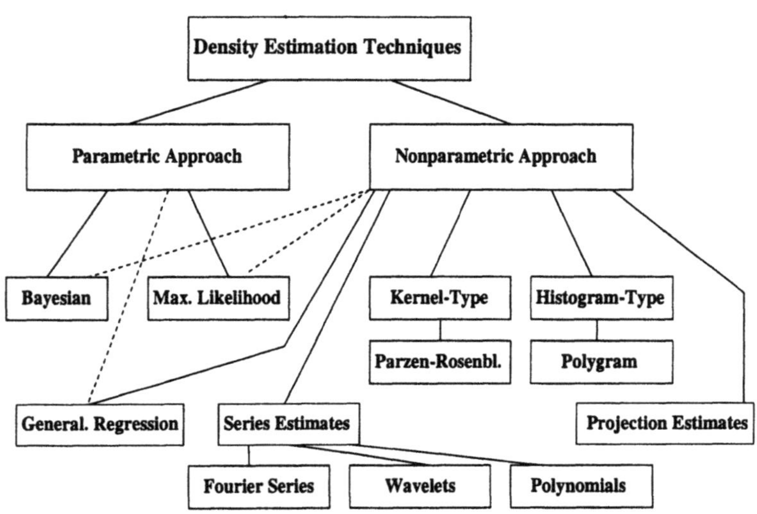

Fig. 1. Density estimation techniques

The paper is organized as follows. In Section 2 the collected traffic measurements, their evaluation methodology and an analysis by parametric models are discussed. In Section 3 we present our nonparametric modeling approach as feasible alternative and estimate the density functions of the corresponding web-traffic characteristics based on real data. Finally, we conclude with a summary of our findings.

2 Measurement and Analysis of Web Traffic

The description of web traffic for analysis, simulation and network design purposes requires simple, universal models capturing different levels of abstraction and different time scales. Here, we first point out the difficulties of a parametric modeling approach. In the next section, we propose a nonparametric estimation technique and illustrate its effectiveness based on data of measured web traffic.

2.1 Data Collection and Evaluation Methodology

In [17] a simplified hierarchical model of web traffic has been derived from measured IP traffic arising from world wide web applications. Responses to web requests are identified as the main part of the transferred data and the time between these responses is used to model the relationship between the responses. The measured

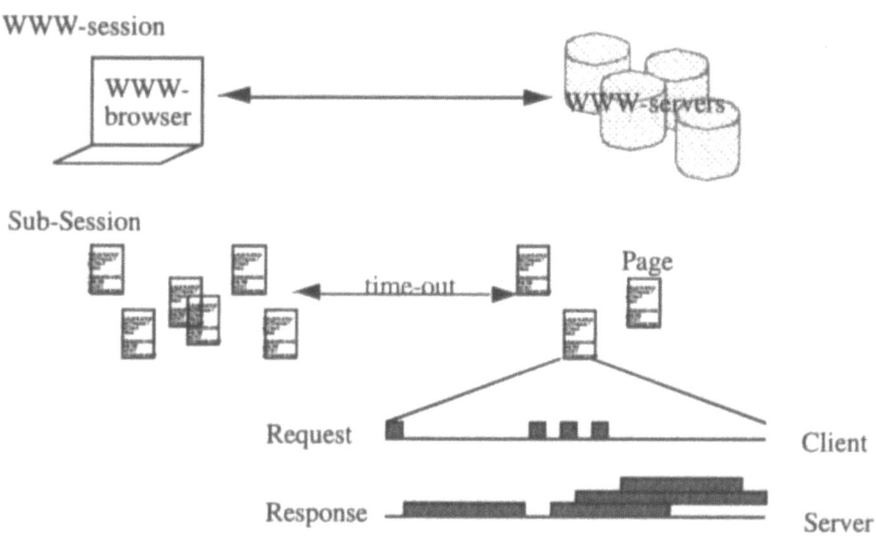

Fig. 2. Hierarchical modeling of web sessions

traffic is described by this hierarchical model distinguishing a session and a page level where the first one is characterized by sub-sessions (see Fig. 2 - see also [14,18]). The levels are modelled by four related random variables, i.e. the characteristics of a sub-session are its size in bytes and its duration in seconds and the characteristics of the transferred web pages are the size of the responses in bytes and the inter-response times in seconds (see Table 1).

Table 1. Characteristics of web sessions

Level	Characteristic	Definition
sub-session	duration	time between beginning and termination of browsing a series of web pages
	size	data volume of visited web pages
page	inter-response time	time between beginning of the old and of the new transfer of pages within a sub-session
	page size	total amount of transferred data (HTML, images, sound, ...)

The data collection has been based on a two weeks measurement of web traffic in an Ethernet segment of the Department of Computer Science at the University of Würzburg. About 20 workstations including 1 file-server and 2 web servers were connected to this segment. A packet trace was captured with the tool TCPDump and analyzed to identify the components of web sessions as depicted in Fig. 2. A timeout mechanism was used to discern sub-sessions (with a timeout of 15 minutes) and web pages.

To model the web sessions in the sketched manner, the statistical identification of the characteristic r.v.s is required. Applying the described heuristics for the data evaluation, the 940 MByte trace shown in Fig. 3 has been analyzed and 373 parallel client sub-sessions were detected (see Fig. 4). Approximately half of the traffic originated from external requests to the web servers within the department.

Considering first the evaluation at the session level, the average sub-session has a size of 1.28 MByte and the coefficient of variation is 3.2. The mean sub-session duration is 29 minutes with a coefficient of variation of 3.0. All measured sessions caused the transmission of 480 MByte of data. About 10% of the traffic (the requests) was directed from the clients to web servers while the main part of the traffic was caused by responses on requests. Therefore, we concentrate our further investigation on the characteristics of this response traffic which generates the major volume.

In [11] the mathematical modeling of the session arrivals by a nonhomogeneous Poisson process and the restoration of its intensity function by a new nonparametric estimation approach are discussed. Based on a characteristic Volterra integral equation the estimation task is formulated as stochastically ill-posed problem and Tikhonov's regularization method is applied as basic solution technique (cf. [16]).

Regarding the characterization of web pages the latter were detected by a timeout mechanism with a detection interval of 3 seconds. The size of a response is defined as the sum of the sizes of all packets which are down-loaded from a web server to the client upon a request. On average one response contains four separate files - the actual web page and further inline objects - and 19.6 web pages are loaded in one sub-session.

The average response size is 54 kByte with a coefficient of variation of 9.1. In the evaluation times between subsequent sub-sessions were not taken into account. The

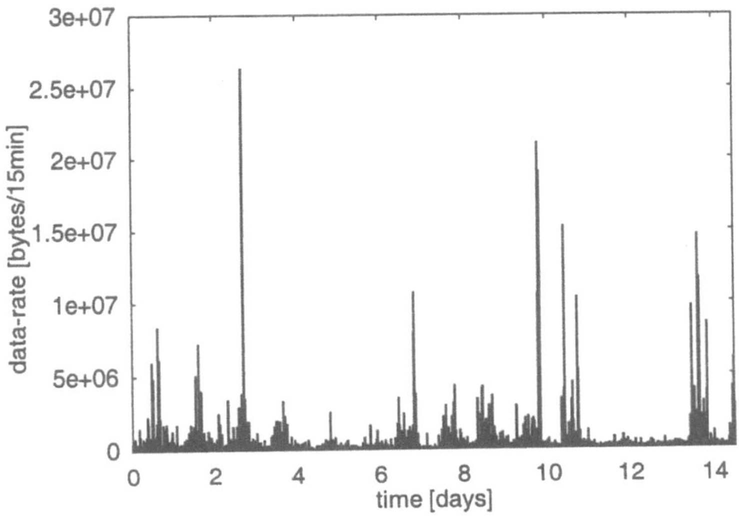

Fig. 3. Measured traffic volume

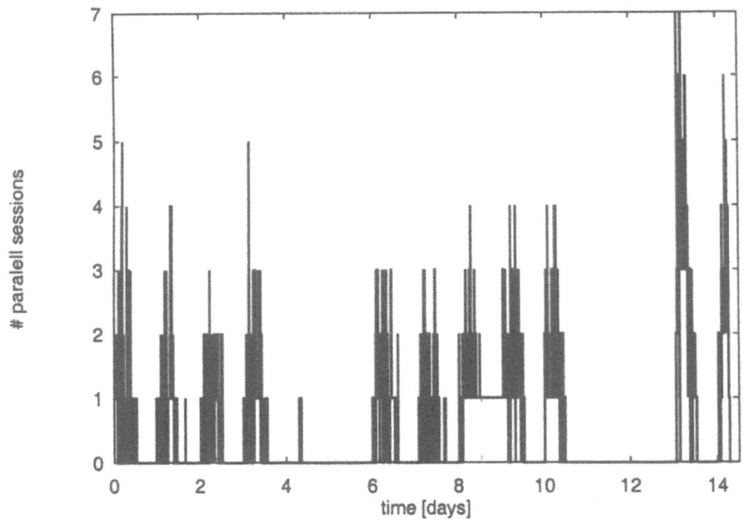

Fig. 4. Detected parallel sessions

mean inter-response time is 81 seconds and the coefficient of variation of the inter-response times is 9.0.

The scatter plot in Fig. 5 shows the dependence between the time to the next response and the size of the current page where the axes are scaled logarithmically. The area covered by the pairs of inter-response time and current response size is quite large. Obviously, no particular relation between large response sizes and large inter-response times or small response size and small inter-response time can be found. The coefficient of covariance of the samples is 0.04. These properties in-

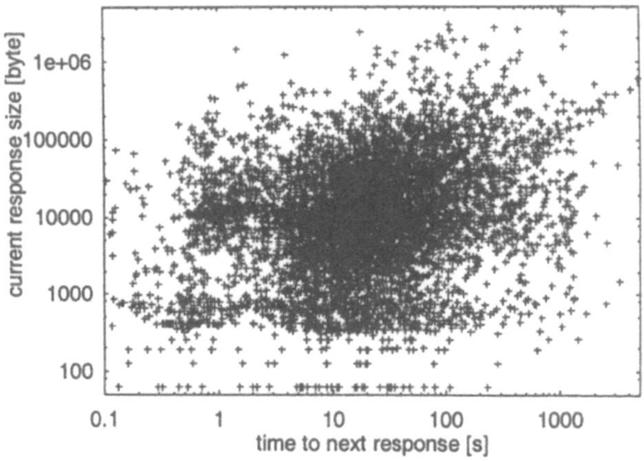

Fig. 5. Dependence of the time to the next response and the current response size

Table 2. The attributes of the data set of the analyzed web traffic

Data		Descriptive Statistics				Sample	Scaling
type	unit	mean	variance	minimum	maximum	size	term
s.s.s.	bytes	$1.283 \cdot 10^6$	$1.664 \cdot 10^{13}$	128	$5.884 \cdot 10^7$	373	10^7
d.s.s.	seconds	$1.728 \cdot 10^3$	$2.71 \cdot 10^7$	2	$9.058 \cdot 10^4$	373	10^3
s.r.	bytes	$3.467 \cdot 10^4$	$2.853 \cdot 10^{10}$	$1 \cdot 10^{-3}$	$4.351 \cdot 10^6$	1000	10^6
i.r.t.	seconds	73.282	$4.184 \cdot 10^4$	0.03	$3.277 \cdot 10^3$	1000	10^3

dicate that the dependence of the response-size and inter-response time could be neglected in the modeling of web sessions.

2.2 Data Analysis by Parametric Models

In the following, a parametric modeling approach is used for a thorough statistical data analysis. The data are described by four r.v.s related to the two basic hierarchical layers of the session and the page level, i.e. the size of a sub-session (s.s.s), the duration of a sub-session (d.s.s.), the size of the response (s.r.) and the inter-response time (i.r.t.). They exhibit a heavy-tail behavior. The data of the page sizes contain the information about 7480 web pages which have been downloaded during 14 days by several TCP/IP connections. To perform the analysis, we have used samples with the reduced sample size $l = 1000$ for the sizes and inter-response times of web requests which have been observed in a shorter period within these two weeks. For simplicity of the calculations, all data were appropriately scaled, i.e. divided by a scaling term.

The attributes of the analyzed data set including the sample size, scaling terms and descriptive statistics, i.e. mean, variance, minimal and maximal values, are depicted in Table 2.

In the following, let $F_l(t) = \frac{1}{l}\sum_{i=1}^{l} \Theta(t - x_i)$, $\Theta(t) = 1, t \geq 0$ and zero otherwise,

denote the empirical distribution function (d.f.).
The statistical analysis of the underlying four r.v.s of these traffic characteristics is similar: we have checked whether an exponential or a Pareto distribution is an appropriate model for the restoration of the related p.d.f.s. First, we tried to fit the data by a thorough statistical technique using exponential and Pareto distributions with the d.f.s $F_{exp}(t,\lambda) = 1 - \exp(-\lambda t)$ for $t > 0$ and $F_p(t,\alpha) = 1 - t_0^{\alpha} t^{-\alpha}$ for $t \geq t_0$, $\alpha > 0$, respectively. We have chosen $t_0 = 10^{-3}$ for each sample. Maximum likelihood estimates were calculated by the formulas

$$\lambda = \left(\frac{1}{l}\sum_{i=1}^{l} x_i\right)^{-1}$$

for the exponential distribution and by

$$\alpha = \left(\frac{1}{l}\sum_{i=1}^{l} \ln(x_i) - \ln(t_0)\right)^{-1}$$

for the Pareto distribution, where $x_1, \ldots, x_l$ denotes the s.s.s., d.s.s., s.r. or i.r.t. sample, respectively. For the s.s.s. sample we obtained $\lambda = 7.795$, $\alpha = 0.305$, for the d.s.s. $\lambda = 0.579$, $\alpha = 0.161$, for the s.r. $\lambda = 28.846$, $\alpha = 0.483$, for i.r.t. $\lambda = 13.646$, $\alpha = 0.344$.
In Figs. 6 - 9 the survival functions $1-F_l(t)$, $1-F_{exp}(t,\lambda)$ and $1-F_p(t,\alpha)$ are shown for s.s.s., d.s.s., s.r. and i.r.t., respectively. The application of the Kolmogorov-Smirnov (K-S) test shows that no sample follows exponential or Pareto distributions despite of the visual similarity of these models. Since the samples contain more than 100 points, the quantiles of the K-S statistic have been estimated by the formula (cf. [2])

$$\tilde{D}_l(Q) = \sqrt{\frac{y}{2l}} - \frac{1}{6l}, \qquad \text{with} \quad y = -\ln(0.005Q),$$

where Q is the confidence level. For $Q = 5$ we get $\tilde{D}_l(Q) = 0.07$ for $l = 373$ and $\tilde{D}_l(Q) = 0.043$ for $l = 1000$. The values of the K-S statistic

$$\frac{D_l}{\sqrt{l}} = \max\{\sup_{0\leq i\leq l-1}(\frac{i+1}{l} - F(x_{(i)})), \sup_{0\leq i\leq l-1}(F(x_{(i)}) - \frac{i}{l})\}$$

calculated by the empirical samples for the exponential and Pareto d.f. $F(x)$ are given by 0.281 and 0.229 for the s.s.s., for the d.s.s. by 0.157 and 0.344, for the s.r. by 0.276 and 0.217, and for the i.r.t. by 0.282 and 0.259, respectively. Since $\frac{D_l}{\sqrt{l}} > \tilde{D}_l(Q)$ holds for all cases, the H_0 hypothesis that the empirical distribution coincides with the selected theoretical one should be rejected.
We conclude that it is difficult to select appropriate parametric models of the r.v.s characterizing the traffic in the extremely dynamic environment determined by world wide web applications.

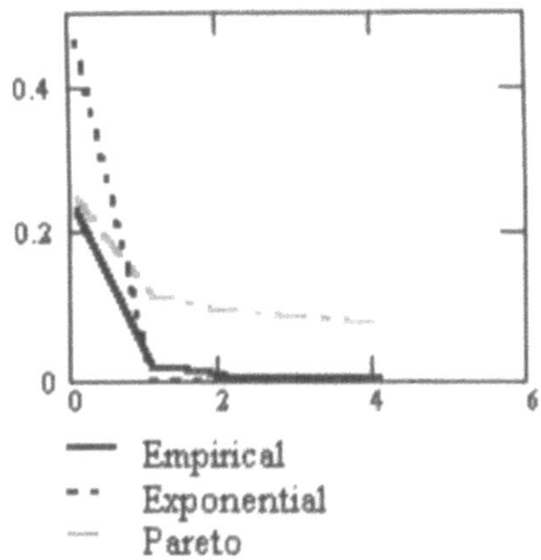

Fig. 6. Estimated survival functions of the size of a sub-session

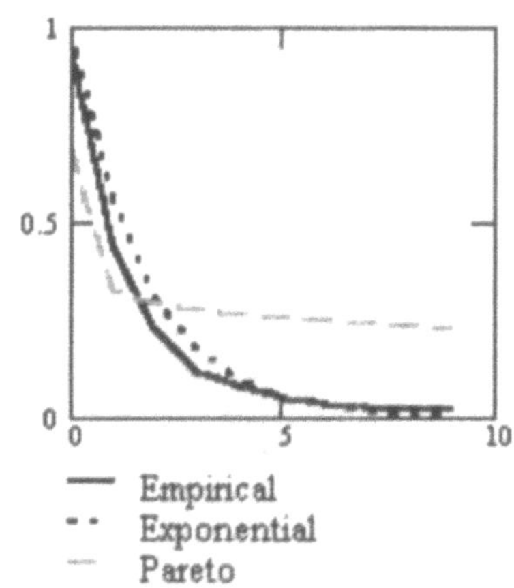

Fig. 7. Estimated survival functions of the duration of a sub-session

3 Nonparametric Estimation of Heavy-Tailed Distributions

The analysis of the existing measurements of web traffic by statistical methods has shown that its characteristics can be described by long-tailed distributions (see [1,4,12,17]). However, so far, mainly parametric approaches have been used to capture this behavior of the basic random variables such as page sizes and inter-response times between requests. Therefore, we present a nonparametric approach as a feasible alternative to estimate the related p.d.f.s.

Standard nonparametric estimates, such as a histogram, projection or Parzen-Rosenblatt (P-R) kernel estimate, cannot describe the behavior of a p.d.f. on the tail due to the lack of information outside the closed interval determined by the

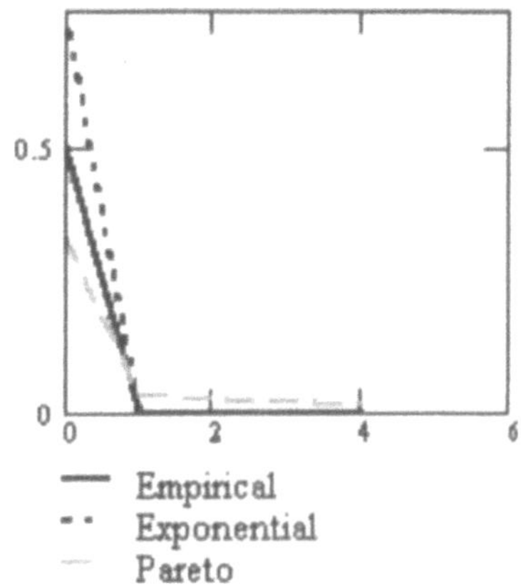

Fig. 8. Estimated survival functions of the size of a response

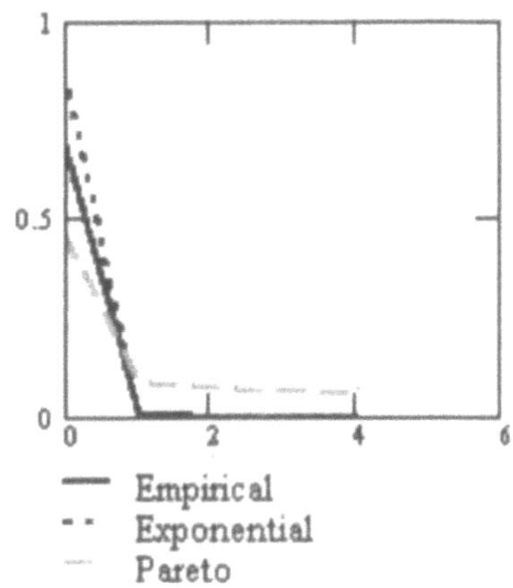

Fig. 9. Estimated survival functions of an inter-response time

range of an empirical sample. They operate just with the empirical samples of limited sizes which are not representative regarding the tails. Hence, parametric tail models and estimates have been developed to describe the tails. Among those Hill's estimate and kernel tail-estimates are very popular (cf. [5,7]). However, parametric estimates need a priori information about the kind of the tail and a large sample size and don't reflect the behavior of the p.d.f. for relatively small values of a r.v.. Nonparametric estimation methods don't require information about the form of the distribution. However, the P-R estimate, which is defined on the whole real axis and may, therefore, be applied to estimate long-tailed p.d.f.s, may be unreliable for heavy-tailed p.d.f.s (see [6,10]). To avoid this lack of reliability of the estimate, we have proposed to use a transformation of an initial r.v. to a new one which has a p.d.f. with a finite support (cf. [10]). The p.d.f. of the latter may be restored by some reliable estimator such as a P-R estimator, a histogram etc. Then the inverse transformation provides a reliable estimator in the sense of the L_1 metric, even in the case of a heavy-tailed p.d.f.. This feature is due to the invariance of the L_1 metric regarding any continuous transformation (cf. [6]). This means that the L_1 estimation error of the p.d.f. with finite support is equivalent to the L_1 error of the estimator of the original p.d.f. with infinite support. Such a nonparametric procedure does not guarantee the accurate restoration of the tail, but it provides the reliable estimation of the original p.d.f. as an entire object in the metric space L_1. Subsequently, we describe this estimation algorithm that is proposed in [10] for the reconstruction of real web-traffic characteristics.

3.1 The Estimation Algorithm

A sample $X^l = (x_1, \dots, x_l)$ of l independent observations of a r.v. with the p.d.f. $f(x)$ and the d.f. $F(x)$ is observed, e.g. the size of web responses. In [10] two basic nonparametric estimates are considered, a P-R estimate with a Gaussian kernel

$$f_{h,l}(t) = \frac{1}{lh\sqrt{2\pi}} \sum_{i=1}^{l} \exp\left(-\frac{1}{2}\left(\frac{t-x_i}{h}\right)^2\right), \tag{1}$$

and a polygram, i.e. a histogram with variable bin width based on statistically equi-probable cells,

$$f_{k,l}(t) = \frac{k}{(l+1)\lambda(\Delta_{rk})} \tag{2}$$

for $t \in \Delta_{rk}$ (cf. [15]). Here, λ is Lebesgue's measure, $\lambda(\Delta_{rk}) \longrightarrow 0$ and $k = o(l)$, where $x_{(1)}, \dots, x_{(l)}$ is the order statistics of the sample X^l and the number of points inside each interval $\Delta_{1k} = [x_{(1)}, x_{(k)}]$, $\Delta_{2k} = (x_{(k)}, x_{(2k)}]$, $\Delta_{3k} = (x_{(2k)}, x_{(3k)}], \dots$ is less or equal to k. The estimate (2) is rewritten in the form

$$\begin{aligned} f_{k,l}(t) = {} & \frac{k}{l+1} \sum_{r=0}^{[\frac{l}{k}]-1} \frac{\Theta(t - x_{(rk+1)}) \cdot \Theta(x_{(k(r+1))} - t)}{x_{(k(r+1))} - x_{(rk+1)}} \\ & + \frac{l - k[\frac{l}{k}]}{l+1} \cdot \mathcal{I}\left(\left[\frac{l}{k}\right] \neq \frac{l}{k}\right) \cdot \psi(t,k) \end{aligned} \tag{3}$$

where $[r]$ denotes the integer part of $r \in \mathbb{R}$ and

$$\mathcal{I}\left([\tfrac{l}{k}] \neq \tfrac{l}{k}\right) = \begin{cases} 1, & [\frac{l}{k}] \neq \frac{l}{k} \\ 0, & [\frac{l}{k}] = \frac{l}{k} \end{cases}, \qquad \Theta(t) = \begin{cases} 1, & t \geq 0 \\ 0, & t < 0 \end{cases}$$

$$\psi(t,k) = \begin{cases} 1, & l - 1 = [\frac{l}{k}]k \\ \dfrac{\Theta(t - x_{([\frac{l}{k}]k+1)}) \cdot \Theta(x_{(l)} - t)}{x_{(l)} - x_{([\frac{l}{k}]k+1)}}, & l - 1 \neq [\frac{l}{k}]k \,. \end{cases}$$

To provide the L_1-consistency of the estimates, a monotone increasing continuous one-to-one transformation function $T : [0, \infty) \rightarrow [0, 1]$,

$$T(x) = \frac{2}{\pi} \arctan x \ , \qquad T'(x) = \frac{2}{\pi} \frac{1}{1+x^2} \tag{4}$$

is applied to the sample X^l by $y_i = T(x_i)$, $Y^l = (y_1, \ldots, y_l)$. It does not depend on the empirical sample X^l and transforms any r.v. x_1 with positive values to a new one y_1 whose p.d.f. $g(t)$ has a compact support, namely the interval $[0, 1]$.
Then the normalized estimate $\tilde{g}_l(x) = \frac{g_l(x)}{\int_0^1 g_l(u)du}$ of $g(t)$ is used to reconstruct the unknown long-tailed p.d.f. $f(x)$ by:

$$\tilde{f}_l(x) = \tilde{g}_l(T(x))T'(x) \tag{5}$$

If the estimate of the p.d.f. $g(t)$ of this new r.v. y_1 is provided by the P-R estimate, one obtains

$$\tilde{f}_{h,l}(x) = \frac{\sqrt{2}}{lh\pi^{\frac{3}{2}} I_{[0,1]}(h)(1+x^2)} \sum_{i=1}^{l} \exp\left(-\frac{1}{2}\left(\frac{\frac{2}{\pi}\arctan(x) - y_i}{h}\right)^2\right) \tag{6}$$

where $I_{[0,1]}(h) = \frac{1}{l}\sum_{i=1}^{l}\left(\Phi\left(\frac{1-y_i}{h}\right) - \Phi\left(-\frac{y_i}{h}\right)\right)$ is the integral of $g_l(x)$ on $[0, 1]$, and $\Phi(x) = \frac{1}{\sqrt{(2\pi)}}\int_{-\infty}^{x} \exp\left(-\frac{u^2}{2}\right) du$ is the Gaussian d.f.. If the polygram is used, one gets

$$\tilde{f}_{k,l}(x) = \frac{2}{\pi(1+x^2)}\, g_{k,l}\left(\frac{2}{\pi}\arctan x\right) \tag{7}$$

where $g_{k,l}(y)$ is derived from (3).
The appropriate selection of the smoothing parameters, i.e. the "window width" h for the P-R estimate and the number of points k in the equi-probable cells for the polygram, is a very important task to provide a reliable estimate. To fit a p.d.f. better in practice, such smoothing parameters must be adapted to the empirical data. We have considered two variants of the discrepancy method, the ω^2- and the D-method, as tools of such an adaptation (cf. [9]).
A practical implementation of these methods is as follows. Let $y_{(1)} \leq y_{(2)} \leq \ldots \leq y_{(l)}$ be the order statistics of the transformed observations. Then, in the case of the ω^2-method, the parameter h is obtained from the equality

$$l\hat{\omega}_l^2 = \delta_1 = 0.05, \tag{8}$$

where $l\hat{\omega}_l^2 = \sum_{i=1}^{l}\left(G^h(y_{(i)}) - \frac{i-0.5}{l}\right)^2 + \frac{1}{12l}$ is the estimator of the Mises-Smirnov statistic $l\omega_l^2 = \sum_{i=1}^{l}\left(G(y_{(i)}) - \frac{i-0.5}{l}\right)^2 + \frac{1}{12l}$, or, in the case of the D-method, from the equality

$$\hat{D}_l = \delta_2 = 0.5 \tag{9}$$

where $\hat{D}_l = \max(\hat{D}_l^+, \hat{D}_l^-)$, and $\hat{D}_l^+ = \sqrt{l}\max_{1\le i\le l}\left(\frac{i}{l} - G^h(y_{(i)})\right)$, $\hat{D}_l^- = \sqrt{l}\max_{1\le i\le l}\left(G^h(y_{(i)}) - \frac{i-1}{l}\right)$ are the estimates of the Kolmogorov-Smirnov statistics $D_l^+ = \sqrt{l}\max_{1\le i\le l}\left(\frac{i}{l} - G(y_{(i)})\right), D_l^- = \sqrt{l}\max_{1\le i\le l}\left(G(y_{(i)}) - \frac{i-1}{l}\right)$. Here

$$G^h(x) = \frac{1}{I_{[0,1]}(h)}\int_0^x g_{h,l}(t)dt = \frac{1}{lI_{[0,1]}(h)}\sum_{i=1}^{l}\left(\Phi(\frac{x-y_i}{h}) - \Phi(-\frac{y_i}{h})\right)$$

is used for the normalized P-R estimate.
For a polygram we have $\hat{D}_l = \sqrt{l}\left(\frac{k}{l+1} - \frac{1}{l}\right)$, and

$$k = [(\frac{\delta_2}{\sqrt{l}} + \frac{1}{l})(l+1)] \tag{10}$$

provides the solution of (9). Here $[x]$ is the smallest integer larger than x.
In conclusion, the proposed algorithm to estimate a long-tailed p.d.f. $f(x)$ reads as follows:

1. The nonparametric estimate of the transformed sample Y^l located on [0,1] is constructed.
2. An optimal smoothing parameter (h or k) for the used estimate is calculated.
3. An inverse transformation (see (5)) is applied to obtain the estimate of the p.d.f. $f(x)$.

In [10] it is shown by a simulation study that a polygram (7) and the P-R estimate (6) are preferable for the application to real data if the true p.d.f. is not available. If one knows that the p.d.f. is heavy-tailed, then a polygram is recommended as simple nonparametric estimate.

3.2 Statistical Analysis by Nonparametric Methods

To illustrate the power of the nonparametric approach, we have applied the sketched estimation algorithm to the data of the measured web traffic.
In Figs. 10 - 13 the polygram and the P-R estimates are presented for the s.s.s., d.s.s., s.r. and i.r.t. samples, respectively. Each figure depicts two graphs to demonstrate better the behavior on the tails and for small values. All graphs were constructed in the points $x_{(1)}, \ldots, x_{(l)}$.
Both estimates were first applied to the samples transformed by the transformation (4), $\{y_i = \frac{2}{\pi}\arctan(x_i), i = 1, \ldots, l\}$, and then the inverse transformation (7) for a polygram and (5) for a P-R estimate were used. The polygrams were calculated by

Table 3. The attributes of the smoothing procedures

Sample	Estimate	Smoothing		
		method	parameter	parameter value
s.s.s.	P-R	ω^2	h	$7.5 \cdot 10^{-4}$
	polygram	D	k	11
d.s.s.	P-R	ω^2	h	$8.1 \cdot 10^{-3}$
		D	h	$3.6 \cdot 10^{-3}$
	polygram	D	k	11
s.r.	P-R	ω^2	h	$1.75 \cdot 10^{-4}$
		D	h	$9.5 \cdot 10^{-5}$
	polygram	D	k	17
i.r.t.	P-R	ω^2	h	$2.3 \cdot 10^{-4}$
		D	h	$2.6 \cdot 10^{-4}$
	polygram	D	k	17

the formulas (3) and (7), the P-R estimates by (6).
The smoothing parameter h of the P-R estimate was computed by the ω^2- and D-methods, i.e. from the equations (8) and (9), called P-R estimate 1 and P-R estimate 2, respectively (see Table 3). The corresponding values h of the ω^2-method for the s.s.s., d.s.s., s.r. and i.r.t. samples are provided for $l\hat{\omega}_l^2 = 0.05$. The corresponding values h of the D-method for the d.s.s., s.r. and i.r.t. samples are provided for $\hat{D}_l = 0.5$. For s.s.s. $\hat{D}_l$ never reaches its maximum likelihood value for any h and we did not apply the D-method. We see that the discrepancy methods ω^2 and D select similar values of h. The parameter k of the polygram was only calculated by the D-method (see (10) and Table 3).
The P-R estimate and the polygram restore the tail of the p.d.f. in a similar manner for each considered r.v. except the s.s.s.. The difference between the estimates occurs for the small values. The maximal values of the polygram and the P-R estimate are given by 165.049 and 48.518 for s.s.s., 14.7 and 2.277 for d.s.s., 999.001 and 196.728 for r.s. and 98.605 and 70.557 for i.r.t., respectively. Due to the small distances between the order statistics near zero the polygrams may have big values. The P-R estimate is smoother. The difference becames smaller for large sample sizes.

4 Conclusions

Considering the characterization of web traffic in the Internet, we have presented a new statistical methodology to analyze measurements of limited size. We have proposed a nonparametric framework to estimate the underlying long-tailed probability densities functions of the relevant random variables such as session sizes and durations as well as response sizes and inter-response times.
Following this nonparametric approach, we have assumed that just general information about the kind of the distributions is available. To implement the proposed

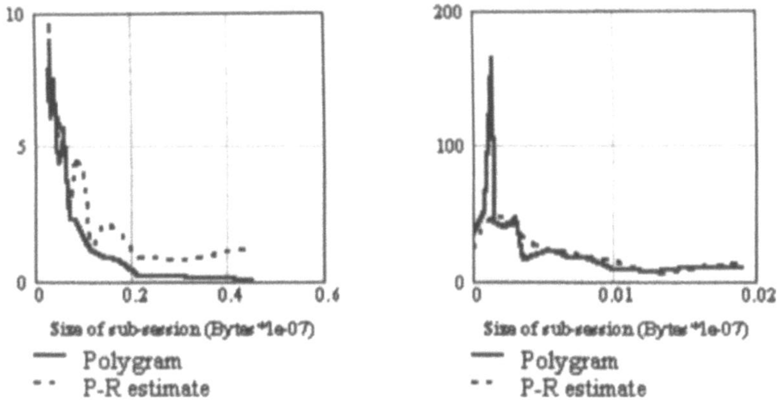

Fig. 10. Estimated probability density functions of the size of a sub-session

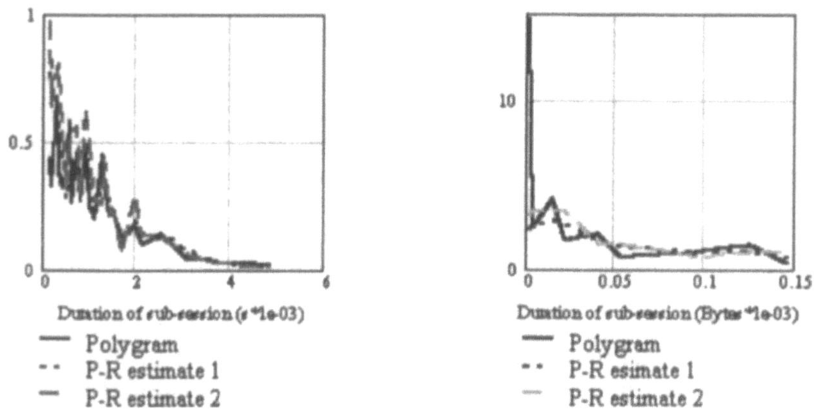

Fig. 11. Estimated probability density functions of the duration of a sub-session

approach, two estimates, a Parzen-Rosenblatt kernel estimate and a histogram with statistically equi-probable cells called a polygram, are selected. To get L_1-consistent estimates for the long-tailed p.d.f.s, the transformation of an initial r.v. to a new one with a finite distribution on the interval $[0, 1]$ is proposed. The introduction of such a transformation allows us to apply apart from the P-R estimate those estimates defined on a closed interval such as a histogram or projection estimates. Finally, an algorithm to construct L_1-consistent p.d.f. estimates has been stated.
From a practical point of view, we are interested in the accuracy of the estimation for empirical samples of limited size. The reliability of the estimates is provided by the selection of corresponding smoothing parameters. In the paper two discrepancy-type methods based on the Kolmogorov-Smirnov and the Mises-Smirnov statistics are used to select the latter parameters. They provide the estimation based on the

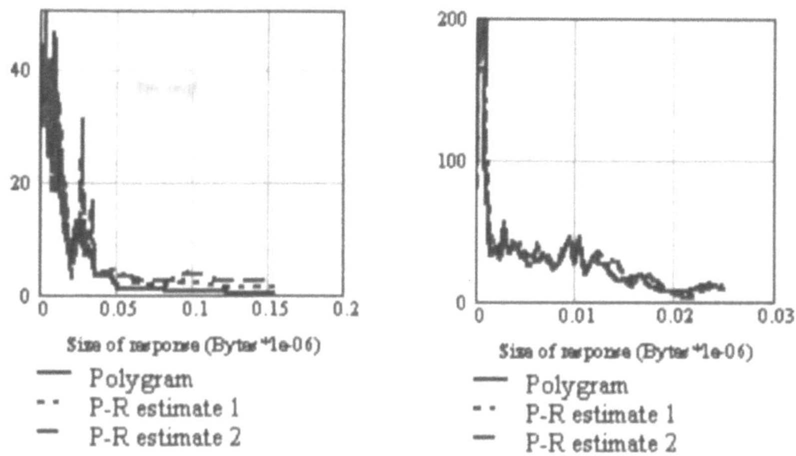

Fig. 12. Estimated probability density functions of the size of a response

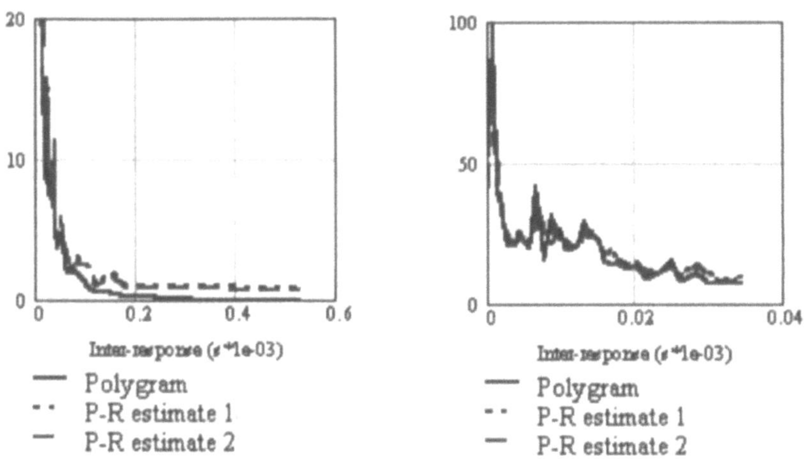

Fig. 13. Estimated probability density functions of an inter-response time

observed ungrouped sample points. The proposed ω^2- and D-methods for the selection of the smoothing parameters provide similar results and are simpler to apply than the well-known cross-validation method (cf. [3]).

To illustrate the power of the proposed estimation approach, we have finally applied it to analyze measurements arising from web traffic. The latter were gathered at the Computer Science Department of the University of Würzburg. Using these real data, the probability density functions of relevant traffic characteristics have been estimated. Furthermore, we have shown that parametric modeling by exponential and Pareto distributions does not provide an adequate description of the related densities.

In conclusion, we have pointed out a new effective way to cope with the thorough statistical data analysis of measured web-traffic characteristics. This sound analysis is the first and one of the most decisive steps towards an effective design of the IP-based transport infrastructure in the extremely variable environment of world wide web applications.

References

1. P. Barford and M.E. Crovella, Measuring Web performance in the wide area, *Performance Evaluation Review*, August (1999).
2. L.N. Bol'shev and N.V. Smirnov, *Tables of Mathematical Statistics*, Nauka, Moscow (In Russian) (1968).
3. Y.-S. Chow, S. Geman and L.-D. Wu, Consistent cross-validated density estimation, *Annals of Statistics*, 11 (1983) 25–38.
4. M.E. Crovella, M.S. Taqqu and A. Bestavros, Heavy-tailed probability distributions in the World Wide Web, in: R. Adler et al., eds., *A Practical Guide to Heavy Tails: Statistical Techniques and Applications*, Birkhauser, Boston, (1998) 3–25.
5. S. Csörgö, Asymptotic Methods in Probability and Statistics, in: B. Szyszkowicz, ed., *A Volume in Honour of Miklos Csörgö*, Elsevier, Amsterdam, (1998) 833-881.
6. L. Devroye and L. Györfi, *Nonparametric Density Estimation, the L_1 View*, John Wiley, New York (1985).
7. B.M. Hill, A simple general approach to inference about the tail of a distribution, *Ann. Statist.*, 3 (1975) 1163–1174.
8. J. L. Jerkins et al.. Operations measurements for engineering support of high-speed networks with self-similar traffic. In P. Key and D. Smith, eds., *Teletraffic Engineering in a Competitive World*, Teletraffic Science and Engineering, vol. 3b, 895–906. Elsevier, Amsterdam, 1999.
9. N.M. Markovich, Experimental analysis of nonparametric probability density estimates and of methods for smoothing them, *Automation and Remote Control*, 50, 7, Part 2 (1989) 941–948.
10. N. M. Markovitch and U. R. Krieger. Nonparametric estimation of long-tailed density functions and its application to the analysis of World Wide Web traffic. Technical Report 257TD(00)13, COST-257, 2000.
11. N. M. Markovitch and U. R. Krieger. Estimating basic characteristics of arrival processes by empirical data. Technical Report 257TD(00)35, COST-257, 2000.
12. M. Nabe, M. Murata and H. Miyahara, Analysis and modelling of World Wide Web traffic for capacity dimensioning of Internet access lines, *Performance Evaluation*, 34 (1998) 249–271.
13. V. Paxson and S. Floyd. Wide area traffic: The failure of Poisson modeling. *IEEE/ACM Transactions on Networking*. Vol. 3, no. 3:236–244, 1995.
14. A. Reyes-Lecuona et al.. A page-oriented WWW traffic model for wireless system simulation. In P. Key, D. Smith, eds., *Teletraffic Engineering in a Competitive World*, Vol. 3b, 1271–1280, Elsevier, Amsterdam, 1999.

15. F.P. Tarasenko, On the evaluation of an unknown probability density function, the direct estimation of the entropy from independent observations of a continuous random variable and the distribution-free test of goodness-of-fit, In *Proceedings IEEE*, 56.-11 (1968) 2052–2053.
16. V.N. Vapnik, *Estimation of Dependences Based on Empirical Data*, Springer, New York (1982).
17. N. Vicari. Measurement and Modeling of WWW-sessions. Technical Report 257TD(97)52, COST-257, 1997.
18. N. Vicari. Models of WWW-Traffic: a Comparison of Pareto and Logarithmic Histogram Models. In *Proceedings of the 5th Polish Teletraffic Symposium / Project COPERNICUS 1463 ATMiN - Closing Seminar*, pp. 2.2-1 – 2.2-12, 1998.

Design and Evaluation of New Communication Control Method to Support the Quality Change of Communication Line

Yusuke Noguchi, Hideo Taniguchi, and Kazuo Ushijima

Graduate School of Information Science and Electrical Engineering, Kyushu University, Hakozaki 6-10-1, Higashi-ku, Fukuoka 812-8581, JAPAN

Abstract. In the Next Generation Network, various environments such as wired network and wireless network coexist. For example, there will be high-speed network and low-speed network, various communication delays, and various communication qualities in it. Additionally, the situation that a computer handles such a various communications network simultaneously will occur. There are many studies about improvement of communication efficiency and QoS guarantee in each fields of communication engineering and computer science. However, because these studies are closed in each field, it is difficult to achieve the higher communication efficiency and to guarantee a QoS certainly in this complicated network environment. Therefore, for the Next Generation Network, we have been studying about a new communication control method in which approaches of communication engineering and of computer science are fused. Especially we focus our attention on wireless network such as satellite communication, which will be utilized broadly.

In this paper, we propose our new communication control method to support the quality change of communication line. First, we describe the communication control method that fit a change of communication quality. Second, we illustrate a flow of communication and formalize the relation between degree to the change of communication quality and communication time. And we formalize the "*optimum window size*" that keeps the increase of communication time in a minimum. Third, the new data transfer protocol that avoids degrading the capacity of communication and an implementation policy are suggested. Further, we implement a communication control mechanism that uses this data transfer protocol and evaluate it on the environment that simulates satellite communication line. And we discuss relations among quality of communication line, communication condition, and communication time. By use of our data transfer protocol, an increase of communication time hardly occurs when it becomes 10^3 times degree of communication error rate.

1 Introduction

Various studies about improvement of communication efficiency and a QoS guarantee have been done until now. As for these studies, there are respects about communication engineering and about computer science. In each field, a lot of researchers are studying actively and many successes are achieved. In communication engineering, Nonnenmacher et al.[1] proposed the party-based loss recovery method for reliable multicast transmission. Ono[2] presented the tradeoffs between delay and throughput with respect to system parameters, and proposed the mechanism to use a satellite channel effectively. Ward et al.[3] focused their attention on size of a communication buffer, and presented the low altitude multiple satellite data link protocol. In computer science, Durst et al.[4] expanded TCP in accordance with nature of satellite communication line. Miyake et al.[5] proposed TCP gateway system for satellite-based Internet access to accelerate the throughput. These studies each make interesting results, but they are closed in each field. There is not a research intending link between fields.

On the other hand, as special feature of one of Next Generation Networks, it has various "transmission rates". In future, in the network where a computer is connected to, there will be networks that are comparatively low-speed as like a telephone wire and very high-speed networks known as "Gigabit network". Additionally, on the Next Generation Networks, we will use a wireless network (e.g. wireless LAN, satellite communication line , etc.) broadly as like traditional wired network that we use. On these situations, the communication control method face various "communication delays" and various "communication quality", because there is the large difference of nature between a wired network and a wireless network. On a wired network, the communication delay is short and the communication quality is stable highly. In reverse, on a wireless network, generally, the communication delay is long and the communication quality is low, and the change is large comparatively. Additionally, the application programs that use these Next Generation Network have various requests. One of these applications demands a real-time property strictly. And another demands higher QoS. The communication control method on the Next Generation Network can adapt itself to various "transmission rates" and various "communication delays" and various "communication quality". To realize this, it is difficult with such closed measures in an individual field as in the past. New communication control method is necessary to achieve the higher communication efficiency, and to guarantee a QoS certainly. We have been studying about a new communication control method in which approaches of communication engineering and of computer science are merged. Especially we focus our attention on a wireless network such as satellite communication[6]-[8] that will be utilized broadly.

In this paper, we propose communication control method that can adapt itself to a quality change of communication line. Nature of communication line is decided from "speed", "delay" and "quality". When we compare them, a change of "quality" is the most intense. In particular, in the wireless network such as satellite communication line, the change of quality is conspicuous. Section 2 describes the basic method that can adapt itself to a quality change of communication line. Section 3 illustrates the flow of communication and formalizes communication time at the environment where communication error occurs. Furthermore, we formalize the window size to make communication time to be shortest. We name this widow size "*the optimum window size*". Section 4 explains protocol specification and commands about the communication control mechanism that we propose. Section 5 presents performance test results of our mechanism in simulated environment and discusses about it. Finally, section 6 concludes with recommends.

2 The basic methods which fits a quality change of communication line

Fig. 1 shows the basic mechanism to fit a quality change of communication line, and we explain brief description of this mechanism to the following.

(Cause) A quality change of communication line varies with noise of electromagnetic wave, obstacle, and so on.
(Phenomenon) By the changing of quality, received signal level, bit error rate, rate of packet retransmissions and communication capability are varied.
(Tackling) In communication control mechanism part, it controls "maximum packet length" and "window size" according to the phenomenon. We name this control **the improvement control**.
(Purpose) We plan maximization of communication capability by using this improvement control.

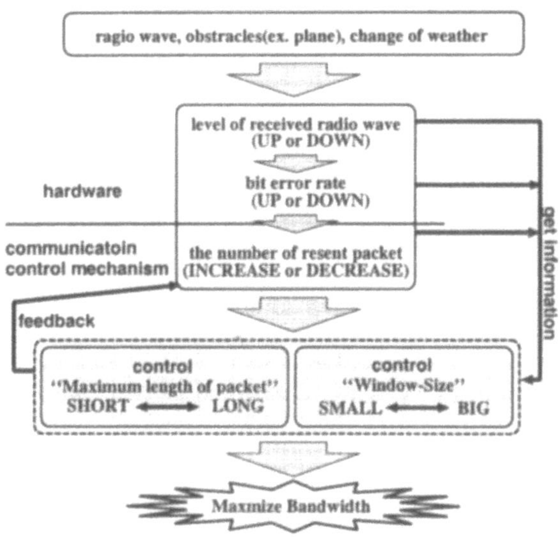

Fig. 1. Basic mechanism

Convention of improvement control is very important to maximize communication capability. In this control, the improvement that expects quality of future communication line is more effective than the improvement in accordance with quality of the present time. So we examined the following points.

- Expectation factors
- Improvement targets
- Expectation means
- Improvement contents

As expectation factors, there are received signal level, bit error rate, rate of packet retransmissions and communication capability.

An improvement target depends on communication control procedure. So by using rate of packet retransmissions and communication capability as expectation factors, it is easy to grasp the change of them and can process to an improvement target immediately.

As shown in Fig. 1, the improvement targets are "maximum packet length" and "window size". Judging from a point of implementation of communication control procedure, modification of window size is easier than modification of maximum packet length. On this account, we adopt the policy that we do modification of window size first, and change maximum packet length when not enough.

To make the improvement control more efficient, relations between expectation factors and communication capability need to be made clear. In next section, we describe formulation of communication time and window size that makes communication time shortest. We name this window size "**optimum window size**".

3 The formulation of communication time and optimum window size

3.1 Variable definisions

In communication control procedure, there are three items to be conscious of quality of a communication line. They are **maximum packet length (*P*)**, **window size (*W*)** and **timeout limit for packet retransmissions (*T*)**. We name these factors **the communication condition.**

Maximum packet length is length of the longest when a packet is sent, and it is expressed in length of bit. **Window size** is the number of packets that can be transmitted consecutively without waiting for response. **Timeout limit** is a border of latency before the control procedure decides to do retransmission of a packet.

Table 1 shows other variable definitions to formulate communication time.

Table 1. Variable definitions

t	:	transmission rate of communication line(bit/sec)
e	:	error rate of communication line
Ids	:	the number of instructions to transmit data packet(instructions)
Idr	:	the number of instructions to receive data packet (instructions)
Ics	:	the number of instructions to transmit command packet(instructions)
Icr	:	the number of instructions to transmit command packet(instructions)
Ps	:	processor performance of sender host (instructions/sec)
Pr	:	processor performance of receiver host (instructions/sec)
d	:	communication delay (sec)
Lc	:	length of packet for command (bit)
D	:	total length of transmitted data (bit)

3.2 Basic expression of communication time

Flow of communication depends on the relation between performance of computer and performance of communication line greatly. Concretely, it depends on relations with data transfer time of communication line ($\frac{\boldsymbol{P}}{t}$) and data transmission processing time of sender host ($\frac{Ids}{Ps}$) and data reception processing time of receiver host ($\frac{Idr}{Pr}$).

In this paper, we name the relations of them with **communication environment**. In this communicationenvironment, flow of communication consists of three cases as below.

(Case 1) $\frac{Ids}{Ps} > \frac{Idr}{Pr} \geq \frac{\boldsymbol{P}}{t}$
(Case 2) $\frac{Idr}{Pr} > \frac{Ids}{Ps} \geq \frac{\boldsymbol{P}}{t}$
(Case 3) $\frac{\boldsymbol{P}}{t} > \frac{Ids}{Ps} \geq \frac{Idr}{Pr}$, $\frac{\boldsymbol{P}}{t} > \frac{Idr}{Pr} \geq \frac{Ids}{Ps}$

In case 1, communication time is decided by data transmission processing time of sender host. In case 2, communication time is decided by data reception processing time of receiver host. In case 3, communication time is decided by data transfer time of communication line. Fig. 2 shows flow of communication in case 1 at assuming bit error does not occur with communication line.

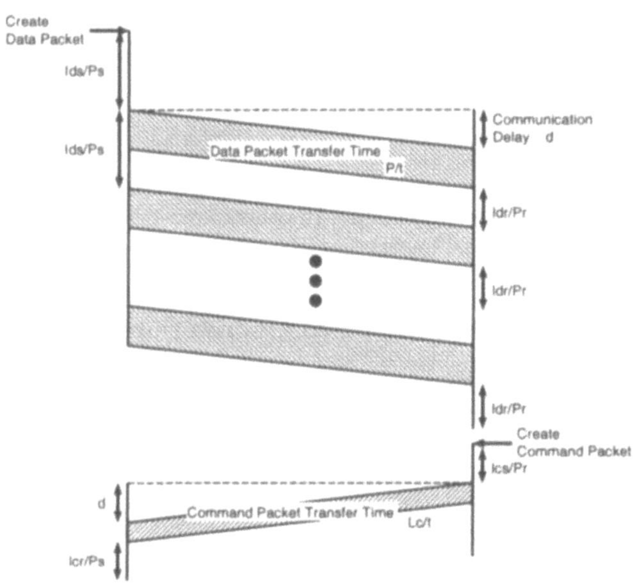

Fig. 2. Flow of communication ($\frac{Ids}{Ps} > \frac{Idr}{Pr} \geq \frac{P}{t}$)

We formulate communication time as below:

Case 1

$$T_{01} = \left(\frac{Ids}{Ps}\boldsymbol{W} + d + \frac{P}{t} + \frac{Idr}{Pr} + \frac{Ics}{Pr} + d + \frac{Lc}{t} + \frac{Icr}{Ps}\right) \times \left(\frac{D}{(P - Lc)\boldsymbol{W}}\right) \tag{1}$$

Case 2

$$T_{02} = \left(\frac{Ids}{Ps} + d + \frac{P}{t} + \frac{Idr}{Pr}\boldsymbol{W} + \frac{Ics}{Pr} + d + \frac{Lc}{t} + \frac{Icr}{Ps}\right) \times \left(\frac{D}{(P - Lc)\boldsymbol{W}}\right) \tag{2}$$

Case 3

$$T_{03} = \left(\frac{Ids}{Ps} + d + \frac{P}{t}\boldsymbol{W} + \frac{Idr}{Pr} + \frac{Ics}{Pr} + d + \frac{Lc}{t} + \frac{Icr}{Ps}\right) \times \left(\frac{D}{(P - Lc)\boldsymbol{W}}\right) \tag{3}$$

And a general expression of communication time, T0, is expressed as below:

$$T_0 = (a\boldsymbol{W} + b)\left(\frac{D}{(P - Lc)W}\right) \tag{4}$$

In this expression, parameter a and b are coefficient decided by each case. And we suppose $\frac{D}{(P-Lc)W}$ to be positive number. In other words, maximum packet length is more always than command packet length large.

3.3 Expression that considered communication error

Next, we discuss about formulating communication time under the environment that communication error occurs.

We introduce new parameter V as variable representing number of the transmitted packet, and suppose that communication error occurs only when a quantity of transmitted data becomes $\boldsymbol{P}V + Lc$. Adversely, communication error never occurs till the length of transmitted data reaches $\boldsymbol{P}V + Lc$.

$$\boldsymbol{P}V + Lc = \frac{1}{e\mu} \tag{5}$$

In this expression, we define parameter μ as **degree of quality decline**.

As degree of quality decline increases, communication error comes to occur with a more little transfer data volume. So e is regarded as initial quality of communication line. In addition, we suppose that $\boldsymbol{P}V \gg Lc$ and bit error occurs only with a data packet. Under this supposition, we consider about the circumstance that initial quality of communication line is the best ($\mu \geq 1$), and will worse. At being $W > V$, communication time T is a total time of T_0 (4) and sending time of retransmitted message and error detection time in a sender host and a receiver host.

The expression that represents communication time T is below:

$$
\begin{aligned}
T &= (a\boldsymbol{W} + \mathrm{b})\left(\frac{V+\boldsymbol{W}}{\boldsymbol{W}}\right)\frac{D}{(\boldsymbol{P} - Lc)\,\boldsymbol{W}\frac{V}{\boldsymbol{W}}} \\
&= (a\boldsymbol{W} + \mathrm{b})\left(\frac{V}{\boldsymbol{W}} + 1\right)\frac{D}{(\boldsymbol{P} - Lc)\,V}
\end{aligned}
\tag{6}
$$

In communication condition, there is an optimum communication condition that provides the data transmissions efficiency with a maximum for current communication quality.

Kenji Ono [2] shows that there is a trade off about system performance between parameters of communication condition (e.g. window size) with communication by satellite channel of time-sharing. From a point implementation of communication processing, it is easy to change window size than to change the maximum packet length. In the rest of this subsection, we formulate the *optimum window size*.

At first we suppose that maximum packet length and a timeout limit are fixed value, and regard communication time T as function of window size. From (6), we pursue the first differentiation about window size by communication time. The *optimum window size* W_0 is expressed in the following expression generally.

$$
W_0 = \sqrt{\frac{b}{a} \times \frac{\frac{1}{e\mu} - Lc}{\boldsymbol{P}}} \tag{7}
$$

From this expression, by detecting initial quality of a communication line and degree of quality decline of it, we get possible to derive the *optimum window size* that maximizes transmission efficiency at one point in time.

3.4 Derivation of expression that removed element of communication delay

However, using of this *optimum window size* (W_0) that derived from (6) and (7) has some problems.

Concretely, it is difficult to get an accurate value of instruction steps for sending or receiving data (Ids, Idr, Ics and Icr) and of communication delay (d). The reasons are as follows:

1. The number of instructions steps for sending or receiving data depend on kinds of processors and/or compilers.
2. Because a result of (7) changes greatly by little changes of a value of communication delay (d), we must get an accurate value of d. If the supposed value of d is different from real value of it, the communication efficiency will be worsened.

So we try to remove in the place (6) and (7) these variables.

Then, on the communication line that transmission rate is t and in the communication environment that the value of window size ($\boldsymbol{W}$) is equal in 1, the communications delay d is the following expression when we need the time $T_{t,W=1}$ in sending of data.

$$d = \frac{\frac{(\boldsymbol{P}-Lc)T_{t,W=1}}{D} - \frac{\boldsymbol{P}}{t} - \frac{Ids}{Ps} - \frac{Idr}{Pr} - \frac{Ics}{Pr} - \frac{Lc}{t} + \frac{Icr}{Ps}}{2} \tag{8}$$

From (1), (2), (3), (6), (7) and (8), the *optimum window size* in each case are below:

(Case 1)

$$W_{01} = \sqrt{\left(\frac{Ps}{Ids} \times \frac{(\boldsymbol{P}-Lc)T_{t,W=1}}{D} - 1\right) \frac{\frac{1}{e\mu} - Lc}{\boldsymbol{P}}} \tag{9}$$

(Case 2)

$$W_{02} = \sqrt{\left(\frac{Pr}{Idr} \times \frac{(\boldsymbol{P}-Lc)T_{t,W=1}}{D} - 1\right) \frac{\frac{1}{e\mu} - Lc}{\boldsymbol{P}}} \tag{10}$$

(Case 3)

$$W_{03} = \sqrt{\left(\frac{t}{\boldsymbol{P}} \times \frac{(\boldsymbol{P}-Lc)T_{t,W=1}}{D} - 1\right) \frac{\frac{1}{e\mu} - Lc}{\boldsymbol{P}}} \tag{11}$$

To use above expression, we must measure $T_{t,W=1}$ beforehand in order to get a value of these expressions. However, we must send equal data to volume D beforehand to get value $T_{t,W=1}$ from (6). This is not realistic in practical use. So, we describe a prediction of $T_{t,W=1}$ from the time for transmitting of a little volume.

In (6), communication time T is in proportion to D and is in inverse proportion to $\boldsymbol{P}$. But $\boldsymbol{P}$ gives influence to number of instruction steps for sending or receiving data(Ids, Idr, Ics, Icr). As $\boldsymbol{P}$ increases, process order number of data copies increases. When we express the smallest $\boldsymbol{P}$ in implementation of a communication control mechanism in $\boldsymbol{P_{base}}$, Ids, Idr, Ics and Icr are expressed with the each following expression.

$$Ids = Ids_{base} \times \frac{\boldsymbol{P}}{\boldsymbol{P_{base}}} \tag{12}$$

$$Idr = Idr_{base} \times \frac{\boldsymbol{P}}{\boldsymbol{P_{base}}} \tag{13}$$

$$Ics = Ics_{base} \times \frac{\boldsymbol{P}}{\boldsymbol{P_{base}}} \tag{14}$$

$$Icr = Icr_{base} \times \frac{\boldsymbol{P}}{\boldsymbol{P_{base}}} \tag{15}$$

In above expression, $Ids_{base}, Idr_{base}, Ics_{base}$ and Icr_{base} each represents the number of orders of sending or receiving process in $\boldsymbol{P_{base}}$.

Table 2. Command list of FCC

Command type	Command name	Semantics
Creation and deletion of path	CPC	Communication Path Create
	CPD	Communication Path Delete
	CPO	Communication Path create/ delete Okay
Transmission of data	I	Information
	PRO	Packet Receive Okay
	PRN	Packet Receive No-good
Suspension and resumption of sending or receiving data	SSD	Send SuspenD
	SRM	Send ResuMe
	RSD	Receive SuspenD
	RRM	Receive ResuMe
Change of communication condition	MCR	Mode Change Request
	MCO	Mode Change Okay
	MCN	Mode Change No-good

4 New Protocol

We designed "FCC(Function of communication Condition Change) communication control protocol". This protocol can support a quality change of communication line. In this section, we describe basic design, a form of packet and specification of FCC.

4.1 Basic design

To support a quality change of communication line, the FCC communication control protocol provides functions as shown below.

1. A facility to change communication condition during existence of communication line.
2. A facility to suspend and resume sending or receiving data during existence of communication line.

4.2 Form of packet

Fig. 3 shows the format of a FCC packet. A FCC packet consists from a part of communication control information and real data. Fig. 3-(A) is structure of the whole packet, and Fig. 3-(B) is the detail of "command part".

The *header* is the number of 8-bit words and is the fixed value of "01111110".

The *destination address* field is the number of 8-bit words. And it contains information representing the receiver computer which packet is sent to.

The *source address* field is the number of 8-bit words. And it contains information representing the sender computer which packet is sent from.

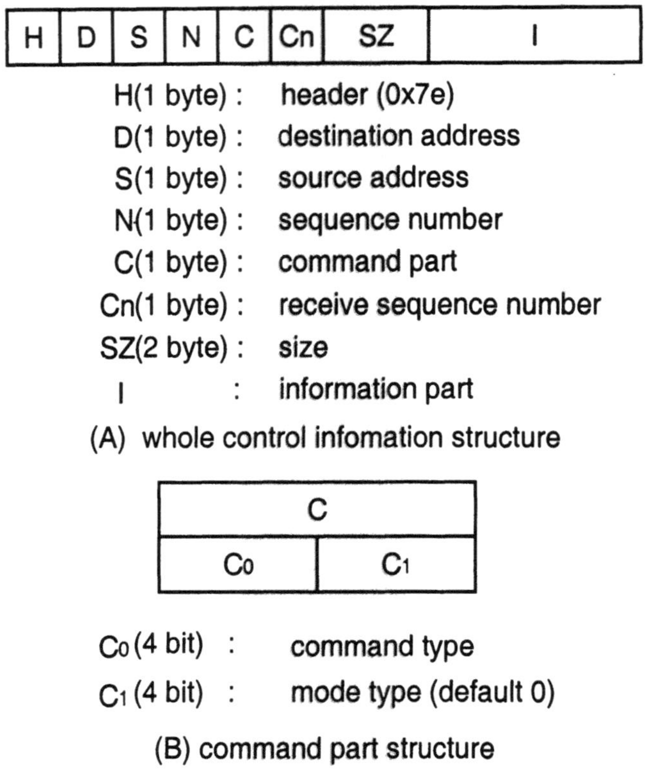

Fig. 3. Format of FCC packet

The *sequence number* field is the number of 8-bit words. And it identifies each packet sent by a host. When the FCC packet is sent, it increases one by one.

The *command part* field consists from *command type* field and *mode type* field. The *command type* field and the *mode type* are each length of 4 bits. The *command type* field keeps information about a command to use by the control of the FCC. The *mode type* field keeps information about communication condition of the FCC.

The *receive sequence number* field is the number of 8-bit words. And it contains the number of the packet received by a host.

The *size* field is the total length of data stored away in the continuing *information part*. Since this is 16-bit field, it limits the information to 2^{16} bytes.

4.3 Mode

The mode stored away in the *command type* field is an identifier of a combination of $\boldsymbol{P}$, $\boldsymbol{W}$ and $\boldsymbol{T}$. The FCC communication control mechanism can judge current communication condition by this mode. At a point in time of communication start, the value of *mode type* equals in 0. And each value of $\boldsymbol{P}$ and $\boldsymbol{W}$ and $\boldsymbol{T}$ are as follows.

- $\boldsymbol{P} = 256$ bytes
- $\boldsymbol{W} = 3$
- $\boldsymbol{T} = 1$ sec

4.4 Command

There are 13 commands of the FCC communication control mechanism. Table 2 shows the 13 types of commands for control. The first column of the table is entry about command type, and the second column is entry about a command name, and the last column is entry about this semantics.

Types of command are classified roughly into four kinds as below:

1. **Establishment and release**
   Commands for creating or deleting communication path and for response of it. For this purpose, there are three commands, `CPC`, `CPD`, `CPO`.
2. **Transmission of data**
   Commands for sending or receiving data and for a notice about failure success of reception. For this purpose, there are three commands, `I`, `PRO`, `PRN`.
3. **Suspension or resumption of sending or receiving data**
   Commands for giving notice of suspension or resumption of data transmission. For this purpose, there are four commands, `SSD`, `SRM`, `RSD`, `RRM`.
4. **Change of communication condition**
   Commands for requiring to change mode and of response for it. For this purpose, there are three commands, `MCR`, `MCO`, `MCN`.

These groups of two latter half are a command to realize this communication control procedure original facility.

4.5 State transition

For a trigger of state transition, there are three kinds as shown below.

- Reception of command that sent from a communication partner
- Request from upper program using this control part
- Notice from a lower system used by this control part

State transition by reception of command is corresponding to 13 commands shown in table Table 2.

Upper programs require nine commands for this control mechanism. These commands are as below.

1. *communication start*
2. *communication close*
3. *sending of data*
4. *sending suspension*
5. *taking data from reception queue*
6. *sending suspension*
7. *sending resumption*
8. *receiving suspention*
9. *receiving resumption*

10. *changing mode*

A notice from a system includes 2 types of the following.

1. *lacking sufficient buffer space*
2. *received a packet with bit error*

5 Measurement and Discussion

We evaluated the mechanism that we proposed under the simulation environment. In this section, we describe simulation environment produced experimentally and the measurement result.

5.1 Pseudo-satellite communication line

Fig. 4 shows positions of each process parts. On communicating each computer, there is a process unit that realizes virtual satellite communication environment. We name this virtual environment "pseudo-satellite communication line". A process that communicates according to FCC control procedure delivers packets to the pseudo-satellite communication line once. The pseudo-satellite process unit takes charge of the real computer communication. In the pseudo-satellite process unit, we can generate communication error and communication delay. There are opportunities for generate communication error as below.

1. **Random :**
   Destroy packet at a venture
2. **The number of appointed packet :**
   Destroy packet in an appointed numerical interval
3. **The time that was appointed :**
   Destroy packet in an interval of appointed time

In this system, three kinds of delay exist. First, the sending time of data from FCC process to the pseudo-satellite process unit is d_i. Second, the real communications delay that occurs between computers is d_r. Third, the delay that occurs inside the pseudo-satellite process unit is d_s. So, the total time of communication delay becomes $d = d_i \times 2 + d_r + d_s$.

5.2 Measurement environment

We measure time at transmission of 10 Mbytes data between two computers under this pseudo-satellite communication environment. We chose Random as a generation opportunity of communication error.

- Computers
  - CPU: Celeron 400MHz
  - OS: BSD/OS Ver3.1
  - I/O board: Efficient ENI-155S-MF-PCI
- Communication line
  - ATM switch: ForRunnerLE 155 ATM switch
  - Maximum transmission rate: 155Mbps

Table 3 shows other parameters that used in (7). In Table 3, each value of Ids and Idr and Ics and Icr are measured in advance by using "CPU clock counter".

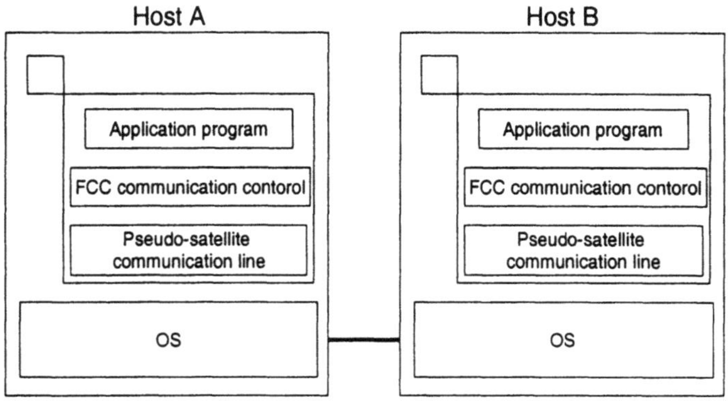

Fig. 4. Positioning of each processes

Table 3. Communication parameters

parameter	:	value	parameter	:	value
$\boldsymbol{P}$	:	1024 (bits)	e	:	9.77×10^{-8} (error/bit)
$\boldsymbol{t}$	:	1 (sec)	Ps	:	4.00×10^{8} (instructions/sec)
Ids	:	23,000 (instructions)	Pr	:	4.00×10^{8} (instructions/sec)
Idr	:	18,000 (instructions)	Lc	:	64 (bits)
Ics	:	5,000 (instructions)	D	:	83,886,080 (bits)
Icr	:	5,000 (instructions)			

5.3 Effects of control by the optimum window size

By the cause of various communication qualities, we changed various kinds of window size with 10, 30 and 50, and transmitted data. Fig. 5 shows the relation between bit error rate (BER) and communication time. In each graph of Fig. 5, "Line" is a result with assuming that processing speed of a computer is faster than a transmission rate. On the other hand, "CPU" is a result by using the *optimum window size* with assuming that a transmission rate is faster than processing speed of a computer. The graph (A), (B) and (C) are cases that processing speeds of a computer are faster than a transmission rate in reality. And the graph (D) and (E) are cases that transmission rates are faster than processing speed of a computer in reality. The following results became clear from graphs.

1. In every condition, while communication quality is high, communication time is comparatively short regardless of size. When communication quality grows than 1.0×10^{-5}, communication time is different every window size greatly. In every case, communication time is shortest by using the *optimum window size* adapted for communication condition. In case of both the situations that processing speed is higher and that transmission rate is higher, as bit error rate of communication line becomes large, the control using the *optimum window size* becomes more effective.

2. In the situation that transmission rate is higher, by the control using the *optimum window size* that assumes high-speed communication line, we can make the increase of communication time very small. We describe the details to the following.
   In graphs (D) and (E), by the control using fixed window size, when BER increases in 9.77×10^{-5} from 9.77×10^{-8}, a minimum of the increase rate of communication time is 90%(150Mbps, $\boldsymbol{W}$=50), and a maximum of it is 153%(100Mbps,$\boldsymbol{W}$=30). By the control using the *optimum window size* with assuming that a transmission rate is higher (`Line`), a maximum of the increase rate of communication time is kept by less than 0.018%. On the other hand, by the control using the *optimum window size* with assuming that processing speed of a computer is higher (`CPU`), a minimum of the increase rate of communication time is 43%(100Mbps), and a maximum of it is 77%(150Mbps). But the rates of the increase are always smaller than the control using fixed window size.
   In graphs (A), (B) and (C), when communication error rate increases in the same way, a minimum of the increase rate of communication time is 33%(1Mbps, $\boldsymbol{W}$=10), and a maximum of it is 202%(2Mbps,$\boldsymbol{W}$=50) by the control using fixed window size. By the control using the *optimum window size* with assuming that processing speed of computers is higher (`CPU`), the increase rate of communication time is kept by less than 23%. On the other hand, with the control with assuming that a transmission rate is higher, a minimum of the increase rate is 60%(1Mbps), and at the maximum case it is 83%(10Mbps).
   From these results, the control using the *optimum window size* that meets communication environment is effective.
3. In graphs (A), (B) and (C), when BER is high and the control procedure uses the *optimum window size* with assuming that transmission rate is higher, communication time is contrary to our expectation and is larger than fixed case. On the other hand, in graph (D) and (E), when it is in the control using the *optimum window size* with assuming that processing speed of a computer is higher, communication time is smaller than fixed case. So, if we want to simplify the control procedure to support the various situations, it is possible to suppress the increase of communication time with assuming that processing speed of a computer is higher.
4. While transmission rate and communication quality is high, communication time with the *optimum window size* is larger than a case without the control. But I am equal to or less than one second even in case of data movement of 10M. But, the difference of time is very small, and it is equal to or less than 1 second even in case of 10Mbytes data transfer.

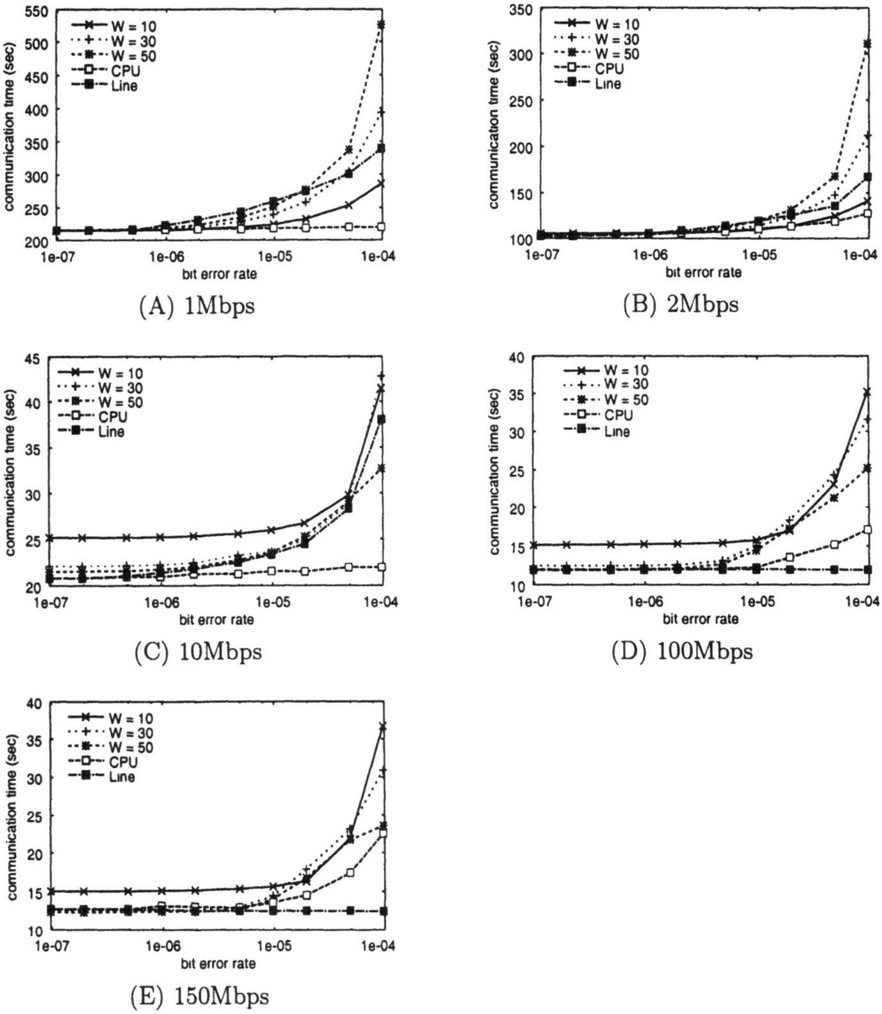

Fig. 5. Relation between BER and communication time

5.4 Changing QoS dynamically

We applied the communication control mechanism that we proposed for the environment that transmission error changed dynamically. As actual data, we took up communication record of a day of localized torrential downpour in Fukuoka of July 2, 1999. We communicated by satellite channel of 64kbps and measured signal fade, the rainfall and transmission error rate. Fig. 6 (A) shows a time change of communication error rate in measurement. In Fig. 6 (A), a x-axis is time in second, and a y-axis is BER. In Fig. 6 (B), a x-axis is time in second, and a y-axis is total amount of transmitted data in byte. But, because the data that we adapted were in the special situation of localized torrential downpour, values of communication error rate are 10^{-3} times larger than real values, and the progress of time is 1.5 times slower than reality. Our communication control mechanism calculated the *optimum window size* from transmission error rate every one second, then changed window size with FCC protocol. In Fig. 6 (A), there is a immediate sharp rise of BER at the last 10 seconds of the measurement period. And in Fig. 6 (B), there is a dull rise of the total amount of traffic using fixed window size ($\boldsymbol{W}$ = 10, 30, 50 and 100). On the other hand, in case of using the *optimum window size* ($\boldsymbol{W}$ = auto), there is hardly the influence. The result of the experiment was that it was able to restrain degradation of communication efficiency by controlling window size according to a quality change of communication line.

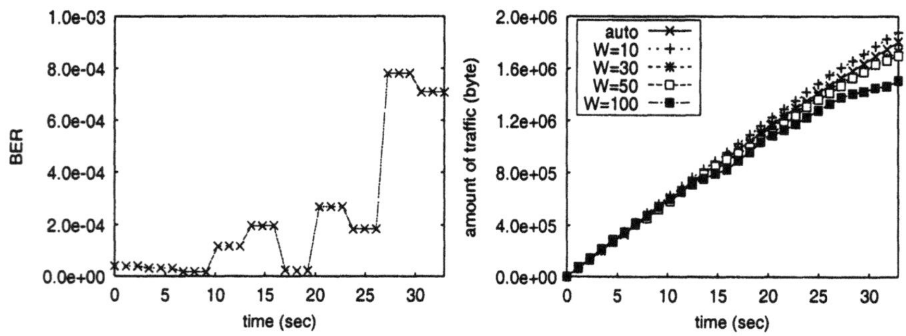

(A) Error Rate (B) Total amount of traffic

Fig. 6. In the case of changing QoS dynamically

6 Conclusion

In the Next Generation Network, there will be various networks that have each different nature. These are various "transmission rates", various "communication delays" and various "communication qualities". However, with traditional communication control mechanism that separates from in each field, it is difficult to achieve the higher communication efficiency, and to guarantee a QoS certainly.

In this paper, we propose our new communication control mechanism, and design new data transfer protocol. Our data transfer protocol has facility to change communication condition during a communication path exists. By using this facility, it can adapt to the quality change of communication line.

From evaluation on the environment that simulates satellite communication line, an increase of communication time hardly occurs when it becomes 10^3 times degree of communication error rate by using of our data transfer protocol.

Acknowledgment

This work was supported by the Research for the Future Program (JSPS-RFTF 96P00603) from the Japan Society for the Promotion of Science.

References

1. J.Nonnenmacher, E.W. Biersack, and D.Towsley, "Parity-Based Loss Recovery for Reliable Multicast Transmission," IEEE/ACM Trans. on Networking vol.6, no.4, pp.349–361 (1998).
2. Kinji ONO, "The Performance Tradeoffs of Periodic Reservation Satellite Channels for Packet Switching Mult-Access System", The Trans. of the IECE of Japan, Vol.E63, No.2, pp.104-111 (1990).
3. C.Ward, C.H.Choi, and T.F.Hain, "A Data Link Control Protocol for LEO Satellite Networks Providing a Reliable Datagram Service," IEEE/ACM Trans. on Networking, vol.3, no.1, pp.91–103 (1995).
4. R.C.Durst, G.J.Miller and E.J.Travis, "TCP extenstions for space communications," ACM The journal of mobile communication, computation and information (Wireless Networks), vol.3, pp.389–403 (1997).
5. Yutaka Miyake, Teruyuki Hasegawa, Toru Hasegawa, and Toshihiko Kato, "A Proposal TCP Gateway System for Satellite-based Internet Ascess", IPSJ-DPS, vol.98, no.55, pp.63–68 (1997).
6. L.S.Golding, "Satellite communication systems move into the twenty-first century", ACM The journal of mobile communication, computation and information (Wireless Networks), vol.4, pp.101–107 (1998).
7. F.Takahata et al., "Satellite Communication Experiments by Universities," The Journal of The Institute of Electronics, Information and Communication Engineers, Vol.80, No.5, pp.435–456 (1997).
8. "Packet Communication Ultra-Small Aperture Terminal System for the Hokkaido Integrated Telecommunication Network," IEEE Trans. MTT, vol.43, no.7, pp. 1692-1698 (1995).

TCP over a multi-state Markovian path

Eitan Altman, Konstantin Avrachenkov*, Chadi Barakat**, and Parijat Dube***

INRIA Sophia Antipolis,
2004, route des Lucioles, B.P.93,
06902, Sophia Antipolis Cedex, France.
Email:{altman,k.avrachenkov,cbarakat,pdube}@sophia.inria.fr

Abstract. In this paper we analyze the performance of a TCP-like flow control mechanism. The transmission rate is considered to increase linearly in time until the receipt of a congestion notification (via loss detection in context of TCP) where the transmission rate is multiplicatively decreased. We introduce a general model based on a multi-state Markov chain for the moments at which the congestion is detected. With this model, we are able to account for correlation and burstiness in congestion moments. Furthermore, we specify several simple versions of our general model and then we identify their parameters from real TCP traces.

1 Introduction

We study in this paper the performance of an additive-increase multiplicative-decrease flow control protocol. This is the kind of control used by TCP, the widely-used transport protocol of the Internet [9]. TCP is used as a reference through the present work, however we anticipate that our results will be also applicable for other flow control mechanisms. A fluid approach is used to model the controlled flow. The transmission rate of the source is assumed to grow linearly at a rate α. In the case of TCP where the flow is controlled via a congestion window, the transmission rate at any instant is equal to the window size divided by the Round Trip Time of the connection. The growth of the transmission rate continues until the source receives a notification of congestion from the network. In case of TCP, the congestion is inferred from the loss of packets. It is an implicit notification compared to the explicit notification used by other flow control protocols as the ABR service in ATM or the ECN proposal in the Internet. We call the moment at which the source reduces its transmission rate a loss moment. Upon detection of a loss, the transmission rate is *scaled down* by a (possibly random) factor $a \in [0, 1]$. The scaling factor depends on many factors as the version of TCP, the number of packet losses in the congestion period and the way with which the loss is detected (e.g. duplicate ACK or Timeout [12]). Note that by choosing in some instants $a = 1$ one can introduce potential loss instants.

* The work of this author was financed by a grant of CNET France-Telecom on flow control in High Speed Networks

** The work of this author was financed by an RNRT "Constellations" project on satellite communications

*** The work of this author was partially financed by the French embassy in India

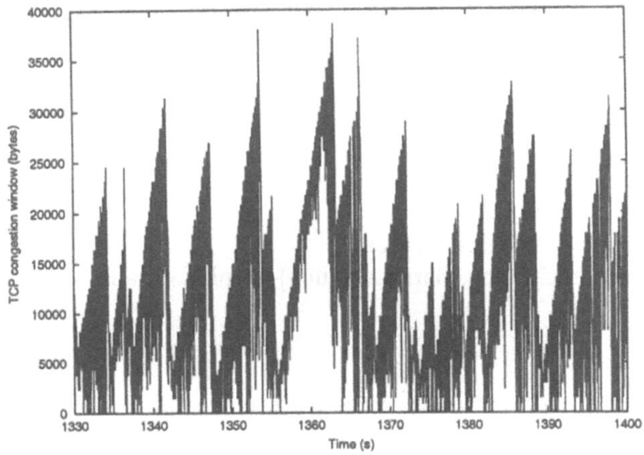

Fig. 1. TCP window evolution

The study of the performance of a flow control mechanism requires a characterization of the moments at which the transmission rate is reduced. These moments can be seen as a point process, where the appearance of a point corresponds to the appearance of a congestion signal or a loss in the context of TCP, causing a reduction in the transmission rate. Simple models as Poisson or iid models may not work in some cases where losses present some kind of burstiness or correlation. For example in Figure 1 one can observe a scenario where the moments of transmission rate reduction are clustered together. This figure corresponds to the window size evolution of a New Reno [5] TCP connection running between two sites at the technology park Sophia Antipolis. Normally in TCP, the window is divided by two upon congestion detection, but we see in this figure a more severe reduction due to multiple consecutive division of the congestion window by two. In a previous paper [1], we present a two-state Markovian model to account for burstiness of losses. In that paper, we considered a lossy path with two states *Good* and *Bad* together with potential loss moments. The transmission rate is reduced upon potential losses. A potential loss can transform into a real loss with probability p_G in the Good state and with probability p_B in the Bad state ($p_G \leq p_B$). The time between potential loss moments is assumed to be independently and identically distributed. Our main contribution in [1] is to show that the throughput of the flow control mechanism increases with the increase in burstiness of losses. However, we validated the model only via simulations, but we did not provide any algorithm for the identification of its parameters from real traces.

The present work is an extension of our previous work [1] to a multi-state Markovian case. Being motivated by some experimentation results (e.g. Figure 1), we allow the path of the connection to be in more than two states. The need for more than two states for describing the channel is also motivated by modelling results from [11,13] on mobile satellite channels, where it was shown that one needs typically at least four states. In [1], the scaling factor a is a random variable equal to either 0.5 (the potential loss becomes a real loss) or 1 (a potential loss is not transformed into a real loss). Here we propose to study the scaling factor with a general distribution that depends on the state of the path. We present then some applications of our general model. These applications can be seen as different ways to infer the parameters of the general model from a real TCP trace. In particular, we provide a method for the parameter identification of our model in [1]. A comparison between the different applications is provided to see which one is the most efficient in predicting TCP performance.

In the following section, we present the general multi-state multi-reduction model for the flow control mechanism. This general model is analyzed in Section 3. In Section 4, we provide several particular cases of the general model as well as their application to TCP modelling. We conclude Section 4 by a comparison between the different particular cases.

2 The model

Let $X(t)$ be the transmission rate at time t. In case of TCP, it is equal to the current window size divided by the Round Trip Time of the connection. Let $K = \{1, 2, \ldots, N\}$ be the set of possible states of the path. We allow losses to occur in any of the N states; the probability of the occurrence of losses in each of these states may be different. To that end, we define a series of potential losses occurring with a certain distribution of times between potential losses. Let T_n denote the time at which the nth potential loss occurs and let X_n denote the transmission rate just prior to T_n. The pair $\{T_n, X_n\}$ can be considered as a marked point process [3]. Let D_n, $n \in \mathbb{Z}$ be a sequence of times between potential losses: $D_n = T_{n+1} - T_n$. D_n are assumed to be i.i.d. with expectation d, second moment $d^{(2)}$ and Laplace Stieltjes Transform $D^*(s) = E[e^{-sD_n}]$. Let Y_n be the state of the channel at the nth potential loss instant. We assume further that the sequences $\{Y_n\}$ and $\{D_n\}$ are independent. We assume that $\{Y_n\}$ is an ergodic Markov chain with the following transition probabilities,

$$p_{ij} = P\{Y_{n+1} = j | Y_n = i\}, \; 1 \leq i, j \leq N$$

Let $P = \{p_{ij}\}_{i,j=1}^{N}$ and let π be the stationary distribution of the Markov chain associated to the path. Next we define N random variables (discrete or continuous), $\{A_n^j; 1 \leq j \leq N\}$, which describe the behavior of the transmission rate when a potential loss occurs: is it reduced and if so by how much. These variables $\{A_n^j; 1 \leq j \leq N\}$ correspond to the N possible states of the model for losses. Each random variable $A_n^j, 1 \leq j \leq N$, takes values in the interval $[0, 1]$. It can take rational or real values within this interval. The choice of the interval $[0, 1]$ stems from the fact that we are *scaling down* the transmission rate at the instant of losses. The set includes 1 since it corresponds to the case when a potential loss is not transformed into a real loss and so the transmission rate is unaltered. $A_n^j, 1 \leq j \leq N$ has a distribution function $F^j(a)$ for all $n \in \mathbb{Z}$. That is, we take the distribution of A_n^j

to be time homogeneous. Denote

$$a_i := \int_0^1 adF^i(a), \quad 1 \le i \le N.$$

We assume that there is at least one i for which $a_i < 1$. The dynamics of the system can be given by the following stochastic recurrent equation

$$X_{n+1} = \sum_{j=1}^{N} A_n^j X_n 1\{Y_n = j\} + \alpha D_n. \tag{1}$$

3 Performance Analysis

First we observe that equation (1) is a particular case of stochastic linear difference equations of type $X_{n+1} = A_n X_n + B_n$, where $\{A_n, B_n\}$ is a stationary and ergodic processes (one can consider the Markov chain $\{Y_n\}$ in the stationary regime). It follows from [4] and [8] that such equations have a stationary solution X_n^* given by,

$$X_n^* = \sum_{k=0}^{\infty} (\prod_{i=n-k}^{n-1} A_i) B_{n-k-1}.$$

The stationary regime exists under the assumption that there is at least one i for which $a_i < 1$. Moreover, for any arbitrary starting point X_0, the sequence $\{X_n\}$ will converge almost surely to this stationary regime, that is

$$\lim_{n \to \infty} |X_n - X_n^*| = 0, \text{ P-a.s.}$$

Therefore, we can assume without loss of generality that the process $\{X_n\}$ is in the stationary regime in order to compute the limit distribution. Next we compute the moments of X_n in this regime. Let us denote,

$$x_i = E[X_n 1\{Y_n = i\}] \; 1 \le i \le N.$$

Obviously, the expectation of X_n is given by,

$$E[X_n] = \sum_{i=1}^{N} x_i.$$

To compute $x_i, 1 \le i \le N$, we use the Laplace Stieltjes Transform approach. Namely, define the following Laplace Stieltjes Transforms:

$$W(s,i) = E\left[e^{-sX_n} 1\{Y_n = i\}\right], 1 \le i \le N,$$

where we assume that X_n is in the stationary regime.

Theorem 1. *The Laplace Stieltjes Transforms* $W(s,j), 1 \le j \le N$, *are solutions of the following implicit equations,*

$$W(s,j) = D^*(\alpha s)\left[\sum_{i=1}^{N} p_{ij}\int_0^1 W(as,i)dF^i(a)\right]\ 1 \le j \le N \tag{2}$$

Proof: We write for any j, $1 \le j \le N$,

$$\begin{aligned}
E[e^{-sX_{n+1}}1\{Y_{n+1}=j\}] &= \sum_{i=1}^{N} E[e^{-sX_{n+1}}1\{Y_{n+1}=j\}|Y_n=i\}P(Y_n=i) \\
&= \sum_{i=1}^{N} E[e^{-sX_{n+1}}|Y_n=i]E[1\{Y_{n+1}=j\}|Y_n=i]P(Y_n=i) \\
&= \sum_{i=1}^{N} E[e^{-s(A_n^iX_n+\alpha D_n)}|Y_n=i]p_{ij}P(Y_n=i) \\
&= D^*(\alpha s)\sum_{i=1}^{N}\int_0^1 E[e^{-saX_n}|Y_n=i]dF^i(a)p_{ij}P(Y_n=i) \\
&= D^*(\alpha s)\sum_{i=1}^{N} p_{ij}\int_0^1 E[e^{-saX_n}1\{Y_n=i\}]dF^i(a)
\end{aligned}$$

This results in the implicit equations (2). □

Although the Laplace Stieltjes Transforms in Theorem 1 are only given as solutions of implicit equations, all moments of $X_n 1\{Y_n = i\}$ for $1 \le i \le N$ (in the stationary regime) can be obtained explicitly. Note that

$$E[X_n^k 1\{Y_n=i\}] = (-1)^k \frac{d^kW(s,i)}{ds^k}\bigg|_{s=0}.$$

We shall now proceed to the calculation of expressions for the first and second moments of $X_n 1\{Y_n = i\}$ for $1 \le i \le N$ from the implicit expressions of the Laplace Stieltjes transforms. Upon differentiating the implicit expressions (2) and using the following relations,

$$W(0,i) = \pi_i, \quad 1 \le i \le N,$$

$$D^*(0) = 1, \quad \frac{dD^*(\alpha s)}{ds}|_{s=0} = -\alpha d,$$

we get N linear equations in N unknowns:

$$x_j = \sum_{i=1}^{N} p_{ij}a_ix_i + \alpha d\pi_j\ 1 \le j \le N. \tag{3}$$

We shall now write the above N equations in matrix notation. Let $x = [x_1, x_2, \dots, x_N]$ and

$$A = \begin{bmatrix} a_1 & 0 & \dots & 0 \\ 0 & a_2 & \dots & 0 \\ \vdots & \vdots & \ddots & \vdots \\ 0 & 0 & \dots & a_N \end{bmatrix}.$$

Then the equations (3) take the form

$$x = xAP + \alpha d\pi \tag{4}$$

Recall that $0 \leq a_i \leq 1$ for all i. Furthermore, we assume that there is at least one i for which $a_i < 1$. The latter guarantees that the matrix AP is substochastic (there is an i for which $\sum_{j=1}^{N} p_{ij} a_i < \sum_{j=1} p_{ij} = 1$). Recall that moduli of all eigenvalues of a substochastic matrix are strictly less than one. Therefore, matrix $I - AP$ has no zero eigenvalue, and consequently, equation (4) has a unique solution. Thus we can state the following result:

Theorem 2. *Let X_n be in the stationary regime. Then $E[X_n]$ is given by*

$$E[X_n] = xe = \alpha d\pi (I - AP)^{-1} e$$

where e is a vector of ones.

To compute the second moment of X_n, we first define

$$x_i^{(2)} = E[X_n^2 1\{Y_n = i\}], \quad 1 \leq i \leq N.$$

Clearly,

$$E[X_n^2] = \sum_{i=1}^{N} x_i^{(2)}.$$

Also let $x^{(2)} = [x_1^{(2)}, x_2^{(2)}, \dots, x_N^{(2)}]$ and

$$A^{(2)} = \begin{bmatrix} a_1^{(2)} & 0 & \dots & 0 \\ 0 & a_2^{(2)} & \dots & 0 \\ \vdots & \vdots & \ddots & \vdots \\ 0 & 0 & \dots & a_N^{(2)} \end{bmatrix},$$

where

$$a_i^{(2)} = \int_0^1 a^2 dF^i(a), \quad 1 \leq i \leq N.$$

Then in the next Theorem we give an explicit expression for $E[X_n^2]$.

Theorem 3. *Let $\{X_n\}$ be in the stationary regime and there is at least one i for which $a_i < 1$. Then $E[X^2]$ is given by*

$$E[X_n^2] = x^{(2)} e = \left(2\alpha d(xAP) + \alpha^2 d^{(2)} \pi\right) (I - A^{(2)} P)^{-1} e.$$

Proof: Differentiating twice the implicit expressions (2), we obtain

$$\frac{d^2W(s,j)}{ds^2} = D^*(\alpha s)\left[\sum_{i=1}^{N} p_{ij}\int_0^1 \frac{d^2W(as,i)}{ds^2}dF^i(a)\right]$$
$$+\frac{d^2D^*(\alpha s)}{ds^2}\left[\sum_{i=1}^{N} p_{ij}\int_0^1 W(as,i)dF^i(a)\right]$$
$$+2\frac{dD^*(\alpha s)}{ds}\left[\sum_{i=1}^{N} p_{ij}\int_0^1 \frac{dW(as,i)}{ds}dF^i(a)\right]$$

Now evaluating the above derivatives at $s = 0$, we get

$$x_j^{(2)} = \sum_{i=1}^{N} p_{ij}a_i^{(2)}x_i^{(2)} + 2\alpha d\sum_{i=1}^{N} p_{ij}a_i x_i + \alpha^2 d^{(2)}\pi_j.$$

Next we rewrite the equations in matrix notation

$$x^{(2)} = x^{(2)}A^{(2)}P + 2\alpha d(xAP) + \alpha^2 d^{(2)}\pi.$$

Solving for $x^{(2)}$, we get

$$x^{(2)} = \left(2\alpha d(xAP) + \alpha^2 d^{(2)}\pi\right)\left(I - A^{(2)}P\right)^{-1}$$

The existence of $\left(I - A^{(2)}P\right)^{-1}$ is guaranteed, because $A^{(2)}P$ is again substochastic as the sum of the elements of the ith row of $A^{(2)}P$ is $\sum_{j=1}^{N} p_{ij}a_j^{(2)} < \sum_{j=1}^{N} p_{ij} = 1$. □

Observe that we computed the expectation of the transmission rate with respect to loss instants. This expectation is also referred to as Palm expectation in the context of marked point processes [3]. Of course, the most interesting is the calculation of the expectation of the transmission rate at an arbitrary time moment. For ergodic processes the latter expectation coincides with the following time average P-a.s.,

$$\bar{x} = \lim_{T\to\infty}\frac{1}{T}\int_0^T X(t)dt$$

This is no other than the throughput of the transfer. It is the total volume of transmitted data over the transfer time. We proceed to evaluate this throughput by employing the concept of Palm probability.

Theorem 4. *The throughput, or the time-average transmission rate, is given by*

$$\bar{x} = E[X(t)] = \sum_{i=1}^{N} a_i x_i + \frac{1}{2}\alpha\frac{d^{(2)}}{d} = \bar{a}x^T + \frac{1}{2}\alpha\frac{d^{(2)}}{d}, \tag{5}$$

where $a = [a_1, a_2, \ldots, a_N]$ and x is given in Theorem 2.

Proof: To compute $E[X(t)]$ one can use the following inversion formula (see e.g., [3] Ch.1 Sec.4)

$$E[X(t)] = \frac{1}{d}E^0[\int_0^{T_1} X(t)dt] \tag{6}$$

where $E^0[.]$ is an expectation associated with Palm distribution. Thus we can write,

$$E[X(t)] = \frac{1}{d}E^0[\int_0^{T_1} (\sum_{i=1}^{N} A_0^i X_0 1\{Y_0 = i\} + \alpha t)dt]$$

Because of the independence of X_n and $\{D_k, k \geq n\}$ and also because of the independence of $\{D_n\}$ and $\{Y_n\}$ we can write,

$$\begin{aligned} E[X(t)] &= \frac{1}{d}[\sum_{i=1}^{N} \left(E^0[A_0^i]E^0[X_0 1\{Y_0 = i\}]\right) E^0[D_0]] + \frac{\alpha}{2d}E^0[D_0^2] \\ &= \sum_{i=1}^{N} a_i x_i + \frac{1}{2}\alpha\frac{d^{(2)}}{d} = ax^T + \frac{1}{2}\alpha\frac{d^{(2)}}{d} \end{aligned}$$

□

In the next theorem we evaluate the second moment of the transmission rate at an arbitrary time instant.

Theorem 5. *Let $d^{(3)}$ be the third moment of the time between potential losses. The second moment of the input rate over a long time interval is equal to:*

$$\begin{aligned} \bar{x}^{(2)} &= \lim_{t\to\infty} \frac{1}{t}\int_0^t X^2(t)dt \\ &= \frac{1}{3}\alpha^2\frac{d^{(3)}}{d} + \frac{1}{d}\alpha d^{(2)} a x^T + a^{(2)} x^{(2)^T} \end{aligned}$$

where $a^{(2)} = [a_1^{(2)}, a_2^{(2)}, \ldots, a_N^{(2)}]$ and $x^{(2)}$ is given in Theorem 3.

Proof: Again by the inversion formula from Palm probability,

$$
\begin{aligned}
E[X^2(t)] &= \frac{1}{d}E^0[\int_0^{T_1} X^2(t)dt] \\
&= \frac{1}{d}E^0\left[\int_0^{T_1}\left(\sum_{i=1}^{N} A_0^i X_0 1\{Y_0 = i\} + \alpha t\right)^2 dt\right] \\
&= \frac{1}{d}E^0\left[\frac{\alpha^2 D_0^3}{3} + \alpha D_0^2 \sum_{i=1}^{N} A_0^i X_0 1\{Y_0 = i\} + \sum_{i=1}^{N} (A_0^i)^2 X_0^2 1\{Y_0 = i\} D_0\right] \\
&= \frac{1}{3}\alpha^2 \frac{d^{(3)}}{d} + \frac{1}{d}\alpha d^{(2)} \sum_{i=1}^{N} a_i x_i + \sum_{i=1}^{N} a_i^{(2)} x_i^2 \\
&= \frac{1}{3}\alpha^2 \frac{d^{(3)}}{d} + \frac{1}{d}\alpha d^{(2)} a x^T + a^{(2)} x^{(2)^T}
\end{aligned}
$$

□

Having obtained the expressions for the general case of N states we shall now focus on some particular cases in the following sections. We show how the parameters of our model can be inferred from a real trace of a TCP connection. Different possible applications of the model to the same trace are presented and the results are then compared to show which method is the most efficient. We will see in the sequel how much the model is general and how multiple sub-models can be derived from it by setting differently the parameters.

4 Specifications of the general model

In this section, we present different ways for the application of our general model to predict the performance of a TCP-like flow control mechanism. We chose to work with real loss processes. From the trace of a TCP connection, we determine the moments of window reduction. We reconstruct then the evolution of TCP congestion window over time under the assumption that the window increases linearly between two consecutive losses. We call this reconstructed window evolution the Exact Fluid Model and we use it below as a reference. We try then to derive simple closed form expressions for the throughput of the exact fluid model, and therefore for the throughput of TCP, using simple versions of our general model.

Our experimentation consists of a long-life New-Reno TCP connection running between `clope.inria.fr` at INRIA and `nessie.essi.fr` at ESSI, both located in the technology park Sophia Antipolis in France. The two machines are connected to the same metropolitan network. The TCP connection is run eleven times for approximately 20 minutes each at the most busy periods (between 10 am and 2 pm). The trace of the connection is captured at the source using the `tcpdump` tool and a program is developed to analyze the traces in order to find the moments at which the congestion window is divided by two. We noticed that most of the time, the loss of packets is detected with the Fast Retransmit algorithm (3 Duplicate ACKs) [12]. We noticed also that the maximum window advertised by the receiver is rarely reached due to working at busy periods. Thus, we can expect that our fluid model approximates correctly the behavior of the congestion window.

4.1 The basic model

We consider here the very simple case where the path has a single state and where the transmission rate is divided once by two at every potential loss moment. We assume that the times between losses are iid. This gives the following expression for the throughput,

$$E[X(t)] = \alpha d + \frac{1}{2}\alpha \frac{d^{(2)}}{d}. \tag{7}$$

Obviously, if times between losses are really iid, this model must give very close result to the throughput of the exact fluid model. And indeed, in our experiments we did not find a significant correlation between inter-loss times. Figure 2 confirms this conclusion. The throughput given by formula (7) follows closely the one given by the exact fluid model. However, to use formula (7) for the throughput calculation, one must know the second moment of inter-loss times. Usually, this quantity is difficult to find since it requires the knowledge of all inter-loss times for the modelled connection. Note that, by contrast, d can be easily calculated by dividing the total time of the connection by the number of losses. The number of losses in turn can be calculated using the packet loss probability. One way to eliminate $d^{(2)}$ is to express it as a function of d. For example, one can assume that inter-loss times form a Poisson process and hence take $d^{(2)} = 2d^2$. The problem with this solution is that it hides the impact of burstiness and expresses the throughput only as a function of the average loss rate. Indeed in Figure 2, the throughput calculated according to the Poisson assumption does not match well the throughput of the exact fluid model. The reason for this mismatch is clearly explained by Figure 3 where we plot the histogram of inter-loss times. This figure shows the deviation of the inter-loss time distribution from the exponential shape. This deviation is caused by the appearance of bursts of losses which causes the pulse of probability around the origin. Indeed, we noticed from the real traces of a TCP connection that the congestion window is divided multiple times by two when a congestion occurs and this due to the loss of packets in multiple consecutive Round Trips (see also Figure 1). However, the important notice we made from Figure 3 is that the time between bursts can still be well approximated by the exponential distribution. Figure 4 shows the distribution of times between losses after the elimination of the pulse around the origin. In the next two sections, we will present two methods to account for this bursty behavior of losses.

4.2 The aggregate loss method

As was noticed in Figure 3, the inter-loss time distribution is a mixture of two distributions, one around the origin represents the time between losses within bursts and another away from the origin represents the time between bursts. This prompts

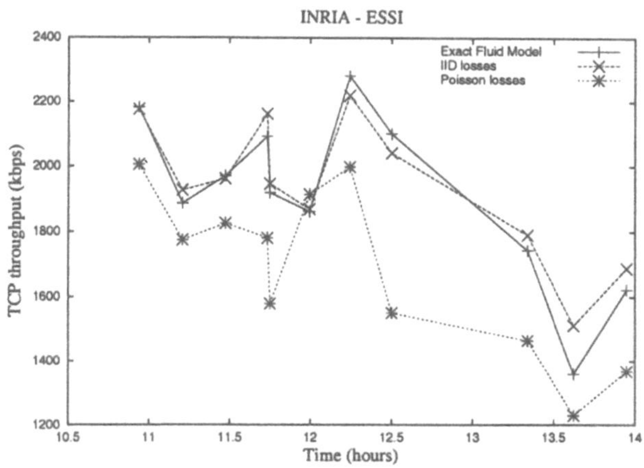

Fig. 2. Comparison of Poisson, iid and exact fluid models

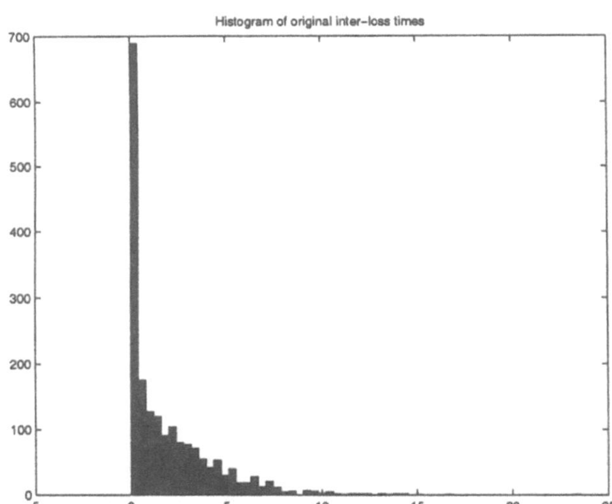

Fig. 3. Histogram of inter-loss times

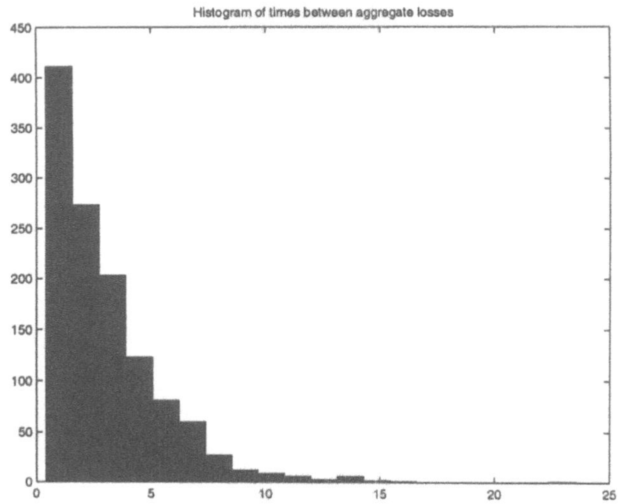

Fig. 4. Histogram of times between bursts

us to aggregate the losses inside a burst into a single loss and to divide the transmission rate upon an aggregate loss occurrence (or a burst occurrence) by two power the number of aggregated losses inside the burst. The aggregate loss process can be considered now as a Poisson process. Upon the arrival of an aggregate loss, the transmission rate is divided by a random factor that can be greater than two. The question that one may ask here is how to characterize a burst, in other words how to decide that two consecutive losses are within the same burst or within two different bursts. In this section we use the following empirical method: we look at the distribution of inter-loss times and to try to find a point which clearly separates the two distributions. We zoom in Figure 5, the distribution of inter-loss times (Figure 3) around the origin. It is clear that two bursts are separated by approximately δ =0.4s. We use this δ for the identification of bursts. In the following, we present two different ways to describe the behavior of the random reduction factor. The first way is to assume that it is iid. The second way is to model it with a Markov chain.

First, let us consider the case of iid reduction factor. The evolution of the transmission rate in this case is given by

$$X_{n+1} = A_n X_n + \alpha D_n,$$

where the reduction factor A_n has a distribution function $F(a)$. D_n is the time between bursts which can be approximated by a Poisson process. Of course, this can be viewed as a particular case of our general model where the path of the connection has only one state. The general results of Section 2 can be specified for the present case as follows,

$$E[X_n] = \frac{\alpha d}{1 - \bar{a}},$$

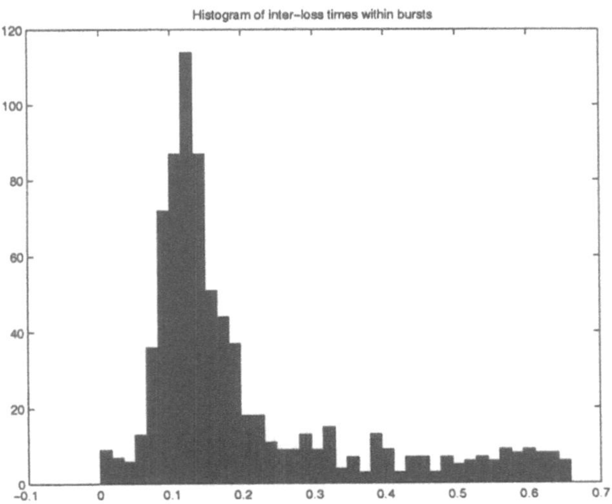

Fig. 5. Histogram of inter-loss times within bursts

$$\bar{x} = E[X(t)] = \frac{\alpha d\bar{a}}{1-\bar{a}} + \frac{1}{2}\alpha\frac{d^{(2)}}{d}, \tag{8}$$

where $\bar{a} = \int_0^1 adF(a)$. Here, the reduction factor A_n is a discrete random variable which takes the values multiple of 1/2. Thus, we calculated $\bar{a}$ as

$$\bar{a} = \sum_{i=1}^{m} \frac{1}{2^i} p_i,$$

where the probabilities p_i are estimated from the TCP connection trace. Let n be the total number of aggregate losses in the trace. We can write

$$p_i = \sum_{k=1}^{n} 1\{a_k = 1/2^i\}/n$$

Note here that the main gain from aggregation, is that the second moment of D_n can now be taken as $2d^2$. Furthermore, from Figure 4, one can see that the distribution of D_n is a shifted exponential distribution given that the time between two aggregate losses is always larger than δ. Thus, a more correct estimation for the second moment is given by

$$d^{(2)} = \delta^2 - 2\delta d + 2d^2.$$

Next we consider the case where the reduction factor is modelled using a Markov chain. We associate a multi-state Markov chain to the path. The transitions of the chain occur upon aggregate loss arrival. The state of the chain when an aggregate loss arrives is equal to the number of losses within the burst. The Markov chain

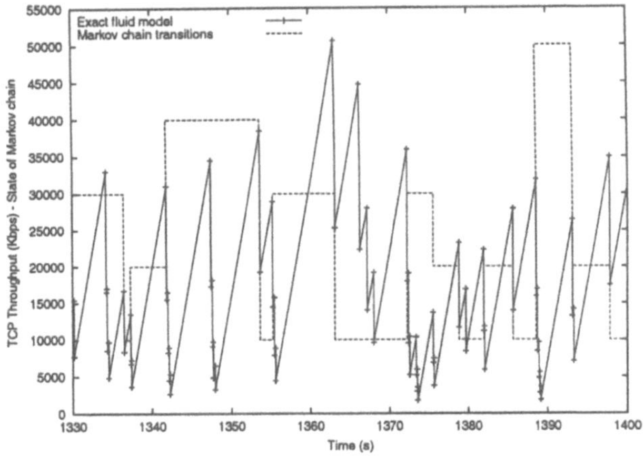

Fig. 6. Transitions of the multi-state Markov chain

determines then how many times the transmission rate is divided by two. Figure 6 explains how the transmission rate and the Markov chain change together. A interval of 0.4s is used to identify the losses belonging to the same burst. The evolution of the transmission rate in this case can be described as follows,

$$X_{n+1} = \sum_{j=1}^{N} a_j 1\{Y_n = j\} X_n + \alpha D_n, \tag{9}$$

where a_j is constant equal to $1/2^j$ and where Y_n is the state of the Markov chain. D_n again represents the time between bursts which can be approximated by a Poisson process. As a corollary of Theorem 3, the throughput can be written as

$$\bar{x} = E[X(t)] = \sum_{j=1}^{N} a_j x_j + \frac{\alpha d^{(2)}}{2d}. \tag{10}$$

The estimations of transition probabilities $\hat{p}_{ij}, i, j = 1, ..., N$, of the Markov chain $\{Y_k\}$ are identified from the trace of the TCP connection as follows,

$$\hat{p}_{ij} = \sum_{k=1}^{n-1} 1\{Y_{k+1} = j | Y_k = i\} / \sum_{k=1}^{n-1} 1\{Y_k = 1\}$$

where the Markov chain state Y_k corresponds to the number of transmission rate reduction at the event of the kth aggregate loss and n is the total number of aggregate loss events. If the number of rate reductions at the aggregate loss moment is greater than N, we assume that the Markov chain is in the state N. Since N is chosen so that it is unlikely to have the rate reduced more than N times during a burst, this assumption should not cause any problem. In the following we take $N = 4$.

Using the maximum distance of 0.4s between losses within a burst (Figure 5), we aggregate in bursts the moments at which the transmission is divided by two. As before, we assume that the resulting aggregate loss process is Poisson. We approximate the throughput of the exact fluid model using equations (8) and (10). Figure 7 shows the results. The iid batch model denotes the first case where the number of losses in a burst is described by an iid random variable. The Markovian batch model denotes the second case where this number is described by a Markov chain. We notice that the two methods give approximately the same result which means that the number of losses within a burst is really iid distributed. The result is closer to that of the exact fluid model than the throughput calculated for the Poisson model. However, it is not as good as we expected. The main reason is that we are ignoring the length of a burst which is here comparable to the time between bursts. Possibly, for other connections where losses are more clustered together, this batch method will have a better performance. One may expect that the Markov version of the batch model will perform better than the iid version on connections where strong correlation exists between burst sizes. In the next subsection, we will present a model that accounts for the time the connection spends during a burst.

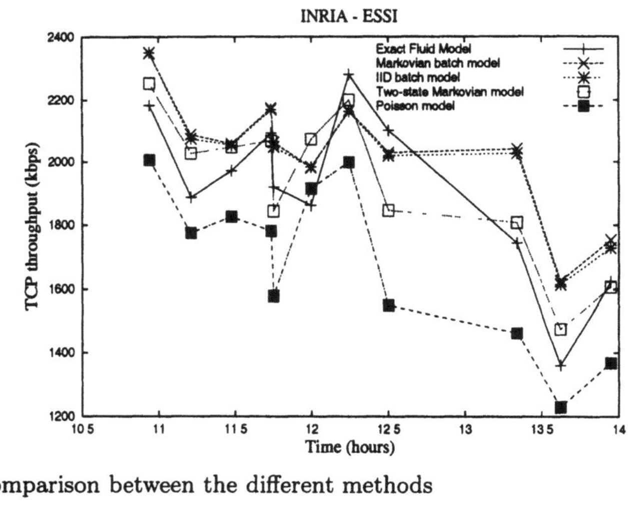

Fig. 7. Comparison between the different methods

4.3 The two-state model

Consider a particular case of our general model where the path switches between two different states. Namely, let $N = 2$ and let the state 1 corresponds to the *Good* state of the path and the state 2 to the *Bad* state. We also denote the transition probabilities of the Markov chain as follows: $p_{11} = g$, $p_{12} = \bar{g} = 1-g$, $p_{21} = \bar{b} = 1-b$

and $p_{22} = b$. The stationary distribution of this chain are equal to,

$$\pi_1 = \frac{\bar{b}}{\bar{b}+\bar{g}}, \qquad \pi_2 = \frac{\bar{g}}{\bar{b}+\bar{g}}$$

The following results can be easily obtained as straightforward corollaries of the theorems for the general N state model.

Corollary 1 *The Laplace Stieltjes Transforms $W(s,i), i = 1,2$, are the solutions of the following implicit equations,*

$$W(s,1) = D^*(\alpha s)[g\int_0^1 W(as,1)dF^1(a)] + D^*(\alpha s)[\bar{b}\int_0^1 W(as,2)dF^2(a)],$$
$$W(s,2) = D^*(\alpha s)[\bar{g}\int_0^1 W(as,1)dF^1(a)] + D^*(\alpha s)[b\int_0^1 W(as,2)dF^2(a)].$$

We shall now proceed to obtain explicit expressions for the first and second moments of the transmission rate at potential loss instants.

Corollary 2 *The first moment of the transmission rate at a potential loss moment is given by*

$$E[X_n] = x_1 + x_2,$$

where

$$x_1 = \alpha d \frac{a_2(\pi_2 - b) + \pi_1}{1 - a_2 b - a_1 g + a_1 a_2 (g + b - 1)} \tag{11}$$

$$x_2 = \alpha d \frac{a_1(\pi_1 - g) + \pi_2}{1 - a_2 b - a_1 g + a_1 a_2 (g + b - 1)} \tag{12}$$

Corollary 3 *The second moment of the transmission rate at a potential loss moment is given by*

$$E[X_n] = x_1^{(2)} + x_1^{(2)},$$

where

$$x_1^{(2)} = \frac{2\alpha d a_1 a_2^{(2)} x_1 (1 - g - b) + 2\alpha d (a_2 x_2 \bar{g} + a_1 x_1 g) + \alpha^2 d^{(2)} (a_2^{(2)} \pi_2 + \pi_1 - b a_2^{(2)})}{\left(1 - g a_1^{(2)} - b a_2^{(2)} - a_1^{(2)} a_2^{(2)} (1 - g - b)\right)} \tag{13}$$

$$x_2^{(2)} = \frac{2\alpha d a_1^{(2)} a_2 x_2 (1-g-b) + 2\alpha d (a_1 x_1 \bar{g} + a_2 x_2 b) + \alpha^2 d^{(2)} (a_1^{(2)} \pi_1 + \pi_2 - g a_1^{(2)})}{\left(1 - g a_1^{(2)} - b a_2^{(2)} - a_1^{(2)} a_2^{(2)} (1-g-b)\right)} \tag{14}$$

Corollary 4 *The throughput, or the time-average of the transmission rate, is given by*

$$E[X(t)] = a_1 x_1 + a_2 x_2 + \frac{1}{2}\alpha \frac{d^{(2)}}{d},$$

where x_1 and x_2 are given in Equations (11) and (12).

Corollary 5 *The second moment of the transmission rate at an arbitrary time instant is given by*

$$E[X^2(t)] = a_1^{(2)} x_1^{(2)} + a_2^{(2)} x_2^{(2)} + \frac{\alpha d^{(2)} (a_1 x_1 + a_2 x_2)}{d} + \frac{1}{3}\alpha^2 \frac{d^{(3)}}{d},$$

where x_1 and x_2 are given in Equations (11) and (12) and $x_1^{(2)}$ and $x_2^{(2)}$ in Equations (13) and (14) respectively.

Next we specialize the model further by taking A_n^j, for $j \in \{1, 2\}$ and $\forall n \geq 0$, to be discrete random variables with values in $\{0.5, 1\}$. Note that $A_n^j = 0.5$ represents the case when a potential loss is transformed into a real loss, namely when it causes a reduction in the transmission rate, whereas $A_n^j = 1$ represents the case when the transmission rate is not reduced at the potential loss moment. We get here the same model as that described in [1]. Note that in [1] we validate via simulation a particular case of this two-state model that corresponds to $p_G = 0$, $p_B = 1$. In the present work, we show how to set the different parameters of the two-state model in its general case. $\{D_n\}$ is the sequence of the times between potential losses. We also denote $p_G := P\{A_n^1 = 0.5\} = 1 - P\{A_n^1 = 1\}$, as the probability of the event when a potential loss is transformed into a real loss in the Good state. Analogously, we define the probability of a potential loss becoming a real loss in the Bad state as $p_B := P\{A_n^2 = 0.5\} = 1 - P\{A_n^2 = 1\}$. We assume that $p_G \leq p_B$. Clearly,

$$a_1 = 1 - \frac{1}{2}p_G \quad \text{and} \quad a_2 = 1 - \frac{1}{2}p_B.$$

Next we demonstrate how the introduced above parameters as well as d and the transition matrix P can be determined from the data in real TCP traces. First, we obtain an estimation of the transition matrix for the Markov chain $\{Y_n\}$. Recall that this is the Markov chain obtained when looking at the state of the channel at potential loss moments. Let $\{S_n\}$ be a sequence of inter-loss times measured from a TCP trace. We need to determine when the path is in the "Good" state and when it is in the "Bad" state. We use the following simple method. Choose some time interval τ. We will explain later how to make this choice. If the inter-loss time S_n is less than τ then the path is in the Bad state, otherwise the path is considered to be in the Good state. If two or more inter-loss times correspond to the same state, we will merge these intervals together and call the new interval L_k^G or L_k^B depending on the state. Note that these new intervals represent the time during which the path of the connection is either in the Good or in the Bad state. Denote n_G (resp. n_B) the number of the time intervals S_k^G (resp. S_k^B) during the time interval that we use for measurement. Then, the evolution of the path of the TCP connection can be described by a two-state continuous time Markov process with the following infinitesimal generator matrix,

$$Q = \begin{bmatrix} -\sigma_G & \sigma_G \\ \sigma_B & -\sigma_B \end{bmatrix} \tag{15}$$

where the rates σ_G and σ_B are calculated as follows:

$$\sigma_G = \frac{1}{E[S_k^G]} \simeq \frac{n_G}{\sum_{k=1}^{n_G} S_k^G}, \quad \sigma_B = \frac{1}{E[S_k^B]} \simeq \frac{n_B}{\sum_{k=1}^{n_B} S_k^B}.$$

Note that on some paths, say a wireless link, this Markov chain is a priori known and can be directly used without the need to look at the trace of the TCP connection. In case it is not known, we need to define it using the parameter τ as described above. We present now two approaches for the determination of τ. The first one is more empirical. We look at the histogram of the inter-loss times (Figure 3) and we choose τ as the time separating the two distributions it encloses (0.4s in the figure). The second method is less empirical and was used in the context of Markov-modulated Poisson processes [10]. In this second approach we define parameter τ as the expectation of the inter-loss times, that is

$$\tau = E[S_k] \simeq \frac{1}{n}\sum_{k=1}^{n} S_k,$$

where n is the total number of inter-loss intervals we get from the trace. Given the continuous time Markov chain associated to the channel, we can now extract the parameters of the discrete time Markov chain embedded at the potential loss moments. We use for this purpose the uniformization technique [14]. Let us choose the potential loss process $\{D_n\}$ as a Poisson process with intensity $1/d$ higher than both σ_G and σ_B. For example, a reasonable choice of d is the estimation of the average Round Trip Time of the connection. According to the uniformization technique [14], the state of the path described by the Markov process (15) and sampled at the moments of potential losses can be equivalently given by a discrete time Markov chain with the following transition matrix,

$$P = \begin{bmatrix} 1 - d\sigma_G & d\sigma_G \\ d\sigma_B & 1 - d\sigma_B \end{bmatrix}.$$

Having chosen d and calculated σ_G and σ_B from the trace, we can easily deduce the parameters b and g of the loss model. Namely, $\bar{g} = d\sigma_G$ and $\bar{b} = d\sigma_B$. Now we determine p_G and p_B. Let ω_k^G (ω_k^B) be the number of real losses in the time interval S_k^G (resp. in S_k^B). Then the probabilities p_G and p_B are given by

$$p_G = \frac{\sum_{k=1}^{n_G} \omega_k^G}{\sum_{k=1}^{n_G} S_k^G / d} = \frac{d\sum_{k=1}^{n_G} \omega_k^G}{\sum_{k=1}^{n_G} S_k^G} = d\lambda_G, \quad p_B = \frac{\sum_{k=1}^{n_B} \omega_k^B}{\sum_{k=1}^{n_B} S_k^B / d} = \frac{d\sum_{k=1}^{n_B} \omega_k^B}{\sum_{k=1}^{n_B} S_k^B} = d\lambda_B.$$

$1/\lambda_G$ and $1/\lambda_B$ represent the average time between window reductions in the Good and in the Bad state respectively. For the same eleven traces obtained in our experiments, we calculated the parameters of the model. We use $\tau = \delta$ =0.4s to separate the Bad state from the Good state. In Figure 7, we compare the result with that of the exact fluid model. A close match is noticed. In addition to the good results and the closed form expression it provides, this model has the advantage of having simple parameters. All what we need to approximate the throughput is the parameters of the two-state Markov chain associated to the path and the intensity of losses in both states. Concerning the parameter d, it is enough to choose in a way that the intensity of potential losses $1/d$ is higher than the intensity of losses in the Bad state λ_B.

5 Concluding Remarks

We considered in this paper a multi-state Markov path for describing the loss process experienced by a connection that has a linear window increase between losses, and multiplicative decrease upon a loss event. The modelling of some channels using a Markov chain with more than two states have long been advocated, see e.g. [11,13].

Using an approach based on the Laplace Stieltjes Transform, we derived explicit expressions for the two first moments of the transmission rate of the connection just prior to losses, as well as the two first moments of the steady state throughput. We note that the expression for the second moment of the throughput could be useful in designing TCP friendly protocols for real time applications [6] in which other parameters of the linear increase and multiplicative decrease are chosen so as to maintain the same expected throughput (as a function of the loss process and of the round-trip time) as the original TCP protocol. (The latter requirement on the expected throughput stmes from fairness arguments.) Such applications (e.g. interactive voice or video connections) typically require a smaller variance of the throughput than the one of the original TCP in order to ensure a reasonable quality of service.

We have recently succeeded also in analysing non Markovian channels [2], and obtain similar performance measures using a completely different approach (that relies on some covariance functions of the interloss times). The approach obtained here, in contrast, leads to formulae that involve only a finite and small number of parameters that can be easily computed. In addition, we proposed here methods for the identification of such parameters.

References

1. E. Altman, K.E. Avrachenkov, and C. Barakat, "TCP in presence of bursty losses", to appear in *Performance Evaluations*. A shorter version appeared in the *Proceedings of ACM SIGMETRICS*, Santa Clara, California, Jun 2000.
2. E. Altman, K.E. Avrachenkov, and C. Barakat, "A stochastic model of TCP-IP with stationary ergodic random losses", ACM SIGCOMM, Aug. 28 - Sept. 1, Stockholm, Sweden, 2000.

3. F. Baccelli and P. Bremaud, "Elements of queueing theory: Palm-Martingale calculus and stochastic recurrences", *Springer-Verlag*, 1994.
4. A. Brandt, "The stochastic equation $Y_{n+1} = A_n Y_n + B_n$ with stationary coefficients", *Adv. Appl. Prob.*, Vol. 18, pp. 211-220, 1986.
5. K. Fall and S. Floyd, "Simulation-based Comparisons of Tahoe, Reno, and SACK TCP", *ACM Computer Communication Review*, Jul 1996.
6. S. Floyd, M. Handley and J. Padhye, "Equation-based congestion control for unicast applications: the extended version", *ACM Sigcomm*, Aug. 28 - Sept. 1, Stockholm, Sweden, 2000.
7. E.N.Gilbert, "Capacity of a burst-noise channel", Bell Systems Technical Journal, Vol. 39, pp. 1253-1265, Sep 1960.
8. P. Glasserman and D.D. Yao, "Stochastic vector difference equations with stationary coefficients", *J. Appl. Prob.*, Vol. 32, pp. 851-866, 1995.
9. V. Jacobson, "Congestion avoidance and control", *ACM SIGCOMM*, Aug 1988.
10. K.S. Meier-Hellstern, "A fitting algorithm for Markov-modulated Poisson processes having two arrival rates", *Euro. J. Oper. Res.*, Vol. 29, pp. 370-377, 1987.
11. M. Rahman, M. Bulmer and M. Wilkinson, "Error models for land mobile satellite channels", *Australian Telecommunication Research*, Vol. 25 No 2, pp. 61-68, 1991.
12. W. Stevens, "TCP Slow Start, Congestion Avoidance, Fast Retransmit, and Fast Recovery Algorithms", *RFC 2001*, Jan 1997.
13. B. Vucetic and J. Du, "Channel modeling and simulation in satellite mobile communication systems", *IEEE J. on Selected Areas in Communnications*, Vol. 10, No. 8, pp. 1209-1218, 1992.
14. J. Walrand, "An introduction to queueing networks", *Prentice Hall*, 1988.

ISP's Internet Backbone Augmentation using Virtual Link Configuration in Link-state Routing

Do-Hoon Kim[1], Soon-Ho Lee[2], and Dong-Wang Tcha[3]

[1] Graduate School of Management, KAIST, 207-43 Cheongryangri-Dong, Dongdaemun-Gu,
Seoul 130-012, Korea, dhkim@telmal.kaist.ac.kr
[2] Graduate School of Management, KAIST, iskra@kgsm.kaist.ac.kr
[3] Graduate School of Management, KAIST, tchadw@sorak.kaist.ac.kr

Abstract: This paper addresses the backbone augmentation problem arising from ISP's hierarchical link-state routing operations. Focusing on Virtual Link (VL) configuration, proposed is an efficient augmentation scheme of increasing redundancy of the original backbone topology. A VL restores hidden information from introducing hierarchy in the topological database of each backbone router, thereby increasing redundancy of the backbone and preventing partition of the backbone even if some links fail. With given potential VL locations and the original backbone topology, we propose a bi-objective efficient VL configuration model that explicitly evaluates the benefit side as well as the cost side of VL configurations, and fully leverages the trade-off between both sides. Furthermore, provided is an efficient branch-and-bound algorithm (implicit enumeration) with good lower and upper bounds so that solution speed can be accelerated in most cases. To our knowledge, this is the first try to attack the VL configuration problem systematically. For an ISP with its own domain, the proposed model and algorithm are expected to relieve network administrators burden of configuring VLs, support making the backbone more tolerable to backbone link failures, and finally, provide a practical vehicle for reliable link-state hierarchical routing indispensable to overall service quality.

1. Introduction

As the number of Internet users grows, scalability issues are emerging as one of the most fundamental network operation problems. Scalability issues exist at every layer of the Internet architecture: at IP (Internet Protocol) routing layer, for example, the explosion of flooded routing information is witnessed as the network size increases. To cope with scalability issues, ISPs, which are intermediaries connecting local users to the global Internet and whose competitive edges are built around routing efficiency, hierarchically divide their own networks (so-called Autonomous Systems (ASs)) into two tiers, local distribution areas and the backbone: thereby, ISPs can limit the range of routing information exchange and resolve the major scalability issue. However, this gain from hierarchy causes side effects. Besides operational inconvenience, the biggest cost of hierarchical configuration is performance degradation due to limitation of available routes, which may not guarantee optimal routing. Furthermore, in hierarchical implementation of link-state routing protocols, such as OSPF (Open Shortest Path First) and IS-IS(Intermediate System-Intermediate

System), the overall performance of the entire AS is very much dependent on the backbone since the routers in the backbone play the core functions to gather, organize, and redistribute routing information across local distribution areas. Thus, sound connectivity of the backbone is critical success factor for running hierarchical link-state networks ([1], [2], [3], [4]). Along this line, an efficient method to increase redundancy of the backbone should be developed as one of the measures to make the backbone robust.

Redundancy is important to prevent the backbone from being disconnected when backbone links or nodes fail. For example, two edge-connectivity should be satisfied between every pair of backbone routers by configuring back-up links so that no single link or node failure may cause disconnection of the backbone. For a hierarchically configured link-state AS, the following two tracks are suggested as basic options to introduce redundancy into the backbone.

- Track 1: to install new point-to-point links
- Track 2: to configure virtual links in the routing databases of the backbone routers

The first track, a general option to increase connectivity of a network not confined to a link-state network, is to build or lease new serial links or an intermediary fast-switched network between the backbone routers. While this option secures good results to enhance the backbone connectivity, it entails lots of monetary cost that may violate budget constraint. Furthermore, considering the rapid growth of Internet users, relatively long time to completely deploy the option may limit the network operational flexibility, which is a key requirement for service operation systems in the flying-demand phase.

Fortunately, many link-state protocols provide an easier mechanism to generate back-up links than track 1. Remark that with hierarchical link-state routing protocols, every backbone router maintains a map(so-called Topological DataBase, TDB) representing the physical backbone topology and may have alternative routes that are not present in its TDB unless the entire AS is disconnected. Virtual Links (VLs) restore such hidden information in the TDB of the backbone routers. Since no monetary expense but a little inconvenience and operational complexity is incurred to configure VLs, track 2 has economic and time-saving advantages over the track 1 in building back-up links within short time and leaves operational flexibility intact. This option becomes more powerful when the AS itself shows ample connectivity.

Even though backbone augmentation by VLs is easier and more economical than by construction of new links, many network administrators still find it difficult to decide where to configure VLs. However, except some practical tips for implementing VLs in the backbone routers, few literatures deal with an efficient method or even a rule-of-thumb to configure VLs in order to enhance the backbone connectivity ([3]). We guess that this is because link-state routing protocols is still rapidly spreading, and believe that VL configuration issues will come to the surface as link-state protocols are attaining the status of de-facto standard for Internet routing. To our knowledge, this paper is the first try to attack the VL configuration problem systematically.

In sum, our goal is to address, analyze, and formulate the optimal VL configuration problem under hierarchically configured link-state networks and develop efficient solution methods within the context of branch-and-bound algorithm. Since there are many backbone routers as well as potential location of VLs in general, the algorithmic complexity spent in determining optimal location of VLs should not be overlooked. Therefore, to develop an efficient branch-and-bound algorithm (implicit enumeration) matters, and such an algorithm may relieve network administrators of the burden of configuring VLs, thereby making the backbone more robust to backbone link failures. In the next section, we first describe the VL back-up system and then analyze the costs and benefits arising from configuring VLs. Section 3 provides a mathematical model for determining optimal locations of VLs in terms of both benefits and costs. We present a solution method based on branch-and-bound algorithm and some numerical examples in section 4. By summarizing important features of the model and the results and presenting future works, we conclude this paper.

2. Virtual Link Configuration Model

2.1 Role of Virtual Links in Hierarchical Link-state Networks

In link-state routing protocol, routers exchange their piecemeal topological information and construct a map (TDB) representing the overall network topology. As the network size increases, the volume of these transactions and TDB size grow exponentially. So does the resource consumption for routing management and control. Because of this scalability issues, large link-state networks are partitioned into relatively small local areas within each of which routing information exchange among member routers is limited. In order to maintain the entire connectivity (i.e., inter-area routing), a backbone connecting local areas should be constructed. That is, a valid backbone must be contiguous (i.e., all backbone routers should be connected to other backbone routers through backbone links only) and ensure direct connection of each local area ([2], [3]). Figure 1 shows an example of a valid hierarchical configuration.

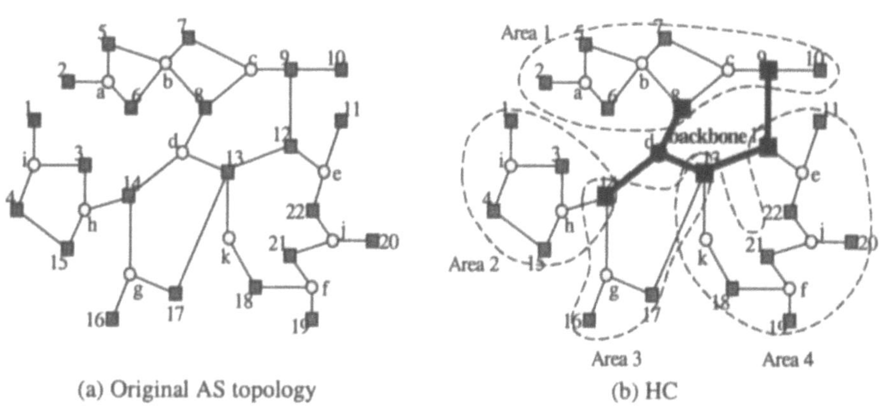

Figure 1: Hierarchically configured link-state network and VL configuration example

The backbone is extremely important in hierarchically configured link-state network. Good practices for backbone design suggest that stability and redundancy are the most important criteria for the good backbone ([1], [2], [3], [4]). To satisfy the stability criterion when designing hierarchical clustering of the AS, placing hosts (workstations, file servers, and so on) in the backbone should be avoided and the backbone should be kept as simple and small as possible ([1], [2], [5]). After hierarchy design, enhancing the backbone configuration in terms of connectivity is accomplished. This augmentation, which can be realized by introducing redundancy with VLs, aims at preventing partition of the backbone when some links fails. That is, VLs work as back-up links to repair the backbone connectivity. Accepted link-state network design practices suggest that the use of VLs should be considered for a backbone poorly designed as a result of unbalanced hierarchical configuration ([2], [3], [6]).

A VL can be configured between separate gate routers that touch the backbone from each side and have a common area. In the case of figure 1-(b), at least three VLs, which connect routers 8 and 9, 12 and 13, 13 and 14 across area 1, area 4, and area 3, respectively, can be configured. Then the VL acts in a similar way as in a tunnel: i.e., a VL creates a path between two gate nodes by using non-backbone links. For example, OSPF routing protocol treats two routers joined by a VL as if they were connected by an unnumbered point-to-point network and registers the VL in TDB of the backbone routers. A VL can only be configured on gate nodes, and in general, only one of the potential VL configurations passing the same local area is configured. The stability of a VL is dependent on the stability of the area it traverses, and the reliability measure (cost metric) of a VL can be defined as that of the weakest link of the links constituting the VL. Also, the amount of the effort required to configure and maintain a VL is proportional to its length (the number of routers on the path corresponding to the VL).

Even though it is desirable to configure sufficient VLs to build back-up links, additional VL configurations required for this purpose cause some burden on the backbone and increase the complexity of the system operation due to the characteristics of the VLs. Therefore, configuring all the potential VLs may lead to an inefficient augmentation of the backbone connectivity. In the following sections, we propose an efficient VL configuration model that fully leverages the trade-off between the benefits and costs of configuring VLs and solution methods to this model.

2.2 Issues in Virtual Link Configuration

Suppose that the required connectivity level for each backbone router is given by the network administrator, then there arise two cases: (1) case 1: all connectivity requirements can be satisfied by using all or some of the potential VLs, (2) case 2: instances other than the case 1. Setting the cost to configure VLs aside temporarily, we can easily check whether case 1 happens or not by applying Gomory-Hu tree method ([7]) on the augmented topology with all the potential VLs. Furthermore, if all the costs to configure VLs were the same, the case 1 would reduce to a simple graph augmentation problem with cardinality objective function, which has a polynomial time algorithms ([8]). However, in the case of heterogeneous VL costs, no polynomial time algorithm but some heuristics has been known so far.

However, the above survivable network design approaches and solution methods are no longer applicable to most practical situations where we should not assume that connectivity requirements of backbone routers are given or available VLs cannot fulfil the given requirements (case 2). For these situations where connectivity requirements have no meaning, the single objective (i.e., minimizing total cost) decision model should be extended in order to explicitly evaluate the benefit side of configuring VLs and to achieve an optimal balance between benefits and costs: benefits from increased connectivity and costs of configuring and maintaining VLs. As a result, the issue of configuring VLs should be multi-objective decision model looking for an augmented backbone topology that maximizes the net gain (= total benefit – total costs) by setting-up VLs.

2.3 A Decision Model for Virtual Link Configuration

The decision problem in this paper is to determine an optimal configuration of VLs with a given set of potential VL locations and the original backbone topology. Within this decision context, a decision alternative is an augmented graph from the original backbone topology by choosing some VLs (decision variables). Thus, in general, there exist finite but many decision alternatives. We will evaluate a decision alternative in terms of two attributes: the aggregated connectivity of the augmented graph and the cost for configuring VLs in the augmented graph. Specifically, compensatory preference (so called utility function of the attributes) approach will be employed for this bi-criteria decision-making problem in order to explicitly reflect the trade-off between attributes and to resolve commensurability issues in order to compare heterogeneous attributes. In this paper, we impose a minimum level of restrictions on the form of utility functions so that the network administrator can have abundant flexibility to develop unique utility functions customized to their own situations.

We are given the original topology of the backbone that can be represented as a simple graph $G = (V, E)$ where V and E represent the set of backbone routers and the set of backbone links, respectively. Also, given is A ($A \cap E = \varnothing$), the pre-defined set of potential VL locations (the set of variables) with assigned installation cost w_e ($e \in A$) which is proportional to the length of VL e. Let $G_Q = G(V, E \cup Q)$ denote an augmented topology by Q ($\subseteq A$), which corresponds to a decision alternative. Note that G_Q is no more a simple graph and may contain parallel links. Define a vector of costs to configure VLs in Q as $\tilde{w} = (w_1, \cdots, w_{|Q|})$. And the nodal connectivity c_v on G means $\min_{u \in V \setminus \{v\}} \gamma(v, u \mid G)$, where $\gamma(v, u \mid G)$ is the maximum number of link-disjoint paths between u and v on G. That is, for a given node v, any failure of (c_v-1) links cannot cause node v isolated from the other part of the backbone. According to this definition, all the nodal connectivity has the same value corresponding to a bottleneck point of the given graph G. This value is, in fact, the cardinality of a minimum cut of G and will be written as $\bar{c}$, which is efficiently computed by Gomory-Hu tree method or others ([7], [9]). Then after determining Q from A, $\bar{c}$ and $\tilde{w}$ represent the attributes of an alternative G_Q.

To complete the decision model for optimal configuration of VLs, utility functions for attributes and total utility function are employed. We first introduce $U_C(G_Q)$ and $\bar{U}_W(G_Q)$, each of which represents the network administrator's evaluation of benefit from connectivity level $\bar{c}$ of G_Q and cost to install and maintain $|Q|$ VLs, respectively. $\bar{U}_W(\text{-})$ can be interpreted as disutility of costs incurred by VL configuration, and is assumed to be $\bar{U}_W(G_Q) = \sum_{e\in Q} \bar{u}(w_e)$ where $\bar{u}(w_e)$ represents disutility in configuring the VL e whose length is w_e. Both $U_C(\text{-})$ and $\bar{u}(\text{-})$ are monotonic non-decreasing function of the connectivity level and VL lengths, respectively(thus, $\bar{U}_W(G_Q)$ is also monotonic non-decreasing to the sum of lengths of VLs in Q). Finally, the network administrator's total utility is described as an additive form of $U(G_Q) = \alpha U_C(G_Q) - \beta \bar{U}_W(G_Q)$. Now we pose the decision model for Optimal VL Configuration (OVLC) as follows:

$$\textbf{[OVLC]}\ \text{Max}_{Q\in A}\ U(G_Q)$$

Figure 2 skeletonizes the decision model explained so far and shows the process to generate and evaluate a decision alternative.

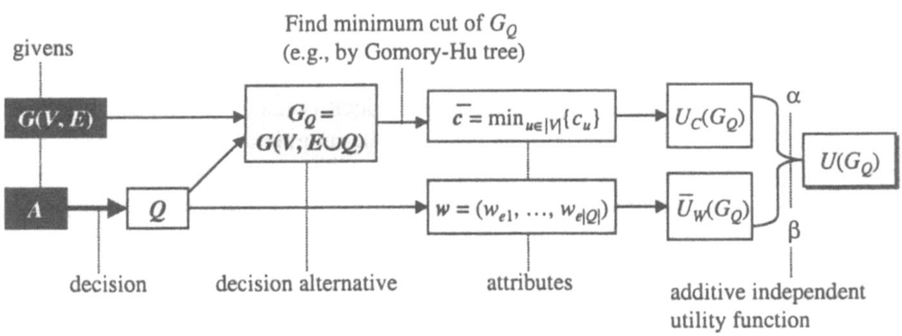

Figure 2: Decision model of optimal VL configuration

3. Solution Method and Examples

In this section, we propose an efficient branch-and-bound algorithm (implicit enumeration). We first demonstrate that [OVLC] is not an easy problem by presenting a counter example that shows a naive solution method based on component-by-component comparison of decision variables cannot succeed. Suppose that $A = \{e_1, e_2\}$ with $w_{e_1} = w_{e_2} = 1$, $U_C(G_Q) = \sum_{v\in V} ac_v$, $\bar{U}_W(G_Q) = b|Q|$, and $U(G_Q) = U_C(G_Q) - \bar{U}_W(G_Q)$. With the original backbone graph shown in the figure 3 and $a \geq (2\ /\ 7)\times b$, no single link (i.e., $Q = \{e_i\}$) seems profitable if it stands alone (its configuration cost exceeds its benefit). However, using e_1 and e_2 together (i.e., $Q = A$) gives a good augmented topology whose benefit far exceeds the cost. This example implies that we have no choice but to examine $2^{|A|}$'s decision alternatives (all possible subsets of A) to find optimal solution at the worst case.

Considering the practical size of the ISP's backbone and the number of potential VL locations, enumerating all the alternatives may discourage field implementation of [OVLC] model. At this point, an efficient enumeration method is requested. We present a branch-and-bound algorithm (implicit enumeration) with good lower and upper bounds so that solution speed can be accelerated.

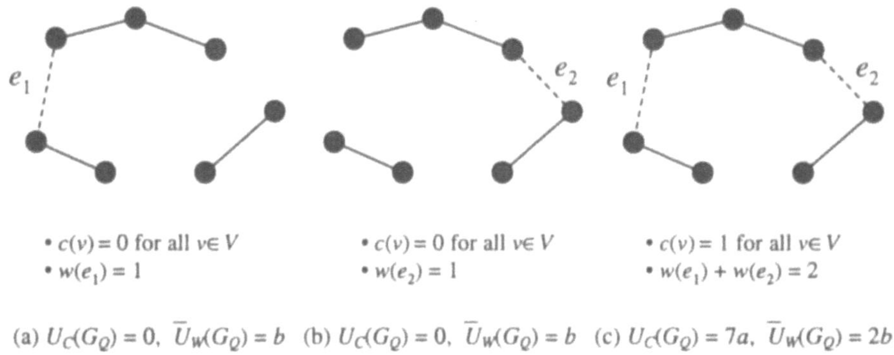

Figure 3: [OVLC] Example and inefficiency of component-by-component comparison

Each Restricted Problem k, [RP^k] in the branch-and-bound tree is characterized as a partition of A (set of potential virtual link locations), $\{A^k, \overline{A}^k, F^k\}$ where A^k and $\overline{A}^k$ represent VL that should be included and excluded in [RP^k], respectively, and $F^k = A-(A^k \cup \overline{A}^k)$, that is, VLs to be determined in [RP^k]. Let a relaxation problem of [RP^k] be denoted by [R-RP^k]. Objective relaxation is employed to compute upper bound to the [RP^k] easily: that is, by ignoring cost attribute (i.e., dropping the term $\overline{U}_W(\cdot)$ from the objective), defined is [R-RP^k]. Since $U_C(\cdot)$ is non-decreasing function of the nodal connectivity level, which is also non-decreasing to the density of the augmented graph, we can easily show that the optimal solution of [R-RP^k] is $Q_R{}^k = A^k \cup F^k$ with its value of $z_R^k = U_C(G_{Q_R^k})$, an upper bound to [RP^k]. Also, let $z(Q_R^k) = U(G_{Q_R^k})$ then $z(Q_R^k)$ becomes a trivial lower bound to [RP^k]. However, the following property shows that we can construct a tighter lower bound from $Q_R{}^k$. Also, the property can be used to choose branching variables or to fix some variables in F^k in the optimal solution to [RP^k], thereby reducing the search space in branch-and-bound algorithm. Before presenting the property, we define $L(e|G_Q) = U_C(G_Q) - U_C(G_{Q-\{e\}})$, which denotes benefit loss from losing a VL e from Q.

<u>**Property**</u>: If the following inequality (1) holds for an element ε in F^k of [RP^k], the solution $\overline{Q} = A^k \cup (F^k - \{\varepsilon\})$ cannot be an optimal solution to [RP^k].

$$L(å \mid G_{Q_R^k}) > \overline{u}(w_\varepsilon) \quad \ldots (1)$$

Proof) For a while, the superscript k will be suppressed. Let Q^* and Q be optimal solutions and any feasible solution to a restricted problem [RP], respectively. Then by definition of Q^* and Q, the following inequality holds:

$$U(G_{Q^*}) = U_C(G_{Q^*}) - \overline{U}_W(G_{Q^*}) \geq U(G_Q) = U_C(G_Q) - \overline{U}_W(G_Q),$$

$$\text{or } U_C(G_Q) - U_C(G_{Q^*}) \leq \overline{U}_W(G_Q) - \overline{U}_W(G_{Q^*}) \quad \ldots (2)$$

Note that (2) defines a necessary condition for Q^* to be an optimal solution to the [RP]. Now, suppose that a solution Q° constructed from Q_R by dropping an element $\varepsilon(\in Q_R)$ satisfying the inequality (1), be an optimal solution to the [RP]: that is, an optimal solution Q° satisfies $L(\varepsilon|G_{Q^0}) = U_C(G_{Q_R}) - U_C(G_{Q_{R-\{e\}}}) = U_C(G_{Q_R}) - U_C(G_{Q^0}) > \bar{u}(w_\varepsilon) = \bar{U}_W(G_{Q_R}) - \bar{U}_W(G_{Q^0})$ (the last equality comes from the definition of $\bar{U}_W(\text{-})$). However, this inequality violates the necessary condition (2), thereby leading to a contradiction. (Q.E.D.)

Inequality (1) means that connectivity loss caused by losing ε on $G_{Q_R^k}$ is large enough to offset-cost saving by uninstalling the VL ε. Thus, $g(e|G_Q) = L(e|G_Q) - \bar{u}(w_e)$ denotes net gain by adding VL e on $G_{Q-\{e\}}$. Remark that we can easily compute or estimate $L(e|G_Q)$, thereupon $g(e|G_Q)$, from a Gomory-Hu tree on G_Q. We can use this property to estimate an upper bound to a successor of [RP^k] when dividing the [RP^k] with ε as a branching variable: that is, an upper bound to a child problem, [RP^{k+1}] defined by excluding ε in consideration (i.e., $\bar{A}^{k+1} = \bar{A}^k \cup \{\varepsilon\}$ and $F^{k+1} = F^k - \{\varepsilon\}$). Since the optimal value of [R-RP^{k+1}] is already known to be worse than that of [R-RP^k], it is likely that deferring to choose [RP^k] as the next exploring candidate will terminate the branch-and-bound process rapidly. In sum, the results from the property may guide branching variables as well as choosing restricted problems in enumeration.

Step 1 (initialization): Set $k = 0$, LIST $= \{(\varnothing, \varnothing, A)\}$, gl(global lower bound) $= -\infty$, and gu(global upper bound) $= \infty$; $k \leftarrow k+1$.
Step 2 (Termination Test): If LIST $= \varnothing$ then the solution Q^* that yields the current gl is optimal solution and exit.
Step 3 ([RP^k] Selection and Relaxation):
(1) Select and delete a restricted problem [RP^k] from LIST.
(2) Solve relaxed [RP^k] ([R-RP^k]); Let z_R^k the optimal value of the [R-RP^k] and Q_R^k the corresponding optimal solution.
Step 4 (Pruning):
(1) If $z_R^k \le gl$ then go to Step 2.
(2) Else if $z(Q_R^k)$ the objective value determined by Q_R^k is greater than gl then update gl and delete from LIST every restricted problem [RP^q] with $z_R^q \le gl$.
(3) Else go to Step 5.
Step 5 (Branching): Let $\{RP^{k_j}\}_{j\in J}$ be a division of [RP^k]; Add $\{RP^{k_j}\}_{j\in J}$ into LIST; Go to Step 2.

Figure 4: Branch-and-bound algorithm for [OVLC]

Also, we build a solution (let say Θ), which consists of only VLs satisfying the inequality (1) at each [RP^k], thereafter comparing the total utility of Θ with that of

Q_R^k. Since those VLs seem the most profitable VLs on $G_{Q_R^k}$, we expect that Θ produces better lower bound than Q_R^k. Even if Θ is no better than Q_R^k, we can improve Θ by sequentially adding the remaining free VLs in the decreasing order of $g(e \mid G_{Q_R^k})$, hoping that a better solution could be achieved in the course of sequential addition. Figure 4 summarizes the branch-and-bound algorithm for [OVLC].

We have tested the proposed framework and branch-and-bound algorithm on some randomly generated problems. However, due to time constraint on this paper, we could not incorporate new lower bounding scheme, but illustrate experiment results for small sample networks. The costs of configuring/maintaining VLs and the potential locations as well as the available number of VLs are randomly chosen in some ranges. Also, for utility of nodal connectivity (u(-)) and disutility of VL cost($\bar{u}$(-)), employed are a concave logarithmic function and a linear function, respectively. Since many literatures on decision analysis (for example, [10]) provide a lot of utility functions used in practical situations, the decision maker of [OVLC], a network administrator, can consult these literatures to choose the best one for his/her own purpose. Finally, given functional forms of u(-) and $\bar{u}$(-), the various levels on coefficients in total utility function U(-) have been analyzed to see how network administrator's relative preference on the backbone connectivity and VL costs affects the results.

Figure 5 shows computational results of the algorithm for [OVLC] on a sample network with 10 nodes, 19 links, and 7 potential VL locations. We used the total utility function with equal weights for connectivity benefits and VL configuration costs (i.e., $\alpha = \beta = 1$). In this case, configuring only two VLs can increase network connectivity from 2 to 3, and the average connectivity of a node pair ($\gamma(u, v)$) increases from 2.97 to 3.47.

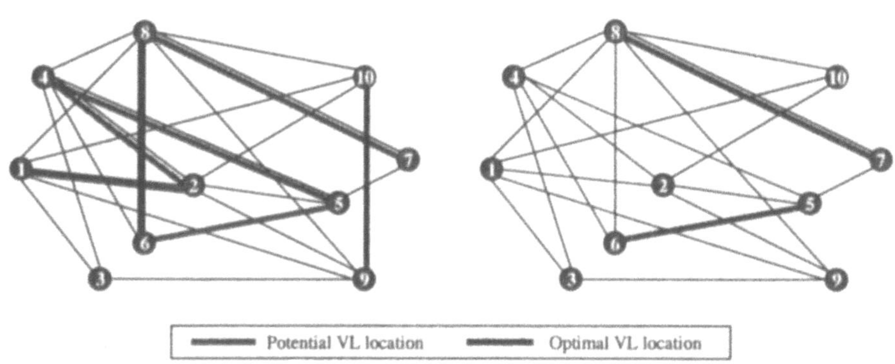

(a) Sample network and potential VL locations (b) Optimal VL locations

Figure 5: Example with $G = (V, E)$ and A where $|V| = 10$, $|E| = 19$, and $|A| = 7$

Also, Figure 6 summarizes sensitivity analysis results by varying weight ratio (α/β) in additive utility function and the size of potential VL locations ($|A|$), respectively. Note that for each instance of varying size of A, we randomly generated ten different sets of potential VL locations in order to see average behaviours of optimal solutions of [OVLC] according to the changes in parameters. As these graphs indicate, the amount of available resources (|A|) seems to differently affect the optimal solutions (the optimal structures of decision alternative, G_Q^*) and the optimal total utility values. The optimal utility values are rather independent of |A|, whereas the number of VLs in optimal solutions tends to increase as |A| increases. However, impacts of utility weight ratio on the number of optimal VLs as well as the optimal total utility look very strong. Furthermore, the result of experiments implies that a certain range of utility weight ratios lead to similar behaviours: for instance, the graphs corresponding to the weight ratio 1 and 1.5 coincide at many levels of A.

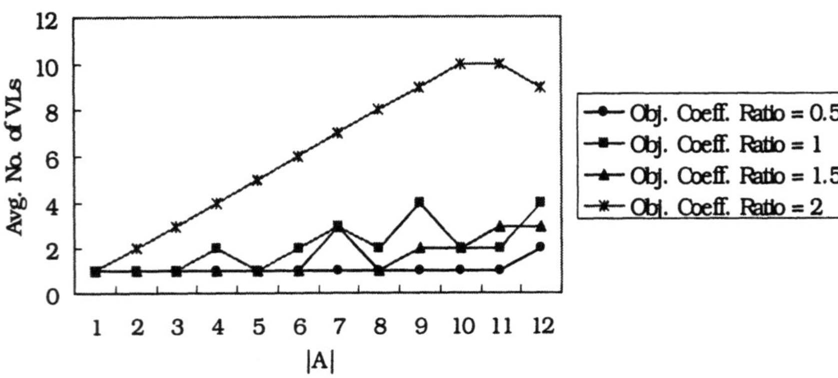

(a) Changes in number of VLs in optimal solutions

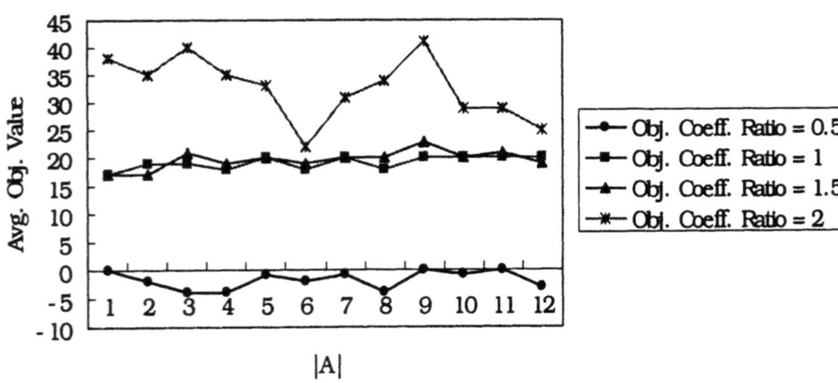

(b) Changes in total utility values

Figure 6: Sensitivity analysis results varying potential VL locations (|A| = 1, ..., 12) and utility weight ratio (α/β= 0.5, 1.0, 1.5, 2.0)

4. Concluding Remarks and Future Works

In this paper, we proposed a framework for enhancing the backbone configuration of ISPs' hierarchical link-state Ass (Autonomous Systems). Good backbone configuration should be accomplished so that some link failures may not cause an isolation of a portion of the backbone. Focusing on Virtual Link (VL) installation, proposed is an efficient augmentation scheme of increasing redundancy of the original backbone topology. With given potential VL locations and the original backbone topology, we propose a bi-objective efficient VL configuration model that explicitly evaluates the benefit side as well as the cost side of VL installation, and fully leverages the trade-off between both sides. Furthermore, provided is an efficient branch-and-bound algorithm (implicit enumeration) with good lower and upper bounds so that solution speed can be accelerated in most cases.

To our knowledge, this is the first try to address the VL configuration problem systematically. For an ISP with its own domain, the proposed model and algorithm are expected to relieve network administrators from the burden of configuring VLs, support making the backbone more tolerable to backbone link failures, and ultimately provide a practical vehicle for reliable link-state hierarchical routing indispensable to overall service quality. For example, the framework is employed to solve short-term network connectivity issues.

Due to unavailability of studies of the same kinds and time limitation, we could not present rich experimental results. However, comprehensive experimental results will soon be reported, performed on several kinds of underlying backbone networks which are randomly generated but devised to reflect the real ISP network characteristics: for example, the remarks shown in [11]. Furthermore, different forms of utility functions will be tested in order to provide and analyze options that a network administrator may take in practice. In particular, we'll develop a set of normalized multi-attribute utility functions, which are not compulsory but employed common practice of multi-criteria decision making ([10]).

Lastly, as an extension of the proposed model, we consider imposing additional constraints on VL configuration: for example, a constraint on the number of VLs that can be installed at the same time.

References

[1] Behrens, J. and Garcia-Luna-Aceves, J.J. (1998) Hierarchical routing using link vectors, *Proceedings of IEEE INFOCOM 98*, 702–710

[2] Moy, J.T. (1998) OSPF: Anatomy of an Internet Routing Protocol, Addison Wesley

[3] Thomas, M.T. II. (1998) OSPF Network Design Solutions, Cisco Press

[4] Tsai, W.T., C.V. Ramamoorthy, W.K. Tsai, and O. Nishiguchi (1989) An adaptive hierarchical routing protocol, *IEEE Transaction on Computers*, 38 1059–1075

[5] Kim, D. and Tcha, D.W., Scalable domain partitioning in Internet OSPF routing, *Telecommunications System*, forth-coming

[6] Halabi, S. (1996) OSPF design guide, Cisco Systems Network Supported Accounts (white paper)

[7] Hu, T.C. (1982) Combinatorial Algorithms, Addison Wesley

[8] Grotschel, M., C.L. Monma, and M. Stoer (1995) Design of survivable networks (in Network Models, eds. M.O. Ball et al.), North-Holland

[9] Padberg, M. and Rinaldi, G. (1990) An efficient algorithm for the minimum capacity cut problem, Mathematical Programming, 47, 19–36

[10] Clemen, R.T. (1996) Making Hard Decisions: an Introduction to Decision Analysis, 2nd ed., Duxbury Press

[11] Zegura, E.W., Calvert, K.L., and Bhattacharjee, S. (1996) How to model an internetwork, Proceedings of IEEE INFOCOM 96, 594–602

Part IV

Mobile Telecommunication I

A Local Anchor Scheme for Mobile IP

Jianzhu Zhang and Jon W. Mark

University of Waterloo, Waterloo, Ontario, Canada N2L 3G1

Abstract. The Mobile IP protocol requires that the mobile computer registers with its home agent through the foreign agent whenever it moves into a new serving domain. If the mobile is far away from its home network, the signaling cost cannot be ignored and the long registration delay would introduce extra traffic cost. The local anchor scheme defines a two-tier registration procedure. When the mobile moves within the coverage of the current anchor, it only needs to register with the current anchor, rather than with its home agent. Thus, the frequency that the mobile needs to register with its home agent is effectively reduced. Comparisons on handoff delays, TCP/UDP costs, and TCP throughput show that the local anchor-based Mobile IP scheme exhibits superior performance over the regular Mobile IP scheme.

1 Introduction to the Local Anchor Scheme

Mobile IP is designed to provide seamless and transparent delivery to a mobile host regardless of its current location. In the regular Mobile IP scheme [1], whenever the mobile moves into a new serving region, it has to register with the home agent. The registration delay depends on the network distance from the mobile to its home agent. Usually when the mobile is far away from its home network, the registration delay cannot be neglected; therefore this leads to handoff performance degradation. When handoff takes place, before the home agent receives the registration request, all the packets destined to the mobile will be delivered to the old foreign agent and be lost. The correspondent host has to re-send those packets via TCP or some upper layer over UDP. This leads to an increase in TCP or UDP cost. The drawback of the regular Mobile IP scheme becomes more apparent if the mobile moves back and forth frequently.

In order to address the locality of the mobile's movement, we introduce local anchoring, first proposed in [2] for personal communications network, to reduce the chance that the mobile has to register with its home agent. This scheme is described as follows.

1. Choose one agent as the focus of an anchoring region and name it as an anchor.
2. When the mobile moves within the anchored region, it does not need to register with its home agent; instead, it registers with the anchor.
3. When the mobile moves out of the anchoring region, it will register with its home agent and the new foreign agent will become the focus of the new anchoring region.

As illustrated in Fig. 1, the packets destined for the mobile will be forwarded to the anchor agent first and, from there, it is forwarded to the foreign agent and then to the mobile.

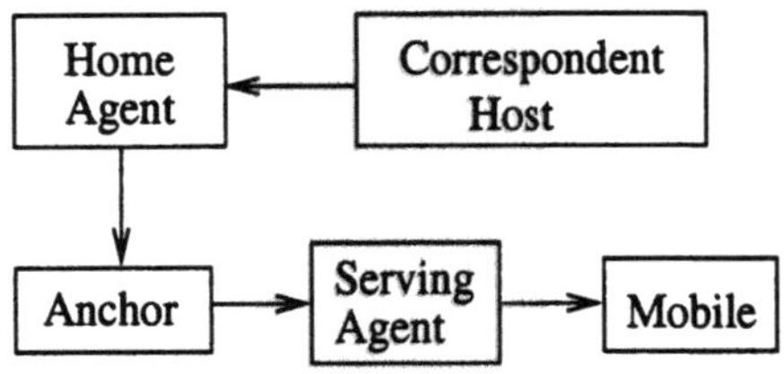

Fig. 1. Packet forwarding in the local anchor scheme

The anchor agent and the new foreign agent are two candidates that can decide whether the mobile should register with its home agent or not. The mobile cannot do this because it does not have the knowledge about the network. We choose the new foreign agent to assume this responsibility. The decision-making process can be based on static or dynamic information. The dynamic approach will use the mobile's past movement information and its current traffic information for decision-making. The static approach makes use of information that is fixed for all time. We use the static approach and the distance from the old anchor agent to the new foreign agent as the criterion to decide whether or not to establish a new anchoring region.

The registration process in the anchor scheme is shown in Fig. 2. There are two cases in the registration procedure, depending on which agent, the home agent or the current anchor, the mobile should register with.

There are 4 steps in the registration procedure:

1. The mobile sends the registration request indicating the current anchor agent and the home agent.
2. Case 1: the new foreign agent decides that the mobile is still in its current anchoring region so it forwards the mobile's registration request to the anchor. Case 2: the new foreign agent decides that the mobile is out of its current anchoring region so it forwards the mobile's registration request to the home agent.
3. The anchor agent or the home agent sends the registration reply back to the foreign agent.
4. The foreign agent returns ACK to the mobile and indicates who, the anchor or the home agent, sends this registration reply. In case 1, the mobile knows that it has not moved out of the current anchoring region and the anchor does not change. In case 2, the foreign agent becomes the focus of the new anchoring region and the mobile will update its anchor agent's IP address for later use.

In order to evaluate the performance of this local anchor scheme against some existing schemes, we need to quantify the problem. Two metrics are used in the evaluation: the average handoff delay and the average cost between two consecutive

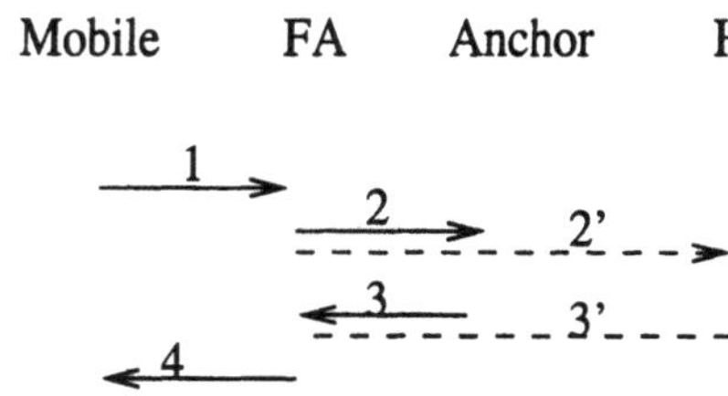

Fig. 2. Registration with anchor scheme

handoffs. It is demonstrated in the sequel that the local anchor-based Mobile IP scheme exhibits much better performance than the regular Mobile IP scheme.

The rest of the paper is organized as follows. Section 2 defines the parameters of the system model. Section 3 discusses and analyzes the handoff delay in the local anchor scheme. Section 4 compares the average UDP cost of the local anchor scheme with the regular Mobile IP scheme. Section 5 compares the average TCP cost of the local anchor scheme with the Indirect TCP (I-TCP) scheme. Numerical results of handoff delay and TCP/UDP costs are presented in section 6. The positive effect of the local anchor scheme on TCP throughput is shown in section 7. Finally, concluding remarks are given in section 8.

2 System Model

It is not easy to precisely model the Internet since the route between any two hosts may change dynamically. To simplify the problem, we assume the route between the involved agents is fixed and the delay on the corresponding route is fixed as well within the mobile's residence time. The delay of a path is proportional to the length of the path or the number of hops on the path. We simply take the delay on a path as the cost of the path.

We only consider the case that the mobile acts as the receiver. Assume that a correspondent host generates data packets destined for the mobile at a fixed rate λ. In the analysis we assume that this packet generation rate is not affected by the mobile's movement and handoff. This assumption is not true for the regular Mobile IP since fast retransmission and congestion control (or slow start in worse case) will occur when the sender detects the packet loss at handoff [3]. This assumption, however, separates the effect of the handoff delay on the TCP cost from other factors which affect the TCP throughput.

2.1 Network model

We use the same network model as proposed in [4] for the handoff delay analysis and the total cost analysis. Fig. 3 shows the network distances between various entities. The network distance is defined by the number of hops between two hosts. The communication cost (delay) is determined by the network distance.

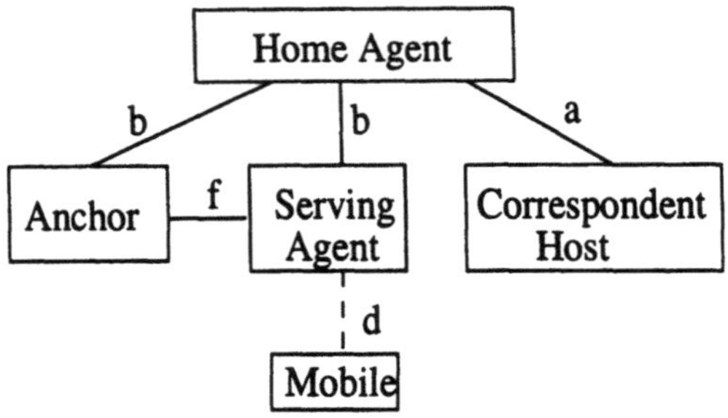

Fig. 3. Network model for cost analysis

Assume a, b, d are fixed but f is variable. In order to quantify f, we assume the relationship between the physical distance and the network distance can be modeled by a bifork tree as shown in Fig. 4. We can see that if any two agents are neighbors in the physical location, they will have a distance of two hops between them; if the physical distance, *Phy_Dist*, between two agent is 2, there will be 4 hops between them in the network ..., therefore we have $f = 2 \times Phy\_Dist$. We can have a similar relationship for the two dimensional case.

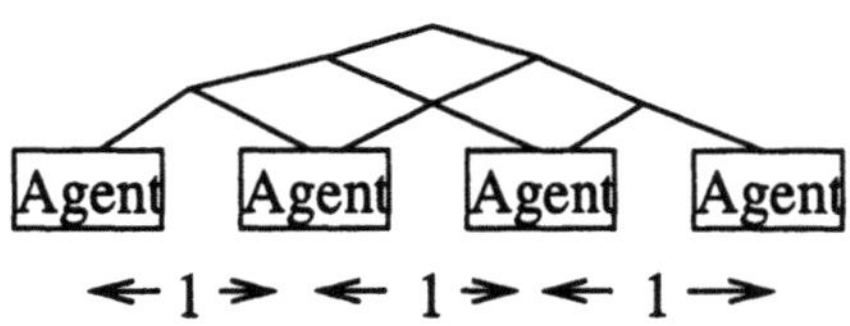

Fig. 4. Relation between network distance and physical distance

In addition to the communication cost (delay), we also need to consider the processing cost (delay) at the agents. We assume the processing cost at each agent is fixed and is denoted by r.

2.2 Mobility model

To facilitate analytical evaluation of system performance as the mobile moves in the anchoring region, we introduce the following mobility model. The residence time of the mobile in each serving area (the interval between two consecutive handoffs) is modeled as an exponential distribution. The assumption of exponential distribution allows us to characterize the mobile's movement by a Markov state transition diagram.

We consider two dimensional movement. In order to investigate the effect of the locality of the mobile on the average handoff delay, we use P_{same}, P_{back} and P_{other}

to describe the probability that the mobile moves in the same direction, in the reverse direction and in any of the other directions relative to the last movement. This mobility model was first proposed in [5].

2.3 Notations

The following notations are used in the ensuing analysis:

- t_{RA}: The registration delay through the anchor agent.
- t_{RH}: The registration delay through the home agent.
- t_{HO_Anchor}: The handoff delay of the local anchor scheme.
- t_{HO_I-TCP}: The handoff delay of the I-TCP scheme.
- t_{Reg_MIP}: The average registration delay of the regular Mobile IP.
- C_{RA}: The registration cost if the foreign agent chooses to register with the anchor agent.
- C_{RH}: The registration cost if the foreign agent chooses to register with the home agent.
- C_{dH}: The packet delivery cost if no anchor is used.
- C_{dA}: The packet delivery cost if the anchor is used.
- C_{dRA}: The total packet delivery cost between consecutive handoffs if the anchor is used.
- C_{dRH}: The total packet delivery cost between consecutive handoffs if no anchor is used.
- C_{dAM}: The cost for a packet to be delivered from the anchor to the mobile.
- C_{Trans}: The cost of transferring TCP states and unacknowledged packets from the old anchor to the new anchor.
- $C_{Anchor-UDP}$: The average UDP cost of the local anchor scheme between consecutive handoffs.
- $C_{Anchor-TCP}$: The average TCP cost of the local anchor scheme between consecutive handoffs.
- C_{MIP}: The average cost of the regular Mobile IP between consecutive handoffs.
- C_{In_Anchor}: The TCP cost of the local anchor scheme when the mobile moves within the local anchoring region.
- C_{Out_Anchor}: The TCP cost of the local anchor scheme when the mobile moves out of the current local anchoring region.

3 Handoff Delay

We define the handoff delay to be the time from when the mobile sends the registration request to the new foreign agent to when the mobile is allowed to send packets to or receive packets from the new foreign agent.

Whether the TCP connection exists or not, when a handoff takes place, it will decide how the handoff will end.

If no TCP connection exists when the handoff takes place, the handoff finishes when the mobile receives a handoff reply from the new foreign agent. We do not expect too much reduction of handoff delay for this case, since the only reduction comes from the shorter registration path, but we do consider it in the UDP cost analysis.

We are more interested in the handoff delay when the TCP connection exists at the time of handoff. We note that TCP can be in an end-to-end connection form, e.g., [6], or in a split connection form, e.g., [7]. A common criticism on split connection is the loss of end-to-end semantics. However, we are interested in studying the effect of combining split connection with the proposed local anchor scheme. Specifically, we want to know the handoff delay when I-TCP [7] is used and the handoff delay when we combine I-TCP with this local anchor scheme.

We already know that in order to improve the throughput of TCP in the wireless/wireline interconnection environment, the air link needs to be dealt with separately. I-TCP was proposed by [7] in which the TCP connection was split into two parts, one for wireline and the other for the air link. It is shown in [7] that I-TCP does improve the throughput but the drawback is that the handoff latency is fairly long (from 265 ms for empty socket buffers to 1430 ms for 32 kilobyte socket buffers).

If we incorporate the local anchor scheme with the I-TCP scheme and make the split happen only at the anchor agent, the only agent which maintains the TCP connection state for the mobile will be confined to the focal agent in the anchoring region. In other words, when the mobile moves within the anchoring region, its new serving agent does not keep track of the TCP connection state for the mobile, so it does not need to copy the mobile's TCP connection state from the old foreign agent, which is done in I-TCP. This will be one source of reduction in handoff delay. Another source of reduction in handoff delay will be the shorter registration path as mentioned earlier.

The registration procedure specified in [1] was not used by the original I-TCP scheme. Therefore we do not refer to the original I-TCP described in [7] when we mention I-TCP in this paper, but instead, we refer to the modified I-TCP which is based on the idea of the original I-TCP but follows the registration procedure of [1]. Details of the modified I-TCP scheme are contained in [8].

Suppose I-TCP uses the same registration procedure as the regular Mobile IP: the registration request and reply have to go through paths b and d in Fig. 3, plus two processing times at the foreign agent and one processing time at the home agent. Therefore the registration delay of I-TCP is $2(b+d)+3r$. After the successful registration, the new foreign agent sends a message to the old foreign agent and requests TCP state transferring. This will incur a delay of $(n+2)(f+r)$ (we will explain why in section 5), where n denotes the TCP buffer size. Therefore the total handoff delay for I-TCP is

$$t_{HO\_I-TCP} = 2(b+d) + 3r + (n+2)(f+r) \tag{1}$$

On the other hand, if the local anchor scheme is used, before the mobile roams out of the serving area of the current anchor, it only needs to register with the anchor, as oppose to the home agent. Thus the registration delay is

$$t_{HO\_Anchor} = 2d + 2f + 3r \tag{2}$$

since the registration request and reply will go through the paths d and f, plus two processing times at the new foreign agent and one processing time at the anchor. If the new foreign anchor decides that the mobile should register with the home agent, as oppose to the anchor, handoff delay should be the same as in the I-TCP scheme, i.e., $2(b+d)+3r+(n+2)(f+r)$.

Because of movement, the mobile will always change its location in the anchor region. Once the mobile moves out of the anchoring region, the new foreign agent will become the focus of the new anchoring region. f is changed when the mobile moves.

Consider the example shown in Fig. 5. The radius of this anchoring region is 2. There are 25 agents in total. The focus (anchor) is labeled c. Eight agents are of distance 1 to the focus and 16 agents are of distance 2 to the focus. For the agents of radius 1, f=2; for the agents of radius 2, f=4; when the mobile moves out of the anchoring region, we use $f = 6$ to compute the TCP state transferring cost and $f = 0$ to compute the packet delivery cost.

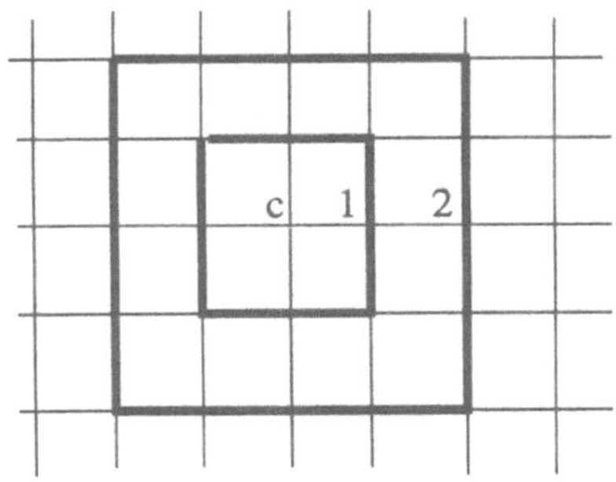

Fig. 5. Anchoring region of radius 2

In order to compute the average handoff delay, we thus need the probability distribution for the mobile to stay at each position. If we know the probabilities that the mobile stays at the different locations, the average handoff delay will be:

$$t_{HO\_Anchor} = \sum_{i=1}^{n} \pi_i t_{HO\_Anchor}(i) \tag{3}$$

where π_i denotes the probability that the mobile stays at location i.

We use Fig. 6 to illustrate how to compute the probability vector $\pi = (\pi_1, \pi_2, \ldots, \pi_n)$ using the mobility model described in subsection 2.2.

For Fig. 6, we use 5 states to describe the mobile's movement and its current location:

- S_1: the mobile is located at position 1;
- S_{21}: the mobile is located at 2 and its last location is 1;
- S_{23}: the mobile is located at 2 and its last location is 3;
- S_3: the mobile is located at 3;
- S_{out}: the mobile moves out of the current anchoring region and needs to register with the home agent.

The above states characterize the Markov chain shown in Fig. 7, where the subscripts "b" stands for "back", "s" for "same", "o" for "other" in the transition

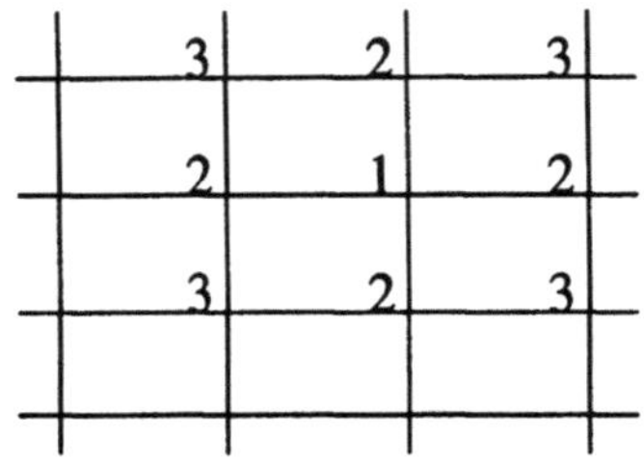

Fig. 6. Anchoring region of radius 1

probabilities. By assigning S_1,S_{21},S_{23},S_3, and S_{out} the positional values 1,2,3,4, and 5 respectively, we can construct the Markov transition matrix, M, for this Markov chain as

$$M = \begin{pmatrix} 0 & 1 & 0 & 0 & 0 \\ P_b & 0 & 0 & 2P_o & P_s \\ P_o & 0 & 0 & P_b + P_s & P_o \\ 0 & 0 & P_b + P_o & 0 & P_s + P_o \\ 0 & 1 & 0 & 0 & 0 \end{pmatrix}$$

It is easy to compute the limiting probability vector $(\pi_1, \pi_2, \dots, \pi_n)$ from this Markov transition matrix.

Fig. 7. Markov chain for r=1

The Markov chain and its transition matrix become significantly bigger when the radius of the anchoring region increases. An alternative approach is to simulate the mobile movements and compute the cost associated with the movements. The simulation results converge to the theoretical value quite well.

4 UDP Cost Analysis

Assume the mobile movement and the packet generation process are independent and stationary. The correspondent host generates packets destined for the mobile at a mean rate λ and the mobile moves between foreign agents at a mean rate μ; then the Packet to Mobility Ratio (PMR) is $\delta = \lambda/\mu$.

Each time when the mobile moves from one subnet to a new subnet, it has to initiate a new registration process. The new foreign agent will take the responsibility to decide whether to register with the home agent or the anchor agent. If the mobile needs to register with the home agent, the new foreign agent will become the new focus; otherwise this foreign agent just contacts the anchor agent and forwards packets to the mobile.

For the regular Mobile IP (MIP) scheme, the registration cost is $C_{RH} = 2(b + d) + 3r$, since the registration request and reply have to go through paths b and d, plus two processing times at the foreign agent and one processing time at the home agent. For the local anchor scheme, if the foreign agent chooses to register with the home agent, the cost will be C_{RH}; if the foreign agent chooses to register with the anchor agent, the cost will be $C_{RA} = 2(d + f) + 3r$. If the mobile is away from its home network, we can assume that b is much greater than f.

Assume that the only cost due to UDP packet loss is that they are delivered to the old foreign agent and have to be resent by the upper layer of the correspondent host. We further assume that retransmission of the lost packets has no effect on λ. If the mobile registers with the home agent, the number of packet loss will be $\lambda * t_{RH}$; if the mobile registers with the anchor agent, this number will be $\lambda * t_{RA}$.

There is some difference in packet delivery cost between the local anchor scheme and the regular Mobile IP. For the regular Mobile IP, the delivery cost is $C_{dH} = (a + b + d) + 2r$; whereas for the local anchor scheme, this will be $C_{dA} = (a + b + f + d) + 3r$. We can see that the packet delivery cost of the local anchor scheme is higher than the regular Mobile IP scheme. Actually the motivation of this scheme is to reduce the registration cost and TCP will benefit most from this scheme because no TCP states need to be transferred when the mobile moves within the anchoring region. UDP will also benefit from the shortened registration path. We expect that the gain on the registration cost can compensate for the loss on the packet delivery cost.

The average residence time for a mobile to stay within an agent's serving range is $1/\mu$. In this period the mobile will receive packets as many as $\lambda \times (t_{RH} + 1/\mu)$ or $\lambda \times (t_{RA} + 1/\mu)$ and the corresponding cost is $C_{dRH} = C_{dH} \times \lambda \times (t_{RH} + 1/\mu)$ or $C_{dRA} = C_{dA} \times \lambda \times (t_{RA} + 1/\mu)$.

Considering the cost at registration, the total cost for a mobile to stay within a particular agent's serving range is $C_{dRA} + C_{RA}$ for the anchor scheme, and $C_{MIP} = C_{dRH} + C_{RH}$ for the regular Mobile IP scheme. The ratio of these two quantifies the gain of the local anchor scheme over the regular Mobile IP.

Because of movement, the mobile will always change its location in the anchor region. Once the mobile moves out of the anchoring region, the new foreign agent will become the focus of the new anchoring region. The costs C_{dRA} and C_{RA} depend on the mobile's location in the anchoring region. When we compute $C_{dRA} + C_{RA}$, we have to average over all the locations.

The limiting probability vector of the Markov chain has nothing to do with the initial condition. Assume we have already got the final state probability vector $\pi = (\pi_1, \pi_2, \ldots, \pi_n)$, then the average cost will be

$$C_{Anchor-UDP} = \sum_{i=1}^{n} \pi_i (C_{dRA}(i) + C_{RA}(i)) \tag{4}$$

Cost ratio is used to show the cost reduction.

$$\frac{C_{Anchor-UDP}}{C_{MIP}} = \frac{\sum_{i=1}^{n} \pi_i((\lambda t_{RA}(i) + \delta)C_{dA}(i) + C_{RA}(i))}{(\lambda t_{RH} + \delta)C_{dH} + C_{RH}} \tag{5}$$

Note that $\pi_n = P_{out}$, the probability that the new foreign agent establishes a new anchoring region.

5 TCP Cost Analysis

Packet loss in the I-TCP scheme is different from UDP at handoff. When handoff takes place, TCP packets will be lost due to delivery to the old foreign agent but the correspondent host does not need to re-send those packets; they will be retransmitted by the anchor agent. In other words, packet loss only happens on the segment from the anchor agent to the old foreign agent. This will be another source of cost reduction by using the local anchor scheme.

Assume the packet generation rate at the correspondent host is not affected by handoff and kept at a fixed rate of λ. When handoff takes place within the anchoring region, the lost packets which are sent to the old foreign agent are retransmitted to the new foreign agent and then forwarded to the mobile. The cost for a packet to be delivered from the anchor to the mobile is denoted as C_{dAM}. C_{dAM} includes the transmission cost in paths f and d and the processing cost at the foreign agent, so we have $C_{dAM} = f + d + r$. Then the cost for TCP within the anchoring region, C_{In_Anchor}, is given by

$$C_{In\_Anchor} = \lambda t_{RA} C_{dAM} + \frac{\lambda}{\mu} C_{dA} + C_{RA} \tag{6}$$

where $t_{RA} = C_{RA} = 2(f + d) + 3r$, is the handoff latency when the mobile registers with the anchor.

When the mobile moves out of the anchoring region, a new anchoring region is set up and the new foreign agent will act as the new anchor. The TCP states maintained at the old anchor are transmitted to the new anchor and this incurs state transferring cost for the local anchor scheme. Therefore the cost when the mobile moves out of the anchoring region, C_{Out_Anchor}, is given by

$$C_{Out\_Anchor} = \lambda t_{RH} C_{dAM} + \frac{\lambda}{\mu} C_{dA} + C_{RH} + C_{Trans} \tag{7}$$

The state transferring cost includes two parts: (1) the new anchor sends a message notifying the old anchor to transfer the mobile's TCP states; (2) the old anchor sends all the buffered packets and the relevant states to the new anchor. Considering the path cost and the agent processing cost, the first part is $f + r$ and the second part is $(f + r) \times packets$.

It is important to decide how many packets have to be transferred from the old agent to the new agent at the time of state transferring. We have proposed a modified TCP/IP stack [8] to support the I-TCP scheme. With this TCP/IP stack, there are two types of packets to be transferred: (1) TCP state of the mobile or specifically, the TCP control block; (2) the unacknowledged packets in the buffer.

The TCP control block is a large structure, occupying 140 bytes [9]. The addresses and ports which identify a socket also need to be transferred — they are not contained in the TCP control block — which include the foreign (the correspondent) IP address, the foreign port, the local (the mobile) IP address and the local port. We need to transfer one set of addresses/ports which is 12 bytes in IPv4 and two TCP control blocks — one for wireless part and one for wireline part — which is 140 bytes/block. Thus we need to transfer $12 + 2 \times 140$ bytes in total. This will form one packet.

The amount of unacknowledged packets which need to be transferred depends on the buffer size of the anchor. If we assume the standard TCP segment size is 1 kilobyte, then a buffer of 8 kilobytes can hold 8 TCP segments, a buffer of 16 kilobytes can hold 16 TCP segments .... Because the wireline bandwidth is higher than the air link, the incoming packets addressed to the mobile will usually be buffered and accumulated at the anchor before they are transmitted to the air link and acknowledged. A simple assumption is that the receiver buffer at the anchor in the direction from the correspondent host to the mobile is always full. Based on the above assumption, the number of unacknowledged packets to be transferred is n for a buffer size of n kilobytes.

So, the total number of packets to be transferred at handoff is $1 + n$ if the anchor's buffer size is n. In addition, another packet is required to inform the old anchor to start the transferring. These messages are exchanged between the old anchor and the new anchor and the network cost is f. According to Fig. 4, $f = 2 \times Phy\_Dist = 2 \times Rd$, where Rd denotes the radius of the anchoring region. Considering the processing cost at the anchor, the total cost for packet transferring is

$$C_{Trans} = (n + 2)(r + 2Rd). \tag{8}$$

The TCP cost for the mobile to move out of the old anchoring region is given by (7). The average TCP cost for the local anchor scheme, therefore, can also be expressed as

$$C_{Anchor-TCP} = \sum_{i=1}^{n} \pi_i (C_{dRA}(i) + C_{RA}(i)) \tag{9}$$

where $C_{dRA} = \lambda t_{RA} C_{dAM} + \frac{\lambda}{\mu} C_{dA}$; but for $i = n$, $C_{RA} = C_{RH} + C_{Trans}$.

For the regular Mobile IP, the TCP cost is always $\lambda(t_{RH} + \frac{1}{\mu})C_{dH} + C_{RH}$.

In terms of TCP, the cost ratio between the local anchor scheme and the regular Mobile IP is

$$\frac{C_{Anchor-TCP}}{C_{MIP}} = \frac{\sum_{i=1}^{n} \pi_i (\lambda t_{RA}(i) C_{dAM} + \delta C_{dA}(i) + C_{RA}(i))}{(\lambda t_{RH} + \delta) C_{dH} + C_{RH}} \tag{10}$$

6 Numerical Results

In the computation we assume:

- the cost of processing a message is equivalent to the cost of communication over a single hop, i.e., $r = 1$.
- the air link incurs more cost (delay) than the wireline, specifically, we assume $d = 5$.
- the distance between the home agent and the correspondent host equals to the distance between the home agent and the foreign agent, and both of them are much greater than the radius of the anchor, specifically, we assume $a = b = 10$.

6.1 Handoff delay comparison with I-TCP

As mentioned earlier, the incorporation of the local anchor scheme into I-TCP can avoid the TCP state transferring when handoff takes place within the current anchor region, and therefore leads to a reduction in handoff delay. In this subsection, we look at how much reduction in handoff delay is achieved.

In Figs. 8-10, the handoff delay ratio means the ratio of the handoff delay of the local anchor scheme to that of the pure I-TCP, i.e.,

$$\text{Handoff delay ratio} = t_{HO\_Anchor}/t_{HO\_I-TCP}.$$

The common point of these figures is that when P_{back} increases, the reduction in handoff delay will also increase. In other words, a mobile which moves with the characteristic of locality will benefit most from the local anchor scheme. This shows that the local anchor scheme does address the locality of movement.

Fig. 8 shows the relationship between the handoff delay reduction and P_{back}, with the radius of the local anchoring region as parameter. Before the anchor radius reaches some critical point, the handoff delay will decrease as the anchor radius increases; when the anchor radius reaches the critical point, the handoff delay will increase as the anchor radius increases. This means that an optimum value exists for the anchor radius. We believe that the optimum radius is relevant to the distance from the home agent to the foreign agent. The explanation is: when the anchor radius is small compared to the distance between the home agent and the foreign agent, the registration delay with the foreign agent is smaller than the registration delay with the home agent; but this is not true when the anchor radius becomes comparable to the distance between the home agent and the foreign agent.

The relationship between the handoff delay reduction and the distance from the home agent to the foreign agent is shown in Fig. 9. When the distance from the home agent to the foreign agent increases, the handoff delay ratio decreases. This is exactly what we have expected. It shows that the shorter registration path in the local anchor scheme does help to reduce the handoff delay.

The relationship between the registration delay reduction and the buffer size at the anchor is shown in Fig. 10. The conclusion derived from Fig. 10 is straightforward: the bigger is the buffer size at the anchor, the greater is the reduction

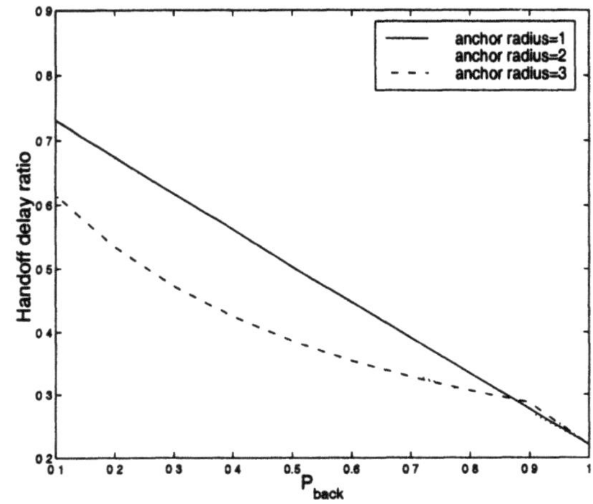

Fig. 8. Effect of the radius of the local anchoring region ($a = b = 10$, $P_{other} = 0$)

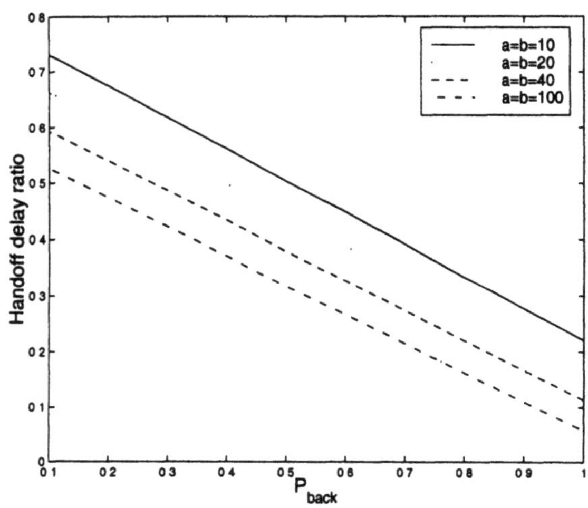

Fig. 9. Effect of the distance to the home agent (anchor radius=1, $P_{other} = 0$)

in handoff delay. This is also what we have expected. By using the local anchor scheme, we actually save the TCP state transferring when the mobile moves within the current anchor region. The bigger is the buffer size, the more we have to transfer in I-TCP at handoff and therefore the more we save by using the local anchor scheme.

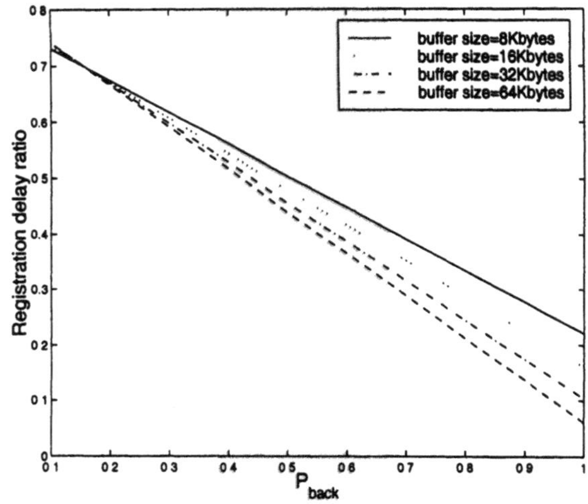

Fig. 10. Effect of the buffer size (anchor radius=1, $P_{other} = 0$)

In Figs. 8-10, we assume $P_{other} = 0$, which actually describes a one dimensional movement. We have the same conclusions if $P_{other} > 0$.

6.2 Numerical result on UDP cost

In this subsection and the next, we use Fig. 6 to illustrate the numerical results on UDP cost reduction and TCP cost reduction. In the computation we assume that the mobile moves in any direction with equal probability, i.e., $P_b = P_s = P_o = 0.25$.

From eqn.(5) we can get the ratio of the average cost of the local anchor scheme to the regular Mobile IP as

$$\frac{C_{Anchor-UDP}}{C_{MIP}} = \frac{568.4\lambda + 28.9\delta + 19.8}{891\lambda + 27\delta + 33} \tag{11}$$

Examining the coefficients in eqn.(11) we can see on average:

- the local anchor scheme has a smaller registration cost (19.8 versus 33);
- the local anchor scheme has a smaller retransmission cost (568.4 versus 891);
- the local anchor scheme has a slightly bigger packet delivery cost (28.9 versus 27).

The relationship between the cost ratio and λ and δ is shown in Fig. 11. Again, we attribute the trends of the curves, which are in favor of the local anchor scheme, to the lower handoff delay of this scheme. If we fix the total number of packets between consecutive handoffs, the cost ratio will decrease when the packet generation rate increases, since a bigger λ is associated with a larger reduction in packet loss by the local anchor scheme at handoff. If we fix λ but decrease δ, we will see that the cost reduction increases. This is because the handoff delay accounts for more in the total cost if the number of packets generated between consecutive handoffs is smaller.

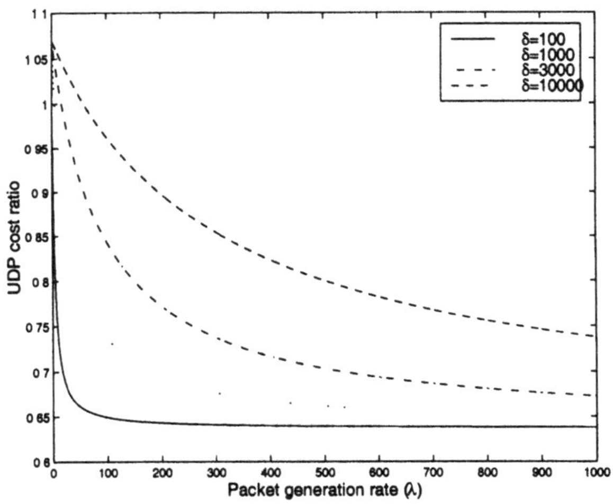

Fig. 11. UDP cost ratio

The effect of P_{back} is shown in Fig. 12. In Fig. 12 we assume $P_{other} = P_{same}$ and $\delta = 1000$. The conclusion is straightforward: the bigger P_{back} is, the more the UDP cost is saved.

6.3 Numerical result on TCP cost

When we compute the TCP state transferring cost, we assume the buffer size at the anchor is 8 kilobytes, i.e., $n = 8$.

From eqn.(10) we can get the ratio of the average cost of the local anchor scheme to the regular Mobile IP as

$$\frac{C_{Anchor-TCP}}{C_{MIP}} = \frac{177.8\lambda + 28.9\delta + 28.7}{891\lambda + 27\delta + 33} \tag{12}$$

Similar to subsection 6.2, here we assume that the mobile moves with equal probability in the four directions when we compute the probability vector π.

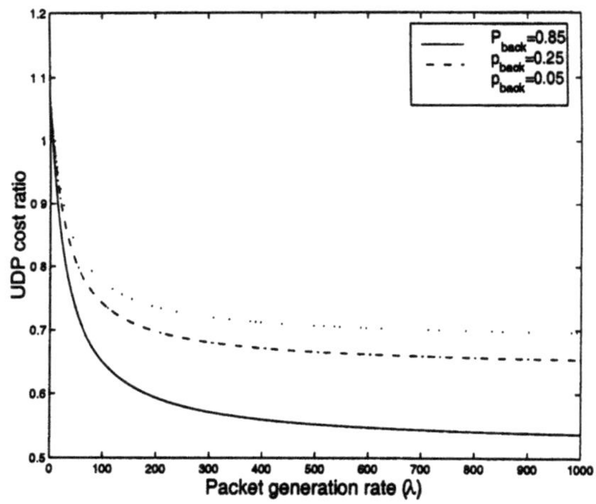

Fig. 12. Effect of P_{back} on UDP cost ratio, $P_{other} = P_{same}$, $\delta = 1000$

Examining the coefficients in eqn.(12) we can see on average:

- the local anchor scheme has a smaller registration cost (28.7 versus 33);
- the local anchor scheme has a much smaller retransmission cost (177.8 versus 891);
- the local anchor scheme has a slightly bigger packet delivery cost (28.9 versus 27).

The relationship between the cost ratio and λ or δ is shown in Fig. 13. We have the same conclusion for TCP cost as for UDP cost. As we have mentioned earlier, the retransmission path for the lost packets at handoff is shorter than regular Mobile IP does, so we can save more on the total cost. This can be seen by comparing the TCP cost ratio with the UDP cost ratio.

The effect of P_{back} is shown in Fig. 14. In Fig. 14 we assume $P_{other} = P_{same}$ and $\delta = 1000$. The conclusion is similar to UDP: the bigger P_{back} is, the more the TCP cost is saved.

7 Effect of Local Anchor Scheme on TCP Throughput

It is shown in our simulation [10] that the local anchor scheme has no effect on TCP throughput if no handoff takes place. In the presence of handoffs, the local anchor scheme improves the TCP throughput significantly. Fig. 15 shows the throughput comparison between the local anchor scheme and the original I-TCP when handoff takes place every 10 seconds. The different throughputs shown in Fig. 15 can be explained from the congestion window variation shown in Fig. 16 and Fig. 17. In Fig. 16, there is only one slow start and the congestion window is kept over 10 kilobytes except for this slow start instant. Most of the time, the congestion window of Fig. 16 is fully opened (64KB). Fig. 17, on the other hand, shows that

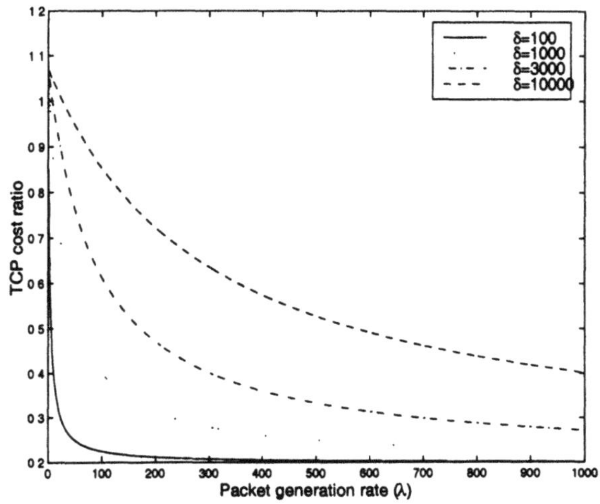

Fig. 13. TCP cost ratio

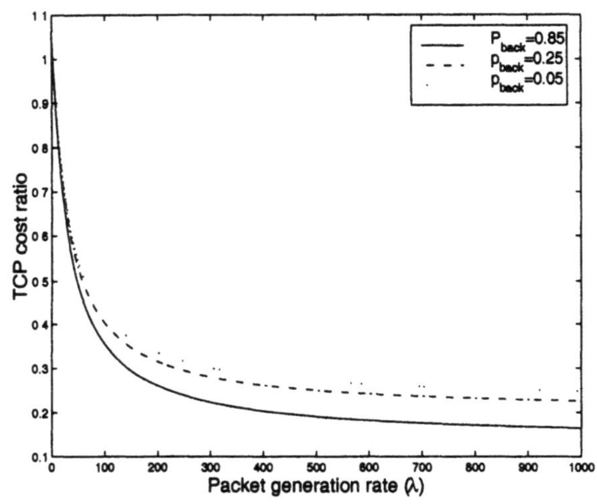

Fig. 14. Effect of P_{back} on TCP cost ratio, $P_{other} = P_{same}$, $\delta = 1000$

the congestion window is less than 20 kilobytes most of the time, and slow start is triggered each time when handoff takes place. After the first handoff, the congestion window in Fig. 17 never goes back to its full size. A possible explanation is that the frequent handoff makes the congestion avoidance threshold [3] very small; therefore the sender enters congestion avoidance state very soon after the slow start. In the congestion avoidance state, the congestion window increases at a slower rate (linearly compared to exponentially in slow start). Since handoff takes place quite often (once every 10 seconds) and the congestion window increases relatively slowly, before the congestion window increases to a large value, the next slow start takes place again. In short, the small congestion window of I-TCP is caused by frequent handoffs and long handoff delay. We conjecture that the local anchor scheme is more suitable for situations of frequent handoff and long handoff delay.

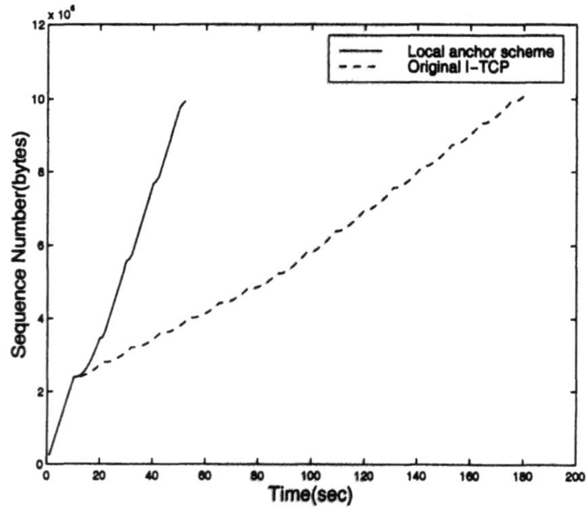

Fig. 15. Comparison of throughput between local anchor scheme and original I-TCP, handoff every 10 seconds, 64KB MAX window

8 Conclusions

We have presented a local anchor scheme for Mobile IP which addresses the locality of mobile movement. Numerical results of the handoff delay analysis show that this scheme can effectively reduce the handoff delay. Compared to the regular Mobile IP scheme, the proposed scheme also leads to reduction in TCP/UDP costs. As well, the local anchor scheme has a positive effect on TCP throughput.

The network distance between the anchor and the new foreign agent is used as the criterion to decide whether to establish a new anchor region or not. In this paper, a static threshold is used. A dynamic scheme may use the information of the mobile's traffic and mobility pattern to update the threshold dynamically. Deployment of a dynamic scheme is currently under investigation.

Acknowledgement

This work has been supported in part by a grant from the Canadian Institute for Telecommunications Research under the NCE program of the Government of

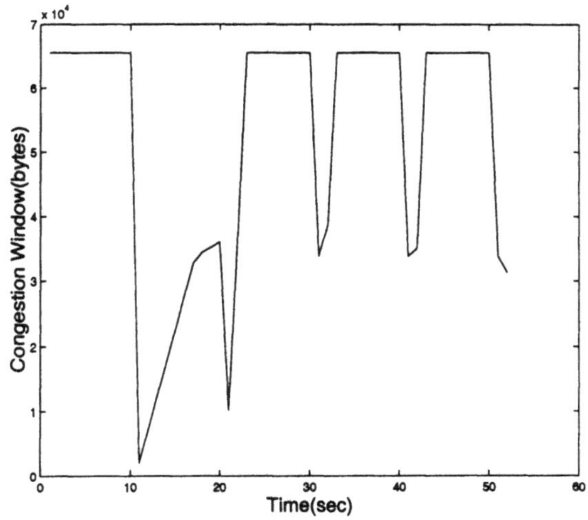

Fig. 16. Congestion window of the local anchor scheme with 64KB MAX window, handoff every 10 seconds

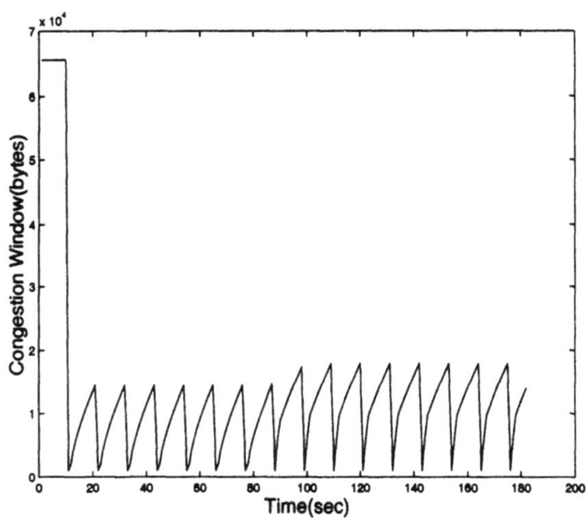

Fig. 17. Congestion window of the original I-TCP with 64KB MAX window, handoff every 10 seconds

Canada, and in part by the Natural Sciences and Engineering Research Council of Canada under grant no. RGPIN7779.

References

1. *RFC 2002, IP Mobility Support.*
2. J. S. Ho and I. F. Akyildiz, "Local anchor scheme for reducing signaling costs in personal communications networks," *IEEE/ACM Transactions on Networking*, vol. 4, pp. 709–725, Oct. 1996.
3. *RFC 2581, TCP Congestion Control.*
4. R. Jain and T. Raleigh, "Mobile Internet access and QoS guarantees using Mobile IP and RSVP with location registers," in *ICC'98*, pp. 1690–1695.
5. J. H. Zhang and J. W. Mark, "A local VLR cluster approach to location management for PCS networks," in *Proc. of WCNC'99*, pp. 21–24, Sept. 1999.
6. R. Caceres and L. Iftod, "The effect of mobility on reliable transport protocols," ICDC'94, 1994.
7. A. Bakre and B. R. Badrinath, "Handoff and system support for Indirect TCP/IP," in *Mobile and Location-Independent Computing Symposium*, pp. 11–24, 1995.
8. J.-Z. Zhang and J. W. Mark, "A modified TCP/IP stack for I-TCP," Tech. Rep., Centre of Wireless Communications, University of Waterloo, Canada, June 1999.
9. G. Wright and R. Stevens, *TCP/IP Illustrated*, vol. 2. Addison-Wesley, 1995.
10. J.-Z. Zhang, "Deployment of Mobile IP in cellular networks," Master's thesis, University of Waterloo, Canada, 1999.

A Distributed Channel Allocation Strategy based on A Threshold Scheme in Mobile Cellular Networks

Yongbing Zhang[1] and Xiaohua Jia[2]

[1] Institute of Policy and Planning Sciences, University of Tsukuba, Ibaraki 305-8573, Japan
[2] Department of Computer Science, City University of Hong Kong, Kowloon, Hong Kong

Abstract. We propose a distributed channel allocation algorithm based on a threshold scheme (D-CAT) for mobile cellular networks. A cell employs two thresholds, a *light* and a *heavy* thresholds, to classify its interference cells in the system into three categories: *light*, *moderate*, and *heavy* cells. Using such a threshold scheme, how many free channels and from where a heavy cell needs to import is clearly determined so that a heavy cell can import free channels as many as possible during one channel acquisition process in order to satisfy its channel demand. This mechanism results in a low overhead cost for message transmission between the cells for acquiring free channels and simplifies the channel allocation mechanism. Simulation experiments and analyses show that D-CAT causes much lower overhead cost of message transmission than the other versions of distributed channel allocation strategies and provides a call blocking probability comparable to an efficient centralized channel allocation strategy.

1 Introduction

The rapid growth in the demand for mobile communications has led the industries into intense research and development efforts towards a new generation of wireless cellular systems. In order to utilize the limited resources (radio spectrum) effectively, the geographical area is divided into cells and the radio spectrum is reused in non-interfering cells. The radio spectrum is divided into channels to serve different calls depending on the various radio technologies: Frequency Division (FD), Time Division (TD), and Code Division (CD) [1]. Many schemes for channel allocation have been proposed in order to use the available channels efficiently and maximize the channel reuse [2–5]. The *performance index* used for measuring the efficiency of a channel allocation scheme is the *call blocking probability*, *i.e.*, the sum of the probabilities of new call blocking and forced termination.

Channel allocation strategies can be mainly classified into two types: *fixed* [6] and *dynamic* [2,7]. A combination of these two classes is also possible [8]. A *fixed* allocation (FA) strategy is to allocate a fixed set of channels to each cell permanently. The same set of channels is reused by another cell at some distance away.

The advantage of the FA strategy is its simplicity. Its disadvantage, however, is that if the number of calls exceeds the number of channels allocated to a cell the excess calls have to be blocked. Variations of FA strategies have been proposed and shown that the performance can be significantly improved [9,10]. A *dynamic* allocation (DA) strategy is to allocate the channels in the system dynamically. Each cell has no channels to itself but requests for free channels if necessary. The system keeps a pool of free channels and any cell can use any channel that does not violate the channel reuse constraint. The DA strategies tend to be more efficient than the FA strategies in conditions of light, non-homogeneous, and time-varying traffic but accompany with high implementation overhead.

In addition to the issue of how to allocate channels among cells, the problem of who plays a key role in making a channel allocation decision is also very important. Most of the algorithms in the literature depend on a mobile switching center (MSC) to accomplish channel allocation, which are referred to as *centralized* channel allocation algorithms [2,5,9,11]. The disadvantage of centralized algorithms is that the centralized scheme may cause the MSC overloaded and the failure of the MSC makes the whole system down. In distributed algorithms, on the other hand, each BS at a cell plays a key role in a channel allocation decision and is capable of running the channel acquisition algorithm if the cell gets overloaded with channel demand. The main advantage of a distributed algorithm is its high reliability and scalability. In a distributed algorithm, a cell collects the information of channel usage by either contacting with other interference cells [3,4,9] or sensing the carriers of the channels only by itself [1,12,13]. The latter method is simple but a channel allocation may not be optimal. The former method, on the other hand, can provide better channel allocation but its implementation cost, *e.g.*, overhead cost for message transmission between cells and resource management, may be high. The overhead cost may increase exponentially in the worst case as the number of cells in the system increases. In this paper, we focus on the algorithms employing the latter information collection scheme and explore how to reduce the overhead cost.

In this paper, we propose a dynamic *distributed channel allocation* algorithm based on a *threshold* scheme called D-CAT. Each cell in D-CAT employs a *heavy* and a *light* thresholds to classify its interference cells in the system according to their states, defined by the number of channels available at a cell, into three categories: *light*, *moderate*, and *heavy* cells. The heavy threshold is a predefined parameter but the light threshold is determined by the states of the other cells within the interference distance. When a cell becomes heavy, the cell (or its BS) triggers the channel allocation algorithm to import free channels. Using such a two-threshold scheme, how many free channels and from where a heavy cell should import is clearly determined. When importing channels a heavy cell takes the light threshold as its target to satisfy its channel demand. The resources (channels) between the cells are therefore balanced and the overhead cost of message transmission between the cells is minimized.

The rest of this paper is organized as follows. Section 2 describes the cellular system model. Section 3 presents the D-CAT algorithm proposed in this paper. Section 4 shows the performance evaluation of D-CAT in comparison with the other algorithms, D-LBSB, D-ES, and CAT, in the literature. The final section presents the conclusion of the paper.

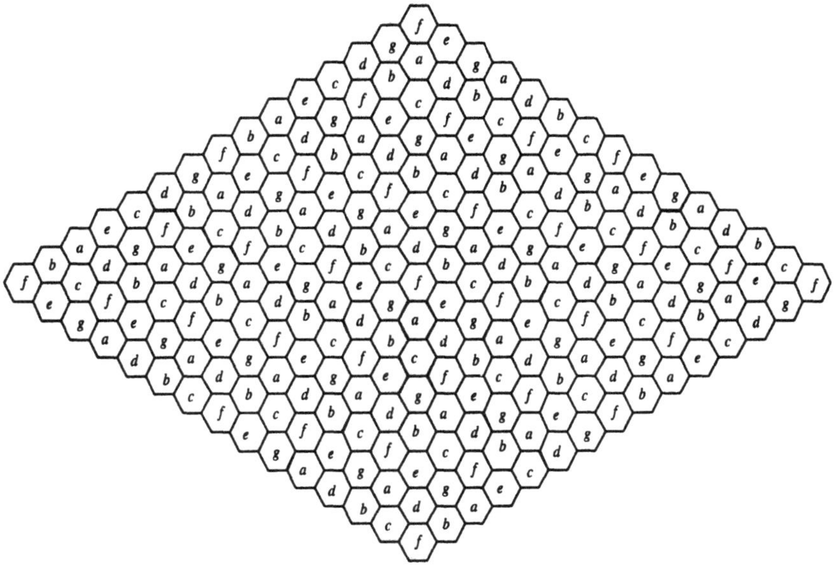

Fig. 1. Cellular System

2 System Model

The geographical area is divided into hexagonal cells in a mobile cellular network as shown in Fig. 1. A base station (BS) and the mobile users communicate through wireless links using radio channels. A number of cells (or BSs) are linked to a mobile switching center (MSC) through dedicated wire-line links. Each MSC is linked to the fixed telephone network again through a wire-line link and acts as a gateway of the cellular network to the fixed backbone network [14].

The system has totally S distinct channels and they are initially assigned to the cells based on a fixed channel assignment scheme. Figure 1 shows an example of the initial state of the system where the alphabets on the cells denotes different sets of channels and the set of cells using distinct sets of channels is 7. It is assumed that a channel is not assigned to a cell permanently but can be reassigned to any cells provided that the same channel is used farther enough than the *minimum reuse distance*, the minimum distance at which the same channel can be reused without interference. A newly incoming call or a handoff call from an adjacent cell will be assigned a free channel immediately if there are any free channels. When an active call traverses the boundary between two cells an *intercell handoff* occurs and the call releases the serving channel in the original cell and is reassigned a new channel at the adjacent destination cell. In order to improve the channel utilization, a cell may also enforce an active call to release its serving channel and reassign it with a new one in the same cell. This process is called the *intracell handoff.*

A set of cells in the system forms a group, N_i, so that each cell in N_i is located within the minimum reuse distance related to the center of the group, cell i, as shown in Figure 2. A set of cells in N_i that all own channel c is denoted by $P_i(c)$.

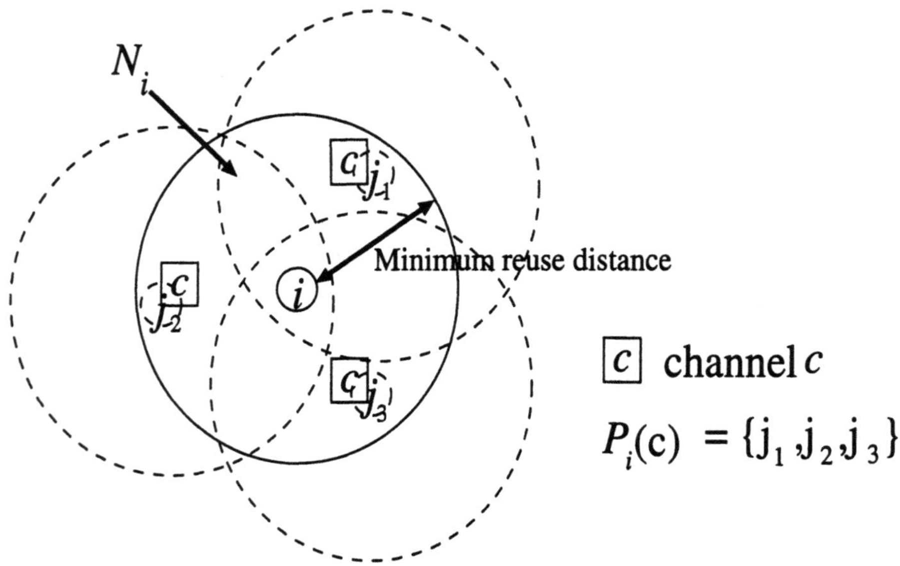

Fig. 2. Interference cells

Channels in the system are not stocked in the global free channel pool but owned by each cell. A cell has the whole right to control the free channels it holds; that is, it can assign or lock any specific channel it owns without negotiation with any other cells.

The cells in N_i are classified into three categories–*heavy, moderate*, and *light* cells–based on a *heavy* and a *light thresholds*, T_h and T_l. If the number of channels available at cell i, v_i, is less than or equal to T_h, that is, $v_i \leq T_h$, cell i is a *heavy* cell. If v_i is equal to or greater than T_l, that is, $v_i \geq T_l$, on the other hand, cell i is a *light* cell. Otherwise, it is a *moderate* cell. A cell intends to import free channels when it becomes heavy. It may, on the other hand, export free channels to the other cells if it has plenty of free channels. It is assumed that cells in the system are cooperative in the sense that they will respond kindly channel requests and do their best to satisfy the channel requests.

The heavy and light thresholds are key parameters in a channel allocation algorithm. An algorithm is triggered by a cell i when v_i becomes equal to or less than the heavy threshold, T_h. The value of T_h are set according to the necessity of the channel allocation policy. If the value of T_h is greater than zero, a cell can still assign a channel to a newly arriving call during the process of channel importing. A cell only attempts to import free channels, on the other hand, if it has no free channels anymore when the value of T_h is zero. A typical value of T_h is therefore 0 or 1. The light threshold, T_l, is the target value that cell i intends to achieve. The value of T_l is determined by the average number of channels available at a cell in N_i as follows.

$$T_l = \left\lceil \frac{\sum_{j \in N_i, j \neq i} v_j + 0.5}{|N_i|} \right\rceil.$$

If $T_l = 0$ then set $T_l = 1$.

The notation used in this paper is listed as follows.

Notation

S set of channels used in the system
C_i set of cells as candidates for channel import
v_i number of channels available at cell i
V_i set of channels available at cell i
u_i number of channels used in cell i
U_i set of channels used in cell i
N_i set of cells within interference distance with cell i
R_{ij} set of channels imported by cell i from cell j
R'_{ij} set of confirmed channels exported from cell j to cell i
T_h heavy threshold, $0 \leq T_h < T_l$
T_l light threshold
r_i number of channels needing to import for cell i, $r_i = T_l - T_h$
$P_i(c)$ set of cells that own channel c in N_i

3 The D-CAT Algorithm

The D-CAT algorithm is distributed and dynamic in the sense that each cell runs the algorithm independently and according to its own state whenever it needs free channels. Unlike some dynamic algorithms, there is no global free channel pool in D-CAT and each cell treats the free channels it holds as its own property. Furthermore, a cell gets some free channels from its light neighbors when it becomes heavy and will keep these channels as its own property. It is therefore not a *borrower-lender* relation between a heavy and a light cells but an *importer-exporter* relation instead.

When D-CAT is activated by a heavy cell i, it asks its interference cells for help and attempts to import sufficient channels to satisfy the channel demand of cell i. The messages transmitted between cell i (importer) and a cell j (exporter) are classified into four kinds of messages as shown as follows.

- *request* message, $request(i)$: Message sent by importer i to all the cells in N_i to request free channels.
- *reply* message, $reply(j, V_j, U_j)$: Message from cell $j, j \in N_i$ responding to the importer i. A reply message contains the identifier of cell j, and the sets of channels available and used in cell j.
- *inform* message, $inform(i, R_{ij})$: Message from importer i to the exporters and the others in N_i to inform them its channel acquisition decision and the end of channel acquisition process. The message also includes the request to the exporters for the reserved channels.

- *confirm* message, $confirm(j, R'_{ij})$: Message from exporter j to importer i to inform it the availability of the requested channels that have been reserved at cell j.

The channel allocation is performed independently at each cell and therefore it need a mechanism to synchronize the time clock between the cells. In this paper, a method proposed by Lamport [15] is used as in [3]. That is, a timestamp scheme is used to synchronize the time of messages transmitted between the cells.

The D-CAT algorithm consists of two components–*channel import component* and *channel export component*. The former component is activated by a cell and works as the client. The latter component, on the other hand, works as the server and waits for the requests from the clients on the other cells.

Channel Import Component

Cell i enters to the "ask-for-help" mode if it becomes heavy and asks its interference neighbors in N_i for acquiring free channels. The channel acquisition algorithm can be described as follows.

1. When cell i becomes heavy, *i.e.*, $v_i \leq T_h$, it sends a *request* message, $request(i)$, for free channels to all of its interference neighbors in N_i. When cell i receives a request from another cell with a larger timestamp, it postpones the response. Otherwise, it replies the request immediately.
2. After cell i has received all the *reply* messages from the cells in N_i, it calculates T_l and r_i. It then determines from where and how many channels it is going to import according to the following steps.
   (a) Seek for the free channels in N_i, *i.e.*, $S - \cup_{j \in N_i} (V_j \cup U_j)$. Stop the algorithm if the request is satisfied. Otherwise, go to the next step.
   (b) Seek for the free channels from the light cells in N_i according to the following steps.
      i. Find a light cell j where $v_j > T_l$ and add it to C_i.
      ii. For each cell j in C_i, find a channel c such that $|P_i(c)| = 1$, and add channel c to the import channel set, R_{ij}. Repeat this process until the request is satisfied, there are no any appropriate channels, or the state of cell j alters.
      iii. For each cell j in N_i, find a channel c such that c belongs to cells $j, j \in P_i(c) \cap C_i$. That is, find channel c that belongs to the light cells in N_i. Add channel c to R_{ij}. Repeat this process until the request is satisfied, there are no any appropriate channels, or the state of cell j alters.
   (c) Find channel c such that c belongs to cells $j, j \in P_i(c)$ and $\max\big(\min_j(v_j), j \in P_i(c)\big) - 1 \geq v_i + |\cup_{j \in N_i} R_{ij}| + 1$. Add channel c to R_{ij}. Repeat this process until the request is satisfied, there are no any appropriate channels, or the state of cell j alters.
3. Mark the channels in R_{ij} that are reserved at cell j and send an *inform* message, $inform(i, R_{ij})$, to each channel exporter j to inform which channels are imported or requested. Wait for the confirmation of the marked channels if necessary but the unmarked channels in R_{ij} are available immediately. Also

send an *inform* message, $inform(i, \emptyset)$, to the other cells in N_i to inform them the end of the channel import process.

4. After receiving the *confirm* messages, $confirm(j, R'_{ij})$, from channel exporter j, make the marked channels available.

Channel Export Component

Requests for free channels arrived at cell j are queued in a request queue based on the timestamps of the requests and processed sequentially. When cell j receives a request from cell i it processes the request according to the following steps.

1. If there are no requests under processing, go to step 2. Otherwise, compare the timestamps of the requests. If the newly incoming request from cell i has a smaller timestamp than the request from cell k under processing then the request from cell k will be aborted and put into the request queue and then go to step 2.
2. Reply cell i with a *reply* message, $reply(j, V_j, U_j)$, and wait until the *inform* message arrives if cell j is not heavy. If cell j is heavy it has no need to wait.
3. When cell j receives an *inform* message, $inform(i, R_{ij})$, it locks the requested channels. If any channels in R_{ij} are the reserved ones but still available, cell j will lock these channels and take the other free channels as reserved instead, and then add these channels to R'_{ij}.
4. Send cell i a *confirm* message, $confirm(j, R'_{ij})$, to inform the availability of the reserved channels.

4 Performance Evaluation

The performance of D-CAT is evaluated with respect to two parameters: the implementation cost and the call blocking probability. For the implementation cost, the number of messages transmitted between the BSs during the process of channel allocation and the delay of message transmission are taken into account. Two distributed algorithms, D-LBSB proposed by Das *et al.* [9], and an enhanced *search* algorithm proposed by Cao *et. al* [3], referred to as D-ES in this paper, and one efficient centralized algorithm proposed by Zhang *et al.* [11], referred to as CAT in this paper, are chosen in comparison with D-CAT.

4.1 Implementation Cost Comparison

The total number of cells in the system is denoted by N. The message delay between the BSs and between a BS and the MSC is fixed to be δ. The postponed response delay is denoted by δ_d. In D-ES, n_p denotes the number of interference primary neighbors of a cell, m denotes conflict rate, and n_u denotes the update message. Since the execution time of a channel allocation algorithm is much shorter than the message delay, it is not taken into account in the comparison. In order not to bring any bias to the comparison results for the chosen algorithms, the case of locking three co-channels for each lender cell for D-LBSB and CAT and the worst case scenario for D-CAT are considered. It is assumed that a heavy cell needs X channels and each channel exporter can offer only one channel. Since in D-ES a heavy cell borrows only one channel during one channel acquisition process, it needs to run the algorithm X times to obtain X channels. The messages transmitted between the BSs and the message delay in D-CAT can be listed as follows.

1. Channel importer i sends a *request* message to each cell j in N_i. The total number of request messages is $|N_i|$ and the delay of message transmission is δ.
2. After receiving the request from importer i, cell j sends back a *reply* message. The total number of the *reply* messages is $|N_i|$ and the delay of message transmission is δ.
3. Importer i sends an *inform* message to each cell in N_i to inform its channel importation decision and notice the end of the channel importation process. The messages to the channel exporters also include the requests for confirmation if it requests any reserved channels from the exporters. The total number of messages is $|N_i|$ and the delay of message transmission is δ.
4. Exporter j sends a *confirm* message to importer i to inform it the availability of the requested channels. Assuming that there are x channels needing to confirm and each channel belongs to a distinct cell, the number of the *confirm* messages is x and the delay of message transmission is δ.

Table 1. Implementation cost comparison of the algorithms.

Scheme	Number of messages	Message delay
CAT	$N + 4X + 1$	2δ
D-LBSB	$2(N-1) + 2(\|N_i\| + 3)X$	$4\delta(1 + 3X)$
D-ES	$(3\|N_i\| + 3n_p m + n_u)X$	$(2(1+m)\delta + \delta_d)X$
D-CAT	$3\|N_i\| + x$	$2(1+\theta)\delta + \delta_d$

The total number of messages transmitted between the cells and the delay for message transmission for channel acquisition is therefore $3|N_i| + x$ and $3\delta + \delta_d$, respectively. Table 1 shows the total number of messages transmitted between the cells and the delay of message transmission for importing X channels for CAT, D-LBSB, D-ES, and D-CAT. In D-CAT, if only all of the imported channels need to be confirmed then $\theta = 1$. Otherwise, $\theta = 0$. Since a cell in D-CAT can import at least one channel in a channel importing operation and therefore in most cases θ will be zero, leading to a message delay of $2\delta + \delta_d$.

It is observed that even though D-CAT shows an implementation cost higher than its centralized counterpart, CAT, but it provides significant improvements on both the message number and the message delay over the other distributed algorithms. It can be seen that by letting each cell have its own channels simplifies the algorithm since the negotiation of a channel allocation between the cells is reduced to only the importer-exporter pair. One possible problem with this scheme, however, is that the identifiers of the channels a cell holds may become disjointed sparsely. To deal with this problem, a priority channel acquisition scheme is used in this paper. The channels a cell is going to import are prioritized according to their identifiers so that a channel with an identifier nearer to the identifiers of the channels initially assigned to the importer cell has a higher priority. For example, cell i is initially assigned with channels, 5, 6, and 7, and now has two channels, 5 and 7. If cell i has found five free channels, 1,4,6,9, and 10, then it attempts to import these channels with a priority of 6, 4, 9, 10, and 1.

4.2 Call Blocking Probability Comparison

Simulation was also used to evaluate the call blocking probability of D-CAT and compare it with D-LBSB and CAT. The results shown in the figures were obtained with 90% confidence interval and within 5% of the sample mean. The simulated cellular system contains 15 × 15 hexagonal cells shown in Figure 1. The letters, $a, b, c, ...$, on the cells in Figure 1 denote distinct sets of channels and the cells with the same letter are assigned with the same set of channels. Each cell is initially assigned 40 channels. Incoming call arrival at each cell is assumed to follow a Poisson process with a mean λ. The holding time of a call is assumed to be distributed based on an exponential distribution with a mean $1/\mu$ of 180 secs (3 mins). The parameters used in CAT are as follows: $\Delta = 2$, $p_s = 20\%$ and $c_{min} = 0.05C$ where C is the number of channels assigned to a cell. The degree of coldness at a cell in LBSB is 0.1.

Figure 3(a) and 3(b) show the call blocking probability of the algorithms under consideration. Two kinds of calling demands, *uniform* and *non-uniform*, was simulated for the algorithms. In the previous case, call arrival at each cell is identical. In the latter case, on the other hand, a cell can get congested from time to time. That is, a cell gets congested from λ to 3λ with a probability of 0.001 and a congested cell will return to the normal state with a probability of 0.01. It is observed that D-CAT outperforms all of other algorithms in terms of call blocking probability. It is observed that there are no large differences in D-CAT between the cases of $T_h = 0$ and $T_h = 1$. This figure has confirmed the potential of distributed algorithms; that is, distributed algorithms are also capable of providing significantly good performance in practical conditions.

Figure 4 shows the average number of channels imported during one channel acquisition operation in D-CAT. The more the number of channels imported in one channel acquisition process, the lower the overhead cost needed for the message transmission. It is observed that D-CAT imports many channels for a heavy cell each time in normal conditions and therefore improves significantly the overhead caused by message transmissions (*e.g.*, near to 6 channels on average at 35 Erlangs).

5 Conclusions

In this paper, a distributed channel allocation algorithm based on a two-threshold scheme, D-CAT, has been proposed and evaluated in mobile cellular networks. It has been shown that D-CAT provides better performance than other centralized and distributed channel allocation algorithms. It has also been shown that the implementation cost, both on message complexity and message delay of D-CAT, is much less than that of either distributed algorithm, D-LBSB or D-ES. Furthermore, letting each cell hold its own channels simplifies the channel allocation procedure and makes a channel allocation algorithm more efficient.

References

1. Ed. J.D. Gibson. *Mobile communications handbook*. CRC Press, 1999.
2. A. Baiocchi, F.D. Priscoli, F. Grilli, and F. Sestini. The geometric dynamic channel allocation as a practical strategy in mobile networks with bursty user mobility. *IEEE Trans. Vehi. Tech.*, 44(1):14–23, February 1995.
3. G. Cao and M. Singhal. Efficient distributed channel allocation for mobile cellular networks. In *Proc. IEEE 7th Int. Conf. Comput. and Commun. Networks*, 1998.
4. G. Cao and M. Singhal. An adaptive distributed channel allocation strategy for mobile cellular networks. In *Proc. 18th IEEE Int. Conf. Performance, Computing, and Commun.*, 1999.
5. G.L. Stuber. *Principles of Mobile Communication*. Kluwer Academic Publisher, Inc., Boston, 1996.
6. M. Zhang and T.S.P. Yum. Comparisons of channel-assignment strategies in cellular mobile telephone systems. *IEEE Trans. Vehi. Tech.*, 38(4):211–215, November 1989.
7. D.C. Cox and D.O. Reudink. Increasing channel occupancy in large scale mobile radio systems: Dynamic channel reassignment. *IEEE Trans. Vehi. Tech.*, VT-22(4):218–222, November 1973.
8. J. Tajima and K. Imamura. A strategy for flexible channel assignment in mobile communication systems. *IEEE Trans. Vehi. Tech.*, 37(2):92–103, May 1988.
9. S.K. Das, S.K. Sen, R. Jayaram, and P. Agrawal. An efficient distributed channel management algorithm for cellular mobile networks. In *Proc. IEEE Int. Conf. Universal Personal Commun.*, pages 646–650, October 1997.
10. H. Jiang and S.S. Rappaport. CBWL: A new channel assignment and sharing method for cellular communication systems. *IEEE Trans. Vehi. Tech.*, 43(2):313–322, May 1994.
11. Y. Zhang and S.K. Das. An efficient load-balancing algorithm based on a two-threshold cell selection scheme in mobile cellular networks. *Comput. Commun.*, 23(5-6):452–461, March 2000.
12. Y. Furuya and Y. Akaiwa. Channel segregation, a distributed adaptive channel allocation scheme for mobile communication systems. *IEICE Trans.*, E74(6):1531–1537, June 1991.
13. H. Furukawa and Y. Akaiwa. Design of underlaid microcells in umbrella cell system. *IEICE Trans. Commun.*, E81-B(4):762–769, April 1998.
14. U. Black. *Mobile and Wireless Networks*. Prentice-Hall PTR, 1996.
15. L. Lamport. Time, clocks, and the ordering of events in a distributed system. *Commun. of the ACM*, 21(7):558–565, July 1978.

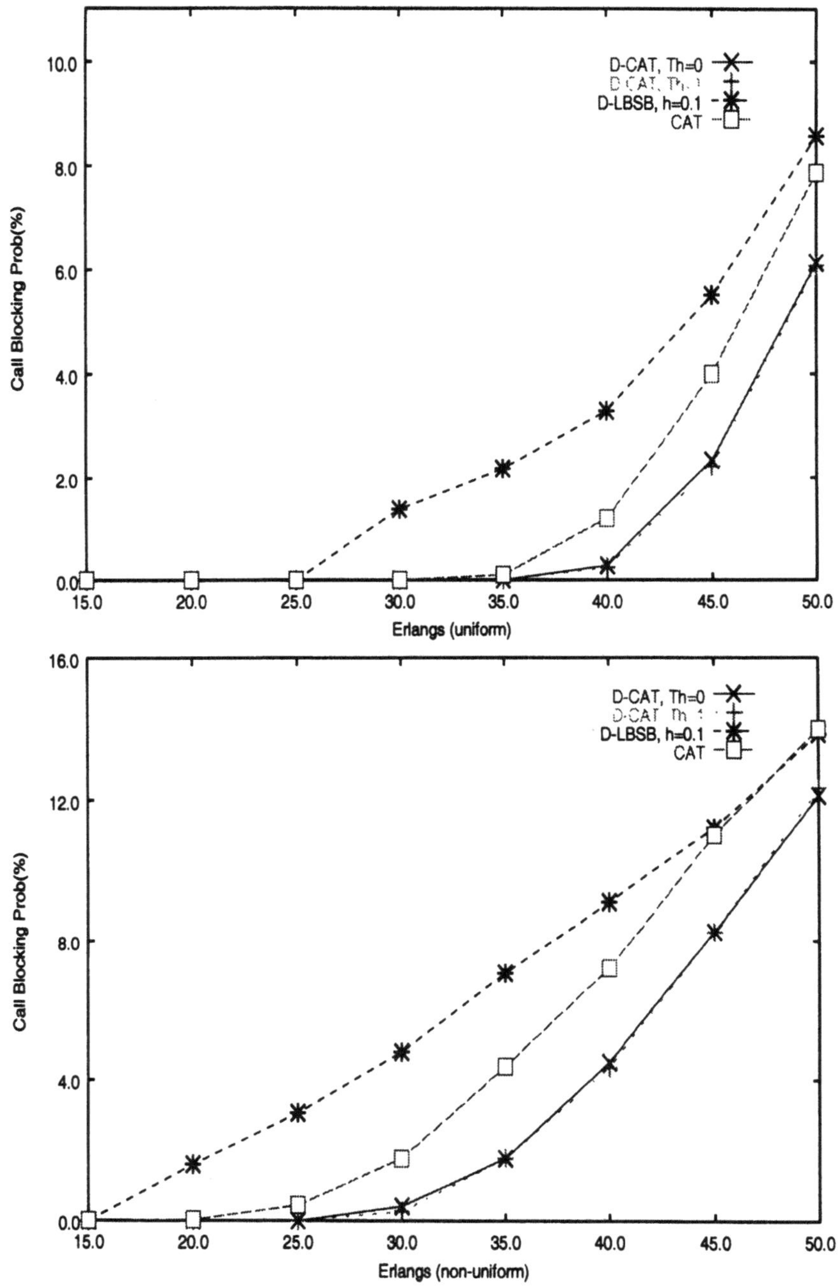

Fig. 3. Comparison of the call blocking probability of the various algorithms.

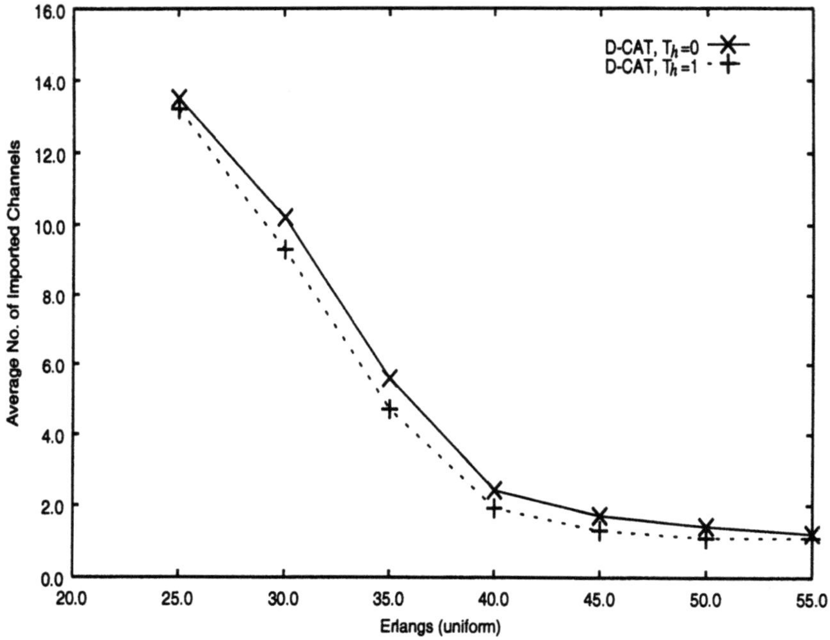

Fig. 4. Average number of channels imported during one channel acquisition process.

Part V

Mobile Telecommunication II

The Feasibility Study on a Spectrum Overlaid System of N-CDMA and W-CDMA

Jie Zhou, Ushio Yamamoto, and Yoshikuni Onozato

Dept. of Computer Science, Faculty of Engineering Gunma University
1-5-1 Tenjin-cho, Kiryu, Gunma, 376-8515 Japan
E-mail: {zhou, kansuke, onozato}@nztl.cs.gunma-u.ac.jp

Abstract. In this paper[0], the performance in terms of signal-to-interference ratio (CIR), of a spectrum overlaid system of N-CDMA over W-CDMA is investigated. The radio channel is composed of the inverse fourth-power path loss law with log-normal shadow fading. The perfect power control is assumed in the estimation. In order to suppress the interference, the other important techniques used in the analysis are the ideal notch filter and the signal level clipper for the transmitters and receivers of W-CDMA system. We propose the concepts of the notch filtering depth and signal level clipping depth. The maximum capacity trade-offs between the two systems and the impact of varying transmission rates on the overlaid system are investigated and quantified. Based on the numerical results, the feasibility of the overlaid system is discussed.

1 Introduction

Wide-band code division multiple access (W-CDMA) is considered as a promising radio access technique for the third generation mobile radio communication systems, which are called $IMT2000$[2]. Especially, this generation mobile system can supply multimedia communication with anyone at any time from anywhere[1],[3]. With the expected wireless revolution in telecommunications, the available spectrum should be used efficiently and flexibly. One step in this direction is the use of spread spectrum overlay. The overlay of systems has significance for adopting some advanced signal processing and interference suppression techniques, such as notch filter[4], and signal level clipper.

In fact, the spectral overlay of a new system to the existing system is not strange but has been studied in Refs.[1],[2],[3],[7] and [9]. However, the spectral overlay system under fading channel employing some interference suppression techniques and also the impact of varying transmission rates on the overlaid system have not been examined until now. For example, Ref.[1] only analyzed the overlaid system of cellular CDMA on AMPS. Refs.[2] and [3] depicted the simple model for N-CDMA/W-CDMA overlaid system without considering any shadow fading

[0] This work is supported in part by the Grants-in-Aid for Science Research No.12680432 and by Research for the Future Program at Japan Society for the Promotion of Science

effects and proposing any interference suppression techniques. The capacity trade-offs between the two systems were only investigated. The feasibility of the overlaid system and the future application have not be discussed.

When W-CDMA is considered as a wireless access technique for $IMT2000$ systems, in order to efficiently make use of a limited radio spectrum, adoption of overlaid system may have many attractions, such as high reuse of bandwidth and high capacity because of using notch filter, and signal clipping techniques. In this paper, we first introduce notch filter, and signal level clipper in the N-CDMA/W-CDMA overlaid system. We propose the concepts of the notch filtering depth and signal level clipping depth and estimate their effects on the overlaid system. The maximum capacity trade-offs and the performance of the system with varying transmission rates are investigated and quantified in terms of signal-to-interference ratio, CIR. We focus on the reverse link in our investigation because the capacity of conventional CDMA cellular system is usually limited by the reverse link, mainly due to the non-orthogonal, asynchronous transmission interference caused by multiple users and the constraint on the mobile station's transmitter power.

The reminder of this paper is organized as follows. The next two sections depict the system model and performance analysis via estimation of interferences. The section 4 describes the N-CDMA/W-CDMA overlaid system under the interference suppressions. Section 5 gives some numerical results and discussions. Finally, the conclusions are given in the section 6.

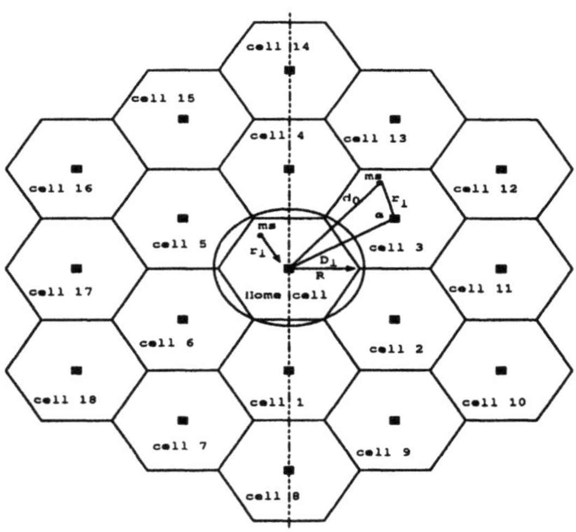

Fig. 1. Standard cellular geometry and interference geometry

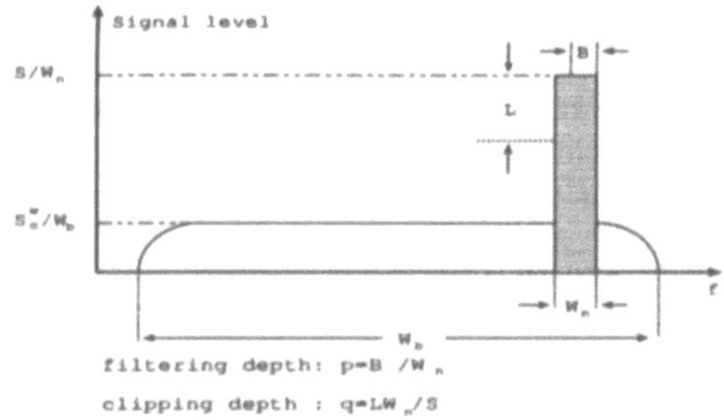

Fig. 2. Filtering depth and clipping depth depicted in the signal level versus frequency plane

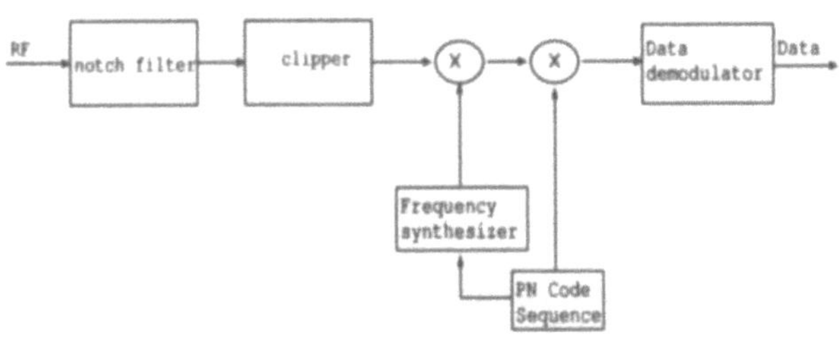

Fig. 3. Block diagram of W-CDMA receiver which employs interference suppression techniques

2 System Models and Interference Suppression

2.1 Overlay Considerations

We consider the N-CDMA/W-CDMA overlaid system employing several interference suppressions which will be introduced in the following subsection.

We model the overlaid system as several CDMA systems with different spreading bandwidths. Its capacity trade-offs for different service purposes are investigated and quantified in detail. Figure 1 shows a typical cellular system where each cell consists of N-CDMA users and W-CDMA users. We can select any one cell as the home cell, in which there are N-CDMA system base-station (BS) and W-CDMA system BS. They are colocated in the center of the cell. For CDMA system, the assignment of the entire channels to each cell, using the pseudo-noise (PN) codes which are uncorrelated, absolutely separates the desired signal.

2.2 Interference Suppression

As shown in Fig.2, where the ordinate is the signal level and the abscissa is the frequency axis, the idea behind the spread spectrum is to transform a desired signal with narrow-bandwidth into a noise-like signal of much wide-bandwidth W_b. As indicated above, one of the most attractive characteristics of adopting spread spectrum in wireless cellular system is the ability to overlay wide-band CDMA system. A wireless access is considered for advanced systems on top of existing narrow-band CDMA. The block diagram of W-CDMA receiver which employs interference suppression techniques[4] is shown in Fig.3. We will estimate their performance in detail.

We know that CDMA systems are relatively sensitive to interference. By the use of a multiplication procedure with the pseudo-noise codes in wide-band, all the signal power, including desired signal and noise are essentially reduced by the processing gain, G as

$$G = \frac{W}{R_b} \tag{1}$$

where W and R_b are the spread bandwidth and information rate, respectively. W may be denoted as the spread bandwidth W_b or W_n for W-CDMA system and N-CDMA system as shown in Fig.2, respectively.

In Fig.2 due to the low-power spectral density, W-CDMA signals cause relatively little interference to N-CDMA system occupying less bandwidth. On the other hand, the N-CDMA system causes high interference to the W-CDMA system without any interference suppressions, which limits the uses of the overlaid system and the system capacity. Then, the interference suppressions must be adopted in the W-CDMA system. For the suppression of the interference, we introduce the notch filter and signal level clipper in the overlaid systems. The notch filter is the limited spectrum filter used in W-CDMA receiver for filtering the part of the overlaid bandwidth of N-CDMA with the higher signal level. The signal level clipper is a kind of signal level limiter used for suppressing the higher signal level of interference. In the design of notch filter and signal level clipper, we propose the filtering depth, p and clipping depth, q defined as follows

$$p = \frac{B}{W_n}, and \quad q = \frac{LW_n}{S} \tag{2}$$

where B and L are the notch filtering bandwidth and clipped signal level, respectively. p represents the tolerance of the collapsed signal in poor shape in W-CDMA decision hardware. q represents the signal clipping depth in W-CDMA receivers. Based on the definition, the deduction ratio of interference of N-CDMA system to W-CDMA system is given by

$$\nu = (1-p)(1-q) \tag{3}$$

3 Performance Analysis Via Estimation of Interference

3.1 Cellular Structure and Propagation Model

Let us consider the overlaid system as shown in Fig.1. We can compute the distance between any users located in the i-th cell and the BS of the home cell as

$$d_0 = \sqrt{r_i^2 + D_i^2 + 2D_i r_i cos(\alpha)} \tag{4}$$

where r_i is the the distance between the BS and the user. D_i is the distance between the home cell BS and the BS of the i-th cell. α is defined in Fig.1.

For the radio channel with the path loss law, κ and shadow fading loss with the random variable λ, which is the log-normal distribution with zero mean and the standard deviation of σ, the received signal power at the receiver S_r can be depicted as[7][9]

$$S_r = \frac{P}{r^\kappa} 10^{\lambda/10} \tag{5}$$

where P is the transmitted signal power at the transmitter. r is the distance between the receiver and transmitter.

3.2 Interference Estimation

In our estimation of the interference, we assumed that both the N-CDMA and W-CDMA systems are interference limited, as is the case in most practical situation. A user receives the interference generated by the other users located in this cell termed as intracell interference, I^{in} and the interference generated by the users located in the surrounding cells termed as intercell interference, I^{out}[9][11]. In the N-CDMA/W-CDMA overlaid system, there are also the cross-interferences among them, such as the intracell W-CDMA interference to N-CDMA and so forth.

In order to delineate the trade-off between the N-CDMA system and the W-CDMA system, CIR constraint of N-CDMA or W-CDMA system is generally expressed as

$$CIR = \frac{S}{I^{in} + I^{out} + N_0} \geq CIR_{req} \tag{6}$$

where S is the desired signal power. N_0 is the background thermal noise. Based on the vast literature on the CDMA cellular systems, such as Ref.[1],[2] and [5], because all the users use the same spread bandwidth in all cells, all the other users' signal power is considered as interference. In contrast with the interference, N_0 is considered as negligibly small in our investigation.

CIR for the overlaid system shown in Fig.1 will be evaluated. In the overlaid system, the following five contributions can be distinguished to the N-CDMA reverse link, namely as[9][11][14];

1) *Desired Signal Power of N-CDMA,* S_c^n *of one user;*

2) *Intracell N-CDMA Interference to N-CDMA,* I_{nn}^{in} *which is generated by the users of N-CDMA system, located in the home cell;*

3) *Intracell W-CDMA Interference to N-CDMA,* I_{wn}^{in} *which is generated by the users of W-CDMA system, located in the home cell;*

4) *Intercell N-CDMA Interference to N-CDMA,* I_{nn}^{out} *which is generated by the users of N-CDMA system, located in the surrounding cells;*

5) *Intercell W-CDMA Interference to N-CDMA,* I_{wn}^{out} *which is generated by the users of W-CDMA system, located in the surrounding cells;*

For the W-CDMA reverse link, the following five contributions can be distinguished namely as[9][11][14];

1) *Desired Signal Power of W-CDMA,* S_c^w *of one user;*

2) *Intracell N-CDMA Interference to W-CDMA,* I_{nw}^{in} *which is generated by the users of N-CDMA system, located in the home cell;*

3) *Intracell W-CDMA Interference to W-CDMA,* I_{ww}^{in} *which is generated by the users of W-CDMA system, located in the home cell;*

4) *Intercell N-CDMA Interference to W-CDMA,* I_{nw}^{out} *which is generated by the users of N-CDMA system, located in the surrounding cells;*

5) *Intercell W-CDMA Interference to W-CDMA,* I_{ww}^{out} *which is generated by the users of W-CDMA system, located in the surrounding cells;*

3.3 Maximum Capacity Trade-offs

For the N-CDMA/W-CDMA overlaid system, it is not easy to estimate the maximum capacity because of the data rates from various traffic sources, different activity factors and the different levels of transmission quality. According to Eq.(6), the maximum capacity trade-off between the two systems can be usually written by the nonlinear function of system parameters as[2][3]

$$N_c^n R_b^n \leq g(CIR_{req}^n, CIR_{req}^w, \sigma, W, N_0, \mu, N_c^w R_b^w) \tag{7}$$

where μ is the activity factor which may be the voice data transmission activity factor, μ_v or high data transmission activity factor, μ_d. CIR_{req}^n and CIR_{req}^w are the requirements of N-CDMA signal transmission quality and the requirement of data transmission quality in W-CDMA system, respectively. N_c^n and N_c^w are the numbers of the users of N-CDMA and W-CDMA systems per cell, respectively. R_b^n and R_b^w are the data rates, respectively. Based on CIR constraints of N-CDMA and W-CDMA, the trade-off between them will be shown in Eqs.(14) and (15), respectively.

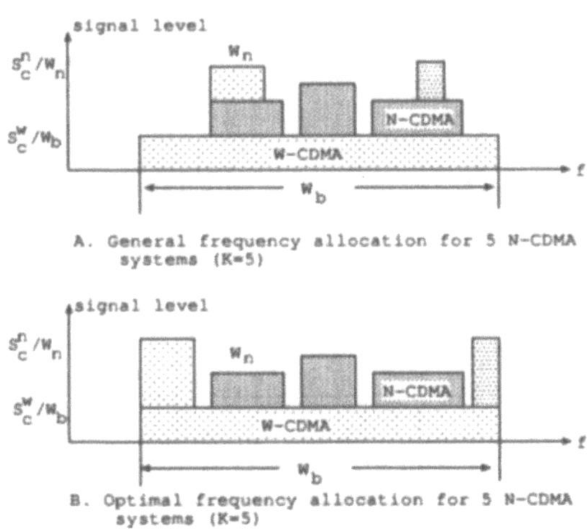

Fig. 4. Overlaid spectrum patterns

4 N-CDMA/W-CDMA Overlaid System

The third generation mobile communication systems will provide not only voice and low data rate services, but also video and higher data rate services. The N-CDMA systems based on IS-95 protocol are mainly asked for voice users under the narrow bandwidth of 1.25MHz. It is difficult to allow video and high data rate services, such as video conference system, with 384kbps or more high data rates, such as 1Mbps or 2Mbps. So recently, there are many new W-CDMA techniques investigated in Refs.[2][3][4] in order to supply multimedia communications with anyone at any time from anywhere. In order to efficiently use bandwidth, the overlaid N-CDMA/W-CDMA systems[2][4] are regarded as one of efficient methods in which some interference suppressions are introduced.

Earlier works of Refs.[1]-[2] identified the possibility of the overlaid systems and estimated some performance of the system. According to those approaches, the mutual exclusion of medium band frequency allocations (FA's) and narrow band FA's like Fig.4, and the capacity trade-offs among all possible overlaid patterns[1][2] were investigated. We model the system structure in the same way as in Ref.[2] in this paper introducing interference suppressions among them. We investigate the capacity of the overlaid system, and obtain the allowed highest data rate under various cellular mobile environments.

Based on Refs.[2] and [3], Fig.4-B shows the optimal frequency allocation between N-CDMA and W-CDMA systems. In the W-CDMA system with W_b, transmitted power density, $\frac{S_c^w}{W_b}$ is shared with K N-CDMA systems of different spread spectrums as shown in Fig.4-B where K is the number of narrow band FA's within the W-CDMA spread bandwidth. All the systems are assumed to use the same base station, otherwise the analysis is very difficult even if the precise transmitted power control is adopted in each system.

4.1 Signal and Interference on the N-CDMA Reverse Link

1)*Desired Signal Power, S_c^n*: In the overlaid system of K N-CDMA systems over W-CDMA system as shown in Fig.4-B, the received desired signal, S_c^n, is assumed as the constant level under the precise perfect power control[9][14] at the $i-th$ N-CDMA base stations for $i = 1, 2, ...K$.
2)*Intracell Interference, I^{in}*: I^{in} is composed of two portions. They are I_{wn}^{in} and I_{nn}^{in} which use the same frequency bandwidth. There is no interference among the N-CDMA systems because there are no frequency overlaid portion among them when we use the optimal allocation of frequency[1][2] of the systems shown in Fig.4-B. I^{in} can be expressed as

$$\begin{aligned} I^{in} &= I_{nn}^{in} + I_{wn}^{in} \\ &= \mu_v N_c^n S_c^n + \mu_d N_c^w S_c^w \frac{W_n}{W_b} \end{aligned} \tag{8}$$

3)*Intercell Interference, I^{out}*: I^{out} is composed of I_{nn}^{out} and I_{wn}^{out}. The interference, I_{nw}^{out} is estimated under the effects of shadow fading, where the approximation has been introduced here that we use the circular cell rather than the true hexagonal cell for simplicity. Notice that R of the circular cell is essentially the same as R of the hexagonal cell. A_i is the area of one cell. I_{nn}^{out} is obtained by the integration of each cell as follows[7][9][12]

$$\begin{aligned} I_{nn}^{out} &= \sum_{i=1}^{m_n} \mu_v \iint_{A_i} S_c^n (r/d_0)^\kappa E[10^{\lambda_i - \lambda_0}] \rho_n dA \\ &= \sum_{i=1}^{m_n} \mu_v S_c^n \Gamma(r/d_0, \lambda_i, \lambda_0, \rho_n) \end{aligned} \tag{9}$$

where

$$\Gamma(r/d_0, \lambda_i, \lambda_0, \rho_n) = \iint_{A_i} (r/d_0)^\kappa E[10^{\lambda_i - \lambda_0}] \rho_n dA \tag{10}$$

where m_n is the total number of outer N-CDMA cells considered in our estimation, ρ_n is the user's distribution density function which is calculated as $\frac{N_c^n}{2\pi R^2}$, $E[10^{\lambda_i - \lambda_0}]$ means the expected value of $[\cdot]$. The derivation of $E[10^{\lambda_i - \lambda_0}]$ is given by C.C.Lee and R.Steel in Ref.[9] or the appendix in Ref.[14].

We can use the similar analytical method above, then I_{wn}^{out} is given by

$$I_{wn}^{out} = \frac{W_n}{W_b} \sum_{i=1}^{m_w} \mu_d S_c^w \Gamma(r/d_0, \lambda_i, \lambda_0, \rho_w) \tag{11}$$

where,

$$\rho_w = \frac{N_c^w}{2\pi R^2}$$

and m_w is the number of outer W-CDMA cells which are considered in the estimation of the overlaid system.

4.2 Signal Power and Interference on the W-CDMA Reverse Link

1)*Desired Signal Power, S_c^w and Intracell Interference, I^{in}*: under the perfect power control, for the W-CDMA reverse link, S_c^w takes the constant level. I^{in} is composed of I_{nw}^{in} and I_{wn}^{out}. It is depicted as

$$\begin{aligned} I^{in} &= I_{nw}^{in} + I_{ww}^{in} \\ &= \mu_d N_c^w S_c^w + \sum_{i=1}^{K} \nu_i \mu_v N_c^n S_c^n \end{aligned} \tag{12}$$

where, ν_i is the deduction ratio of interference to the W-CDMA system generated by i-th N-CDMA system for $i = 1, 2, ...K$.
2)*Intercell Interference, I^{out}*: For the W-CDMA system, there is significant interference to its link because there are many N-CDMA systems which overlay their spread bandwidth within the W-CDMA system bandwidth. I^{out} which is composed of I_{nw}^{out} and I_{ww}^{out} can be summarized as

$$\begin{aligned} I^{out} &= I_{nw}^{out} + I_{ww}^{out} \\ &= \sum_{i=1}^{K} \nu_i \sum_{i=1}^{m_n} \mu_v S_c^n \Gamma(r/d_0, \lambda_i, \lambda_0, \rho_n) \\ &\quad + (1 - \frac{W_n}{W_b} \sum_{i=1}^{K} p_i) \sum_{i=1}^{m_w} \mu_d S_c^w \Gamma(r/d_0, \lambda_i, \lambda_0, \rho_w) \end{aligned} \tag{13}$$

where, p_i the filtering depth of bandwidth of i-th N-CDMA system for $i = 1, 2, ...K$.

4.3 N-CDMA and W-CDMA Constraints

In order to maintain adequate transmission quality in N-CDMA and W-CDMA systems, respectively, we must find the constraint relationship satisfying the requirement that the bit energy-to-noise density ratio of each received desired signal should be greater than or equal to the certain target value for its requirement among the aggregate data of N-CDMA systems and W-CDMA system under various transmission data rates. Based on Eq.(6), we can obtain the trade-off based on the N-CDMA constraint as

$$\frac{W_n}{N_c^n R_b^n \left(\frac{E_b}{N_0}\right)_{req}^n} \geq \mu_v + \mu_d \frac{N_c^w}{N_c^n} \frac{S_c^w}{S_c^n} \frac{W_n}{W_b} + \sum_{i=1}^{m_n} \mu_v \Gamma(r/d_0, \lambda_i, \lambda_0, \rho_n)$$
$$+ \frac{W_n}{W_b} \sum_{i=1}^{m_w} \mu_d \frac{S_c^w}{S_c^n} \Gamma(r/d_0, \lambda_i, \lambda_0, \rho_w) \tag{14}$$

and the trade-off based on the W-CDMA constraint as

$$\frac{W_b}{N_c^w R_b^w \left(\frac{E_b}{N_0}\right)_{req}^w} \geq \mu_d + \sum_{i=1}^{K} \nu_i \mu_v \frac{N_c^n}{N_c^w} \frac{S_c^n}{S_c^w} + \sum_{i=1}^{K} \nu_i \sum_{i=1}^{m_n} \mu_v \frac{S_c^n}{S_c^w} \Gamma(r/d_0, \lambda_i, \lambda_0, \rho_n)$$
$$+ \left(1 - \frac{W_n}{W_b} \sum_{i=1}^{K} p_i\right) \sum_{i=1}^{m_w} \mu_d \Gamma(r/d_0, \lambda_i, \lambda_0, \rho_w) \tag{15}$$

where in Eqs.(14) and (15), $\left(\frac{E_b}{N_0}\right)_{req}^w$ and $\left(\frac{E_b}{N_0}\right)_{req}^n$ are the requirements to maintain adequate transmission quality for W-CDMA and N-CDMA systems, respectively.

5 Numerical Results

In order to derive some numerical results, more specific model assumptions have to be made. We consider the N-CDMA/W-CDMA overlaid systems of the following parameters:

1) $W_b = 25MHz$, $W_n = 1.25MHz$;
2) $0 \leq p \leq 1$, $q \leq 1$, $0 \leq \nu \leq 1$;
3) $\mu_v = 0.375$, $\mu_d = 0.375$;
4) $\kappa = 4.0$;
5) $\sigma = 6, 7, 8, 9dB$;
6) $m_n = m_w = 18$;
7) $S_c^w = 1.00S_c^n, 0.50S_c^n, 0.03S_c^n$;
8) R_b^n = 9.6kbps, R_b^w = 9.6, 56, 128, 384kbps;
9) R = 10 km, $K = 5$;
10) Although, the required bit energy-to-noise density ratio for W-CDMA is generally greater than that for N-CDMA because data requires the lower bit error ratio. The retransmission schemes can enforce the transmission quality as shown in

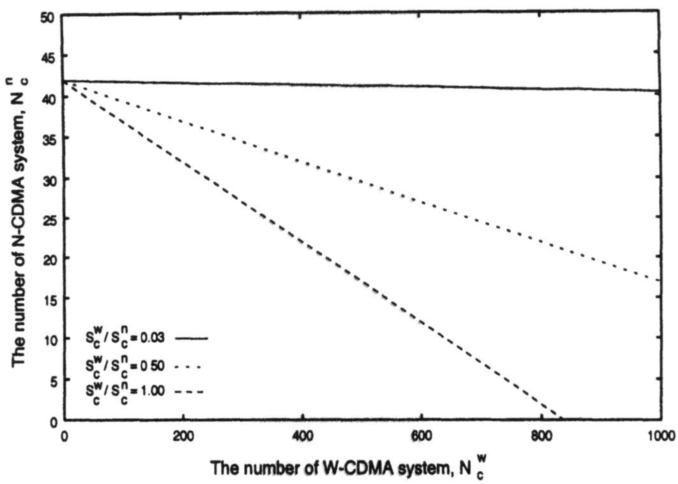

Fig. 5. The number of N-CDMA users due to the number of W-CDMA users with the various ratio of $\frac{S_c^w}{S_c^n}$ under N-CDMA constraint of Eq.(14) ($W_b = 25MHz$, $W_n = 1.25MHz$, $\sigma = 8dB$ $\mu_d = 0.375$, $\mu_v = 0.375$)

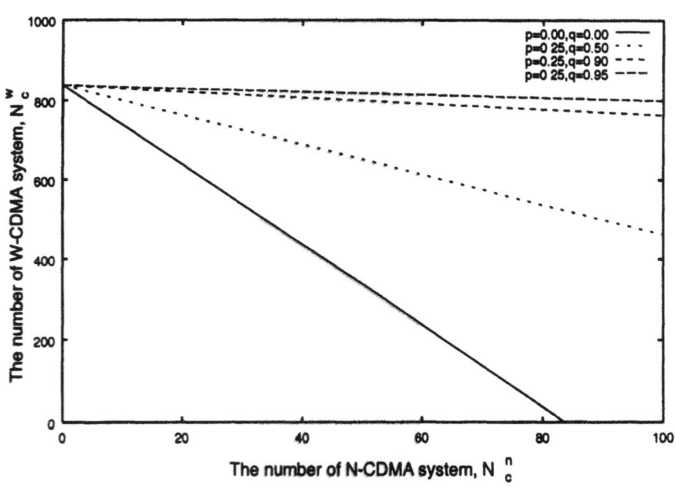

Fig. 6. The number of W-CDMA users due to the number of N-CDMA users with the various notch filtering depth, p and signal clipping depth, q under W-CDMA constraint of Eq.(15) ($W_b = 25MHz$, $W_n = 1.25MHz$, $\sigma = 8dB$ $\mu_d = 0.375$, $\mu_v = 0.375$, $R_b^w = 9.6kbps$, $\frac{S_c^w}{S_c^n} = 0.5$)

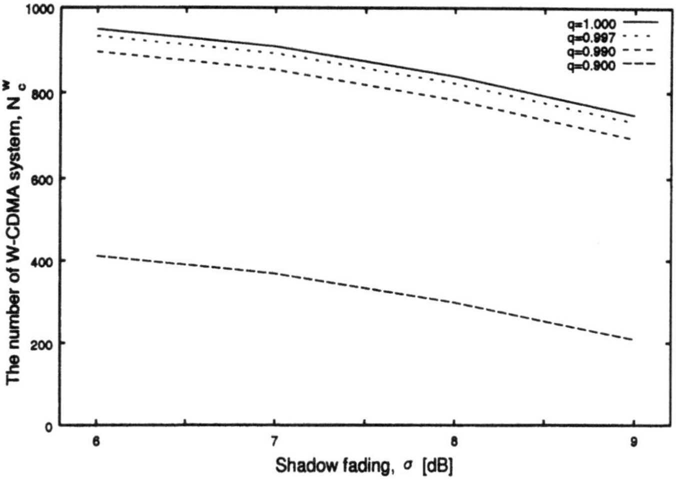

Fig. 7. The number of W-CDMA users due to the various shadow fading, σ for various values of q ($W_b = 25MHz$, $W_n = 1.25MHz$, $\mu_d = 0.375$, $p = 0.25$, $S_c^w = 0.03S_c^n$, $R_b^w = 9.6kbps$, $N_c^n = 43$)

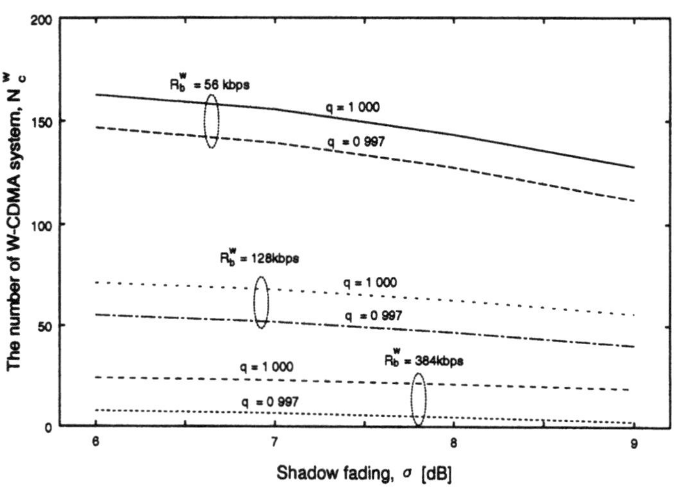

Fig. 8. The number of W-CDMA users due to the various shadow fading, σ with various data rate, R_b^w ($W_b = 25MHz$, $W_b = 1.25MHz$, $\mu_d = 0.375$, $p = 0.25$, $S_c^w = 0.03S_c^n$, $N_c^n = 43$)

Refs.[1] and [2], such as $(\frac{E_b}{N_0})^{w}_{req} \leq 7dB$. We also assume them as the worst situation as

$$(\frac{E_b}{N_0})^{n}_{req} = 7dB, \quad and \quad (\frac{E_b}{N_0})^{w}_{req} = 7dB$$

Figures 5 and 6 illustrate the system capacity trade-offs between the two systems. They are the N-CDMA constraint and W-CDMA constraint, respectively. Figure 5 shows that as the S_c^w/S_c^n ratio decreases, the slope of the capacity curves becomes smoother. In order to ensure no rapid decrease of the two system capacities, we must adopt that $0.03 \leq S_c^w/S_c^n \leq 0.50$. Figure 6 shows the effect of N_c^n on N_c^w varying notch filtering depth and signal clipping depth. We see the overlaid system with $p = 0.25$ and $q \geq 0.5$, the two system capacities can be higher, otherwise the overlaid system has not any advantages for the practical use. If $p > 0.25$, the better results will be achieved.

Figure 7 illustrates the W-CDMA capacity in the presence of N_c^n versus shadow fading, σ for various values of q. From the results, whether the system is in any wireless environments, the more q reaches one, the higher the overlaid system capacity is achieved for commercial applications because we can obtain the increase of total system capacity under the same spread bandwidth. Then, the high reuse of bandwidth can be achieved.

In W-CDMA, for the varying transmission rates, Fig.8 shows the data rates supported when $q = 1$ and $q = 0.997$, respectively. We see the overlaid system can also support the highest data rate at 384 kbps in the case of $q = 1$. For $q \leq 0.997$, the system can not support this data rate. Below 384 kbps, the overlaid system can be supplied in the various wireless environments.

6 Conclusions

The co-channel interference issues of sharing spread bandwidth among N-CDMA and W-CDMA systems have been investigated. We introduced the notch filter and signal clipper in the estimation of interference. The numerical results show the followings:

1) Without precise notch filter and signal clipper, any overlaid systems by employing bandwidth sharing can not be used in practical situation because of significant co-channel interference problems in the overlaid system.

2) CIR in the N-CDMA system was not degraded significantly by the presence of W-CDMA system because of the smaller power spectrum density of the W-CDMA system. On the other hand, CIR of W-CDMA system was degraded greatly, then the W-CDMA receiver and transmitter must be paid attentions carefully.

The important issue on the feasibility of the N-CDMA/W-CDMA overlaid system was investigated for the reverse link. The analytical method depicted in this paper could be easily extended to treat the forward link and additional performance in the overlaid systems which have potentials for the future wireless applications.

Acknowledgment

The authors would like to thank Professor F.Adachi, Tohoku University for helpful discussions.

References

1. H.H.Hmimy and S.C.Gupta, " Overlay of Cellular CDMA on AMPS Forward and Reverse Link Analysis", *IEEE Trans. on Veh. Technol.*, Vol.45, No.1, pp.51-56, Feb. 1996
2. D.G.Jeong, I.G.Kim and D.W.Kim, "Capacity Analysis of Spectrally Overlaid Multiband CDMA Mobile Networks", *IEEE Trans. on Veh. Technol.*, Vol.47, No.3, pp.798-807, Aug. 1998
3. D.W.Kim, I.G.Kim and D.G.Jeong, "Capacity Analysis of Spectrally overlaid Narrowband and Wideband CDMA Systems for Future Mobile Communications Services", *IEICE Trans. on Commun.*, Vol.E82-B, No.8, pp.1334-1342, Aug. 1999
4. R.L.Pickholtz, L.B.Milstein and D.L.Schilling, " Spread Spectrum for Mobile Communications", *IEEE Trans. on Veh. Technol.*, Vol.40, No.2, pp.313-321, May 1991
5. J.S.Wu, J.K.Chung and Y.C.Yang, "Performance Study for a Microcell Hot Spot Embedded in CDMA Macrocell Systems", *IEEE Trans. on Veh. Technol.*, Vol.48, No.1, pp.47-59, Jan. 1999
6. S.W.Wang and L.B.Milstein, "CDMA Overlay Situation for Microcellular Mobile Communication", *IEEE Trans. Commun.*, Vol.43, pp.603-613, Feb. 1995
7. C.Lin I, L.J.Greenstein and R.D.Gitlin, "A Microcell/Macrocell Cellular Architecture for Low-and High-Mobility Wireless Users", *IEEE J. Select. Areas Commun.*, Vol.11, No.6, pp.885-891, Aug. 1993
8. V.K.Garg and J.E.Wilkes, *Wireless and Personal Communications System*, IEEE Press, 445 Hoes Lane, Piscataway, NJ, 1996
9. C.C.Lee and R.Steele, "Effect of Soft and Softer Handoffs on CDMA System Capacity", *IEEE Trans. on Veh. Technol.*, Vol.47, No.3, pp.830-841, Aug. 1998
10. J.Shapira, "Microcell Engineering in CDMA Cellular Networks", *IEEE Trans. on Veh. Technol.*, Vol.43, No.4, pp.817-825, Nov. 1994
11. R.Coombs and R.Steele, "Introducing Microcells into Macrocell Networks: A Case Study", *IEEE Trans. on Commun.*, Vol.47, No.4, pp.568-576, 1999
12. J.S.Wu, J.K.Chung and Y.C.Yang, "Performance Study for a Microcell Hot Spot Embedded in CDMA Mcrocell System", *IEEE Trans. on Veh. Technol.*, Vol.48, No.1, pp.47-59, Jan. 1999
13. L.C.Wang, G.L.Stuber and C.T.Lea, "Architecture Design, Frequency Planning, and Performance Analysis for a Microcell/Macrocell Overlaying System", *IEEE Trans. on Veh. Technol.*, Vol.46, No.4, pp.836-848, Nov. 1997
14. J.Zhou, Y.Onozato and U.Yamamoto, "On the Capacity and Outage Probability of A CDMA Hierarchical Mobile System with Perfect/Imperfect Power Control and Sectorization", *IEICE Trans. on Fundamentals*, Vol.E82-A, No.7, pp.1161-1171, July 1999.

Mobility and Traffic Analysis for WCDMA Networks

Szabolcs Malomsoky and Árpád Szlávik

Traffic Analysis and Network Performance Laboratory, Ericsson Hungary Ltd.
H-1037 Budapest, Laborc u. 1., Hungary
{Szabolcs.Malomsoky,Arpad.Szlavik}@eth.ericsson.se

Abstract. In cellular mobile communication systems, the mobility of vehicles affects some important parameters, such as handover rates, channel occupancy times and blocking probabilities. The present work is based on a model proposed in [1] that suggests a convenient and practical approach to build an analytic traffic model, which also includes the effect of vehicle mobility.

This model is extended for CDMA networks, where the handover is of the so-called "soft handover" type. Performance measures like soft handover rates, soft handover type distributions and the offered communication traffic per cell, etc. are obtained. It is also explained how the model could be used for design and analysis of 3rd Generation WCDMA Systems.

1 Introduction

As today's cellular operators move to increase the number of services they offer to subscribers – e.g. by integrating wireless access to the Internet – new technologies are required in their systems. 3rd generation cellular networks [2] are being developed and standardized currently offering:

- increased capacity within their existing spectrum allocation,
- higher capacities and lower system design costs per subscriber (lower infrastructure costs), which will lead to a lower cost per subscriber,
- new subscriber features and integrated (voice and data) services, which will help the operators to increase their market penetration.

Code Division Multiple Access (CDMA) is regarded as the most suitable multiple access technology to fulfill the above requirements, and Wideband CDMA (WCDMA) [3] is capable to serve the new, high data rate wireless multimedia demand.

The present work is based on a model presented in [1]. This model offers a convenient and practical approach to build an analytic traffic model, which also includes the effect of vehicle mobility. In [1], the method was presented for cellular networks, in which the handover is of the so-called "hard handover" type.

In this work, the above model is extended to a CDMA cellular network, where the significant part of the handovers is of the "soft handover" type. Soft handover means that a mobile terminal can communicate with more than one base station at the same time. (For a thorough discussion of the advantages and disadvantages of soft handover see [4].) The consequence of this is that the traffic load on the fixed access network will not only (or mostly) depend on the call intensity, but the distribution of the number of legs a mobile is connected to the network with will also have a significant effect. This is because for example a call that is served by three base stations for some time will be (during that time) carried over three separate connections in the fixed network up to the so-called Diversity Handover unit (DHO), which combines the information stream on the three connections into one single connection. In the model, a road system is defined and covered with

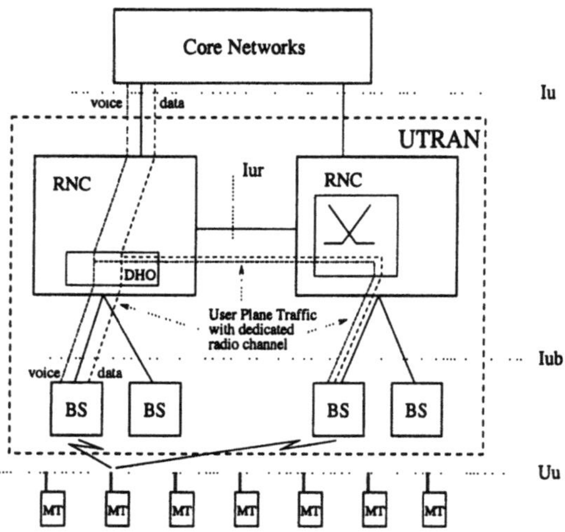

Fig. 1. UTRAN Architecture

overlapping CDMA cells. The vehicle traffic on this road system is described by a vehicle traffic matrix. The vehicle traffic is routed over the road system, and also the vehicle traffic load and the vehicle speed on different streets are calculated. Calls are generated to/from the vehicles as a Poisson process. Different call classes are defined in the model, which makes possible to investigate integrated-services WCDMA systems. Soft handover rates, the offered communication traffic per cell, blocking probabilities and also the distribution of the offered traffic with different number of legs is obtained. The method also gives a possibility to estimate the offered user-plane traffic between Radio Network Controllers (RNCs) in UMTS access networks (i.e. on the Iur interface, see *Fig. 1*).

1.1 Overview of the Literature

Several papers have been written about mobility modelling and spatial traffic distribution in cellular networks. First we give an overview of the literature that is most relevant for our work, then the objectives, goals and limitations of our work are discussed.

In [5], many aspects of mobility modelling in third generation mobile systems are considered, giving a good overview on mobility modelling and a good starting point for the interested reader. In [6] and [7], cells are divided into soft handover regions and calls are uniformly generated in a cell. As a result, the probability distribution function of total sojourn times in the soft handover region of a given cell during a channel holding time is obtained by analytical calculations. In these works, the mobile can move up, down, left and right, changing its speed M times during a call (where M is a geometrically distributed random variable) according to a uniform distribution over $[0, Vmax]$. Applying these distributions, it is difficult to find parameters such that they fit to real measurements. Especially, vehicular traffic is difficult to model, since the effects of a road map (directions, speeds, vehicular traffic hot spots, etc...) are difficult to take into account. A similar approach is presented in [8]. Here, an analytic traffic model is proposed to estimate the soft handover rate in CDMA cellular systems. As a limitation, a mobile can have at most two simultaneous connections (to two base stations) at a time. The call generation rate in the different regions (handover or non-handover) depends on the ratio of the areas of regions and cells. The main result is the analysis of the sensitivity of the handover rate when varying the cell radius, the speed of mobiles and the activity of the users. In [9], the authors present a mathematical formulation for systematic tracking of the random movement of a mobile station in a cellular environment. Based on this detailed formulation, a computer simulation is developed to obtain the behavior of different mobility-related parameters (e.g. the handover rate). An analysis of data obtained by simulation shows that the generalized gamma distribution function is a good approximation for the cell residence time distribution. However, the results in [10] indicate that it is not the residence time distribution, but it is its mean that influences teletraffic results. Therefore, classical Markovian methodology (applied also in our paper) has a chance to be valid and useful in practical analysis. In [11], it is emphasized that spatial teletraffic characterization is essential for planning and dimensioning of mobile communication systems. A geographic traffic model is presented, that makes it possible to involve demographical and geographical factors into teletraffic modelling. However, the evaluation of the performance measures related to handovers is not incorporated in this model.

1.2 Our Work

In this work, we extend the model presented in [1] for CDMA networks. The model is applied to problems arising in CDMA networks (soft handover modelling), as well as problems arising specifically in WCDMA networks (user-plane traffic estimation on the Iur interface). Our contribution is that we allow overlapping cells, and we consider soft handover instead of hard handover. This model may directly take input from a large scale mobility model, e.g. a gravity model [5].

The paper is organized as follows. In *Section 2.1*, a road network model and a method for estimation of the vehicular traffic volume on the road network are explained. *Section 2.2* gives a summary of all the simplifying assumptions we consider in the model. In *Section 2.3*, closed form solutions of the call arrival rate and the residence time in a soft handover region (SHR) are given and the relations between these parameters are shown. In *Section 2.4*, it is shown how the parameters achieved that far can be used to estimate inter RNC traffic. The probability distribution of the channel occupancy time in a cell is given in *Section 2.5*. By assigning capacity to each cell, in *Section 2.5*, we obtain the blocking probability and the offered traffic load on each cell in each call class by applying a recursive method. The handover intensity in a cell is also given in *Section 2.5*. *Section 2.6* presents a simple numerical example. Finally, conclusions are drawn in *Section 3*.

2 Description of the Model

In [1], the authors proposed a method to analyze the mobile communication traffic on a road systems model in which the road network is covered by non-overlapping hexagonal omnicells. This cell layout assumes adjacent cells with common boundary, therefore it enables to analyze only the handovers of the so-called hard handover type. In our work, we changed this cell structure by applying circle-shaped and overlapping cells, where the significant part of the handovers is of the soft handover type. By modelling soft handover operation, we are able to obtain some important performance measures of WCDMA cellular systems.

2.1 Road Systems Model and Traffic Flow Estimation

From the viewpoint of the mobility, the road network characterizes the geographical area under study. In our model, we consider only one transport mode, let us say vehicular traffic that flows over the specified road network between different traffic sources and absorption points. Pedestrian mobility is not considered, but it seems to be straightforward to include.

The road system can be modelled by a graph, where the links of the graph represent the *streets* and the nodes of the graph can be *junctions* (representing crosses in the road system) or the so-called *centroids* (representing the origin and the destination of the traffic flows on the road system). *Fig. 2* shows an example.

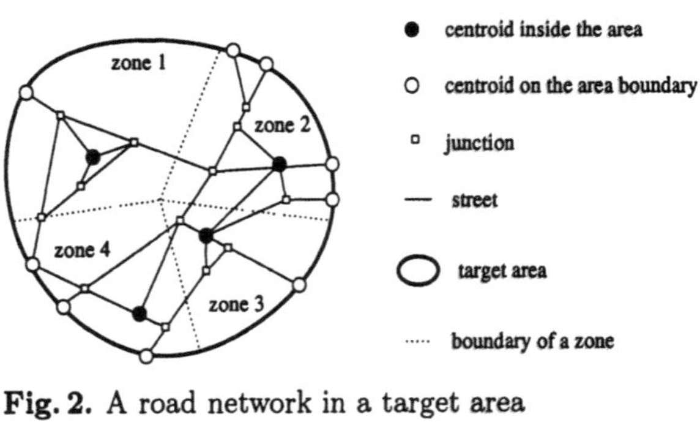

Fig. 2. A road network in a target area

Traffic flow on the road network of the target area is surveyed in advance. At the beginning of the survey the target area is divided into *zones*. Since the division is based on geographic, population and transport characteristics, these zones usually differ from radio zones such as cells in cellular systems. The traffic volume flowing from a zone to another zone is measured for each pair of zones. The results of this survey are usually represented by *Origin-Destination tables* (O-D tables) – *Table 1* for example. Note that this table also can be exploited as a result from a mobility model for large-scale areas that is on a higher abstraction level. We distinguish two kind of centroids:

- each zone of the target area has its own centroid placed inside the zone, which represents the origin and the destination of the zone itself (see *Fig. 2*),
- other centroids, which exist on the edge of the target area, represent traffic flows from/to the zones out of the target area (in *Fig. 2* the target area consists of 4 zones).

Table 1. An O-D table for the road network in Fig. 4

Vehicles/hour	c0	c1	c2	c3
c0	0	1000	250	50
c1	1000	0	50	250
c2	250	50	0	1000
c3	50	250	1000	0

Each element OD_{ij} in the O-D matrix (see *Table 1*) is the vehicular traffic volume defined as the number of vehicles moving from centroid ci to centroid cj during a unit time (e.g. rush hour). To determine which routes these vehicles move along towards their destination, we use an *incremental traffic assignment method* (details can be found in *Appendix A)*, but other algorithms could also be used. With this algorithm, the vehicular traffic is routed over the road network and the vehicular traffic load on all the streets are calculated.

The road network is covered by intersecting circle-shaped cells. The intersection of the circles and the links of the graph will be called *imaginary nodes.* We thus consider a new graph representing our road system, where the nodes of the graph can be centroids, junctions or imaginary nodes and the links of the graph are some parts of the original streets (links between junctions and/or centroids) divided by the imaginary nodes. Since the cells are overlapping, the overlapped regions are the so-called *soft handover regions* (SHR), where the soft handover calls can occur. We assume that a mobile under soft handover can communicate with *maximum three base stations* at the same time (a soft handover region can be in the intersection of one, two or three cells), see *Fig. 3.*

2.2 Notation and Assumptions

Some Preliminary Notation. With some routing algorithm used on the O-D table (see *Appendix A*) we can get the *routes* $r = 1, \ldots, R$ and the Q_r traffic volumes for each route r.

On route r we have *soft handover regions*

$$SHR_1^r, \ldots, SHR_j^r, \ldots, SHR_{J_r}^r,$$

listed in order of appearance (some soft handover regions can appear in the list more than once depending on the line of the route).

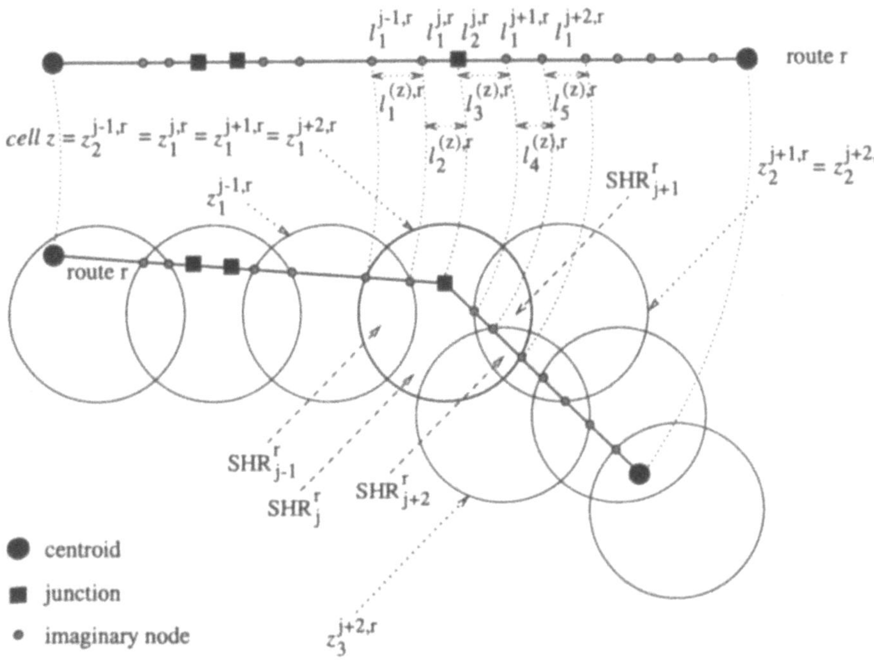

Fig. 3. Hierarchy of sections on route r

For each SHR_j^r we define the *characteristic set* as

$$\mathcal{Z}_{j,r} = \{\text{cells containing } SHR_j^r\} = \{z_1^{j,r}, \ldots, z_{K_{j,r}}^{j,r}\},$$

where SHR_j^r is overlapped exactly by cells $z_1^{j,r}, \ldots, z_{K_{j,r}}^{j,r}$, and therefore the number of those cells is $1 \leq K_{j,r} \leq 3$.

Finally, the smallest undivided section of route r is a link, which can begin and end in an imaginary node, a junction or a centroid. On the route r SHR_j^r contains the *links*

$$l_1^{j,r}, \ldots, l_i^{j,r}, \ldots, l_{I_{j,r}}^{j,r}.$$

(See *Fig. 3* for the better understanding of all the above notation.)

We will use the *indicator function* I throughout the paper: $I(expression)$ equals 1, if the *expression* is true and $I(expression)$ equals 0, if the *expression* is false.

From theory of sets: $x \in X$ ($X \ni x$) means that element x is in set X (set X owns element x), $Y \subseteq X$ ($Y \supseteq X$) means that Y is a subset of set X (set Y contains set X as a subset).

Assumptions. The model parameters and the assumptions are the following:

- The road network in the target area is covered by *circle-shaped* and intersecting *cells* with radius R_0. *(A1)*
- The centers of the adjacent cells are placed as the vertices of a *regular triangle.* The distance between the centers of the adjacent cells (the *base stations* are placed in the middle of the cell and have omnidirectional antennas) is D. *(A2)*
- Consider that we have N_{cell} cells and cell z ($z = 1, \dots, N_{cell}$) has a *capacity* of S_z channels. *(A3)*
- One transportation mode is considered (vehicular traffic) and the *traffic volume* flowing from a centroid to another is given in form of the *O-D table. (A4)*
- The speed of the vehicles on the streets is given by the *load-speed profile* (see *Fig. 5*) and is assumed to be *constant* on a street $\big(v(x)$ stands for the speed on street $x\big)$. *(A5)*
- The length of each street and each link is given in advance $\big(L(x)$ stands for the length of street or link $x\big)$. *(A6)*
- The traffic volumes and the speed of the mobile terminals on the links connected by an imaginary node are equal. *(A7)*
- The distribution of vehicles on a street is *uniform. (A8)*
- We distinguish C different *call classes.* Call class c ($c = 1, \dots, C$) is characterised by the *call arrival rate* λ_c, the *holding time* h_c and the *bandwidth* b_c. *(A9)*
- If there are n vehicles in a cell, then the arrivals of calls of class c form a *Poisson process* with the originating rate $\lambda_c n$ (the number of vehicles in the rush hour in a cell is sufficiently larger than the number of channels in the cell). *(A10)*
- The holding time of a call of class c is an *exponential* random variable with mean h_c. *(A11)*
- The arrival rate of calls originating *outside* the target area and then entering the target area are given in advance at the centroids on the border $\big(\lambda_c^0(p)$ for call class c for the centroid $p\big)$. *(A12)*
- The handover operation is of *soft handover* type (if an active mobile crosses a soft handover region border, we will say its call is "handed over" to the next soft handover region). A soft handover call is not *blocked*, if at least one of its *soft handover legs* is not blocked. *(A13)*

2.3 Soft Handover Region Parameters

From [1] we know that the probability that a new call of class c generated on $l_i^{j,r}$ succesfully reaches the soft handover region boundary between SHR_j^r and SHR_{j+1}^r is $p_c^{new}(SHR_{j+1}^r \mid l_i^{j,r}) \stackrel{def}{=} \frac{h_c}{t_i^{j,r}} e^{-\frac{T^{j,r}}{h_c}}(1 - e^{-\frac{t_i^{j,r}}{h_c}})$, for $1 \le j < J_r$, where $t_i^{j,r} \stackrel{def}{=}$

$\frac{L(l_i^{j,r})}{v(l_i^{j,r})}$ is the whole travelling time on $l_i^{j,r}$ $\big(L(l_i^{j,r})$ is the length of link $l_i^{j,r}$ and $v(l_i^{j,r})$ is the speed on link $l_i^{j,r}$ that equals the speed on the whole street containing this link$\big)$ and $T_i^{j,r} \stackrel{def}{=} \sum_{h=i+1}^{I_{j,r}} t_h^{j,r}$ is the travelling time from the boundary of $l_i^{j,r}$ and $l_{i+1}^{j,r}$ to the boundary of SHR_j^r and SHR_{j+1}^r.

$Q()$ [*vehicles/hour*] $= K()$ [*vehicles/km*] $v()$ [*km/hour*] for vehicular traffic, where Q is the *traffic volume*, K is the *traffic density* and v is the *traffic speed*. Therefore, the proportion of the arrival rate of new calls of class c on $l_i^{j,r}$ that gets connected to the set of base stations $Z \subseteq \mathcal{Z}_{j,r}$ is

${}^{Z}\lambda_c^{new}(l_i^{j,r}) = \lambda_c \frac{Q_r}{v(l_i^{j,r})} L(l_i^{j,r}) \; {}^{Z}P_c(SHR_j^r) = \lambda_c \; Q_r \; t_i^{j,r} \; {}^{Z}P_c(SHR_j^r)$, where Q_r is the traffic volume on route r and

$${}^{Z}P_c(SHR_j^r) \stackrel{def}{=} \prod_{z \in Z} (1 - B_c(z)) \prod_{s \in \mathcal{Z}_{j,r} \setminus Z} B_c(s),$$

where $B_c(z)$ is the *blocking probability* of c-type calls in cell z (the proportion of new call intensities ${}^{\varnothing}\lambda_c^{new}(l_i^{j,r})$ correspond to the blocked portion of calls, $Z = \varnothing$), see *Section 2.5* for the calculation of the blocking probabilities.

Remark: The arrival rate of new calls of class c on $l_i^{j,r}$ is $\lambda_c^{new}(l_i^{j,r}) \stackrel{def}{=} \sum_{Z \subseteq \mathcal{Z}_{j,r}} {}^{Z}\lambda_c^{new}(l_i^{j,r})$. Note that $\lambda_c^{new}(l_i^{j,r})$ is the call arrival rate corresponding to the infinite cell capacity case like in [1] (no blocking, $B_c(z) = 0$ for all cell $z = 1, \ldots, N_{cell}$) and also notice that ${}^{Z}\lambda_c^{new}(l_i^{j,r}) = \lambda_c^{new}(l_i^{j,r}) \; {}^{Z}P_c(SHR_j^r)$.

Suppose that a new call of class c originates in SHR_j^r from a vehicle moving along route r. Then the probability that this new call originates on $l_i^{j,r}$ is $p_c(l_i^{j,r}) \stackrel{def}{=} \frac{\lambda_c^{new}(l_i^{j,r})}{\sum_{h=1}^{I_{j,r}} \lambda_c^{new}(l_h^{j,r})} = \frac{\lambda_c \; Q_r \; t_i^{j,r}}{\sum_{h=1}^{I_{j,r}} \lambda_c \; Q_r \; t_h^{j,r}} = \frac{t_i^{j,r}}{T_0^{j,r}}$, like in [1], where $T_0^{j,r} \stackrel{def}{=} \sum_{h=1}^{I_{j,r}} t_h^{j,r}$ is the travelling time through SHR_j^r on route r.

This results in the *soft handover probability of new calls* of class c from SHR_j^r (the probability that a call of class c originating in SHR_j^r successfully reaches the boundary of SHR_j^r and SHR_{j+1}^r) to be

$$p_c^{new}(SHR_{j+1}^r \mid SHR_j^r) = \sum_{i=1}^{I_{j,r}} p_c^{new}(SHR_{j+1}^r \mid l_i^{j,r}) \; p_c(l_i^{j,r}) =$$

$$= \frac{h_c}{T_0^{j,r}} \sum_{i=1}^{I_{j,r}} e^{-\frac{T_i^{j,r}}{h_c}} (1 - e^{-\frac{t_i^{j,r}}{h_c}}).$$

Notice that the above expression on the right is a *telescopic sum* (an expounded sum, where the terms have alternating signs following each other and with the exception of the first and the last term, the others sum up to zero – the middle neighbouring terms cancel each other out) with $T_{i-1}^{j,r} = T_i^{j,r} + t_i^{j,r}$ and with $T_{I_{j,r}}^{j,r} = 0$, thus we get

$$p_c^{new}(SHR_{j+1}^r \mid SHR_j^r) = \frac{h_c}{T_0^{j,r}}(1 - e^{-\frac{T_0^{j,r}}{h_c}}), \quad \text{for } 1 \leq j < J_r. \tag{1}$$

Next consider the *soft handover probability of soft handover calls* of class c from SHR_j^r, i.e. the probability that while moving on route r the call of class c enters SHR_j^r (so it is "handed over" from SHR_{j-1}^r to SHR_j^r) and succesfully reaches the boundary of SHR_j^r and SHR_{j+1}^r (it is "handed over" to SHR_{j+1}^r) and that is

$$p_c^{ho}(SHR_{j+1}^r \mid SHR_j^r) = e^{-\frac{T_0^{j,r}}{h_c}}, \quad \text{for } 1 < j < J_r \tag{2}$$

because of the assumptions (A9),(A10),(A11).

We still need the boundary conditions for the soft handover probabilities: Consider an ongoing call that enters the target area at the centroid in SHR_1^r and moves from this centroid to the destination along route r. The problem is that the centroid is usually found inside the area rather than on the boundary of the cells of $\mathcal{Z}_{1,r}$, thus the boundary conditions for the soft handover calls entering these cell are not straight-forward. Therefore, considering these types of handover calls, we assume that the first link of route r is "lengthened backwards" to reach the boundary of one of the cells of $\mathcal{Z}_{1,r}$. If the lengthened part is d_r long, the travelling time of the new first street is $T_{0'}^{1,r} \stackrel{def}{=} \frac{d_r}{v(l_1^{1,r})} + T_0^{1,r}$, so we have the boundary condition

$$p_c^{ho}(SHR_2^r \mid SHR_1^r) = e^{-\frac{T_{0'}^{1,r}}{h_c}}. \tag{3}$$

On the other hand, the calls in $SHR^r_{J_r}$ are not "handed over" to further soft handover regions, therefore

$$p_c^{new}(SHR^r_{J_r+1} \mid SHR^r_{J_r}) = 0, \tag{4}$$
$$p_c^{ho}(SHR^r_{J_r+1} \mid SHR^r_{J_r}) = 0, \tag{5}$$

and thus the formulas for the soft handover probabilities are completed.

Let we define the *Z-set call arrival rate of new calls* of class c in SHR^r_j on route r as the proportion of the call arrival rate of new calls of class c in SHR^r_j on route r that get connected to the base station set $Z \subseteq \mathcal{Z}_{j,r}$. It can be calculated as ${}^Z\lambda_c^{new}(SHR^r_j) = \sum_{i=1}^{I_{j,r}} \lambda_c \, Q_r \, t_i^{j,r} \, {}^Z P_c(SHR^r_j)$, thus

$$ {}^Z\lambda_c^{new}(SHR^r_j) = \lambda_c \, Q_r \, T_0^{j,r} \, {}^Z P_c(SHR^r_j), \quad \text{for each } 1 \le j \le J_r. \tag{6}$$

The *call arrival rate of new calls* of class c in SHR^r_j is $\lambda_c^{new}(SHR^r_j) \stackrel{def}{=} \sum_{Z \subseteq \mathcal{Z}_{j,r}} {}^Z\lambda_c^{new}(SHR^r_j) = \lambda_c \, Q_r \, T_0^{j,r}$. It coincides with the call arrival rate of the infinite cell-capacity case (see [1]).

Remark: We have the following simple relationship for the Z-set call intensities of the newly initiated calls:

$$ {}^Z\lambda_c^{new}(SHR^r_j) = \lambda_c^{new}(SHR^r_j) \, {}^Z P_c(SHR^r_j). \tag{7}$$

Let we define the *Z-set call arrival rate of soft handover calls* of class c entering SHR^r_j as the proportion of the call arrival rate of soft handover calls of class c entering SHR^r_j that get connected to the base station set $Z \subseteq \mathcal{Z}_{j,r}$. It consists of two parts, namely the newly initiated calls of class c in SHR^r_{j-1} and then "handed over" to SHR^r_j and the calls of class c "handed over" from the previous soft handover region to SHR^r_{j-1} and then "handed over" further to SHR^r_j for $1 < j \le J_r$.

For $Z \nsubseteq \mathcal{Z}_{j,r} \setminus \mathcal{Z}_{j-1,r}$, if $\mathcal{Z}_{j-1,r} \subseteq \mathcal{Z}_{j,r}$ this means that

$$\begin{aligned} {}^{Z}\lambda_c^{ho}(SHR_j^r) = \; & p_c^{new}(SHR_j^r \mid SHR_{j-1}^r)\ {}^{Z \bigcap \mathcal{Z}_{j-1,r}}\lambda_c^{new}(SHR_{j-1}^r) \quad \times \\ \times & \prod_{z \in \mathcal{Z}_{j,r} \setminus \mathcal{Z}_{j-1,r}} \left\{ \left(1 - B_c(z)\right) I(z \in Z) + B_c(z) I(z \notin Z) \right\} + \\ + \; & p_c^{ho}(SHR_j^r \mid SHR_{j-1}^r)\ {}^{Z \bigcap \mathcal{Z}_{j-1,r}}\lambda_c^{ho}(SHR_{j-1}^r) \quad \times \\ \times & \prod_{z \in \mathcal{Z}_{j,r} \setminus \mathcal{Z}_{j-1,r}} \left\{ \left(1 - B_c(z)\right) I(z \in Z) + B_c(z) I(z \notin Z) \right\} \end{aligned} \tag{8}$$

because for the $Z \subseteq \mathcal{Z}_{j,r}$ that contains a cell from the tighter characteristic set $\mathcal{Z}_{j-1,r}$, the Z-set soft handover calls consist of exactly the $Z \bigcap \mathcal{Z}_{j-1,r}$-set newly initiated and soft handover calls of SHR_{j-1}^r that successfully reach SHR_j^r and get blocked exactly by the required new cells in the looser characteristic set $\mathcal{Z}_{j,r}$.

For $Z \subseteq \mathcal{Z}_{j,r} \setminus \mathcal{Z}_{j-1,r}$, if $\mathcal{Z}_{j-1,r} \subseteq \mathcal{Z}_{j,r}$, we get

$$ {}^{Z}\lambda_c^{ho}(SHR_j^r) = 0 \tag{9}$$

because for the $Z \subseteq \mathcal{Z}_{j,r}$ that does not contain any cell from the tighter characteristic set $\mathcal{Z}_{j-1,r}$, the Z-set soft handover calls does not exist (there are no calls in the previous SHR that can be "handed over" to SHR_j^r in such way).

For $Z \neq \varnothing$, if $\mathcal{Z}_{j-1,r} \supseteq \mathcal{Z}_{j,r}$, we get

$$\begin{aligned} {}^{Z}\lambda_c^{ho}(SHR_j^r) = \; & p_c^{new}(SHR_j^r \mid SHR_{j-1}^r) \sum_{Z \subseteq S \subseteq \mathcal{Z}_{j-1,r}} {}^{S}\lambda_c^{new}(SHR_{j-1}^r) + \\ + \; & p_c^{ho}(SHR_j^r \mid SHR_{j-1}^r) \sum_{Z \subseteq S \subseteq \mathcal{Z}_{j-1,r}} {}^{S}\lambda_c^{ho}(SHR_{j-1}^r) \end{aligned} \tag{10}$$

because for the $Z \subseteq \mathcal{Z}_{j,r}$ not empty sets, the Z-set soft handover calls consist of all the S-set newly initiated and soft handover calls of the previous SHR (of looser characteristic set $\mathcal{Z}_{j-1,r}$) for which S is looser than Z that successfully reach the boundary of SHR_{j-1}^r and SHR_j^r (no blocking can take place for calls "handed over" to SHR_j^r this way).

For $Z = \varnothing$, if $\mathcal{Z}_{j-1,r} \supseteq \mathcal{Z}_{j,r}$, we get

$$
\begin{aligned}
{}^{\varnothing}\lambda_c^{ho}(SHR_j^r) &= p_c^{new}(SHR_j^r \mid SHR_{j-1}^r) \times \\
\times \sum_{\varnothing \neq S \subseteq \mathcal{Z}_{j-1,r} \setminus \mathcal{Z}_{j,r}} & {}^{S}\lambda_c^{new}(SHR_{j-1}^r) + p_c^{ho}(SHR_j^r \mid SHR_{j-1}^r) \times \\
& \times \sum_{\varnothing \neq S \subseteq \mathcal{Z}_{j-1,r} \setminus \mathcal{Z}_{j,r}} {}^{S}\lambda_c^{ho}(SHR_{j-1}^r) \qquad (11)
\end{aligned}
$$

because for the $Z = \varnothing$, the blocked soft handover calls consist of all those non-blocked S-set newly initiated and soft handover calls of the previous SHR that S does not contain a cell from the tighter characteristic set $\mathcal{Z}_{j,r}$ and that successfully reach SHR_j^r (no new blocking again).

For the completion of this recursive formula we need to give the initial condition and that can be

$$
{}^{Z}\lambda_c^{ho}(SHR_1^r) = {}^{Z}\lambda_c^{0}(SHR_1^r) \frac{Q_r}{\sum\limits_{\text{route } h \text{ starts in } SHR_1^r} Q_h}, \qquad (12)
$$

for $Z \subseteq \mathcal{Z}_{1,r}$, where ${}^{Z}\lambda_c^{0}(SHR_1^r) \stackrel{def}{=} \lambda_c^{0}(p_1^r)\, {}^{Z}P_c(SHR_1^r)$ (here p_1^r is the first centroid of route r), if we assume that the call arrival rate of calls coming from outside the target area to SHR_1^r is divided among the routes starting in the particular soft handover region in the proportion of the traffic volumes on these routes.

Remark: The *call arrival rate of soft handover calls* of class c entering SHR_j^r is $\lambda_c^{ho}(SHR_j^r) \stackrel{def}{=} \sum\limits_{Z \subseteq \mathcal{Z}_{j,r}} {}^{Z}\lambda_c^{ho}(SHR_j^r)$ does not provide us such a simple relation like equation (7) for the newly initiated call intensities (an analogous to equation (7) for the soft handover call intensities does not hold, for example because of equation (9)).

Remark: We derive the Z-set call arrival rates to be able to calculate inter RNC traffic (we need the 2-leg and 3-leg proportion of the call arrival intensities in the soft handover regions to calculate the user-plane traffic on the Iur interface, see *Section 2.4*). We also need to know the proportion of the blocked call arrival rates in some soft handover regions when calculating the offered traffic load for a cell, see *Section 2.5.*

We define the *mean SHR residence time of call class c* as the mean value of the holding time of the call of class c being in a specific SHR until the call is terminated or until it reaches the border of (is "handed over" to) the next SHR.

The mean SHR residence time of call class c originated in SHR_j^r (from [1]) is $h_c^{new}(SHR_j^r) = h_c - \frac{h_c^2}{T_0^{j,r}} \sum_{i=1}^{I_{j,r}} e^{-\frac{T_i^{j,r}}{h_c}} (1 - e^{-\frac{t_i^{j,r}}{h_c}})$, and since it includes a telescopic sum with $T_{i-1}^{j,r} = T_i^{j,r} + t_i^{j,r}$, we get

$$h_c^{new}(SHR_j^r) = h_c - \frac{h_c^2}{T_0^{j,r}}(1 - e^{-\frac{T_0^{j,r}}{h_c}}), \quad \text{for } 1 \leq j < J_r. \tag{13}$$

The *mean SHR residence time of call class c* "handed over" from SHR_{j-1}^r to SHR_j^r is

$$h_c^{ho}(SHR_j^r) = h_c(1 - e^{-\frac{T_0^{j,r}}{h_c}}), \quad \text{for } 1 < j < J_r. \tag{14}$$

The boundary conditions are

$$h_c^{ho}(SHR_1^r) = h_c(1 - e^{-\frac{T_{0'}^{1,r}}{h_c}}) \tag{15}$$

because of the "backward lengthening" in the first cells of the route r and

$$h_c^{new}(SHR_{J_r}^r) = h_c, \tag{16}$$
$$h_c^{ho}(SHR_{J_r}^r) = h_c \tag{17}$$

because the calls in the last cell of the route are not "handed over" to further SHRs.

Remark: Note that there is a simple relationship between the soft handover probability and the mean SHR residence time

$$h_c^{new}(SHR_j^r) = h_c\big(1 - p_c^{new}(SHR_{j+1}^r \mid SHR_j^r)\big), \tag{18}$$
$$h_c^{ho}(SHR_j^r) = h_c\big(1 - p_c^{ho}(SHR_{j+1}^r \mid SHR_j^r)\big), \tag{19}$$

where equation (18) stands because of equations (1) and (13); (4) and (16) and equation (19) stands because of equations (2) and (14); (3) and (15); (5) and (17) respectively.

We can calculate the *Z-set traffic load* for $Z \subseteq \mathcal{Z}_{j,r}$ for the call class c for each soft handover region SHR_j^r (defined as the proportion of the traffic load that is induced by the calls that get connected to the base station set $Z \subseteq \mathcal{Z}_{j,r}$) by

$$\begin{aligned} {}^{Z}Load_c(SHR_j^r) = {}^{Z}\lambda_c^{new}(SHR_j^r)\; h_c^{new}(SHR_j^r) + \\ + \; {}^{Z}\lambda_c^{ho}(SHR_j^r)\; h_c^{ho}(SHR_j^r), \end{aligned} \tag{20}$$

and thus *the l-leg offered traffic load for any area* (some soft handover regions together) can be calculated as the sum of those Z-set offered traffic loads of soft handover regions in the area for which $|Z| = l$, for $l = 0, 1, 2, 3$.

2.4 User-Plane Traffic Estimation on the Iur Interface

Let the cell sets $RNC1 \subseteq \{1, \ldots, N_{cell}\}$ and $RNC2 \subseteq \{1, \ldots, N_{cell}\}$ represent two RNC-s ($RNC1 \bigcap RNC2 = \emptyset$ and the corresponding cells are connected to RNC1 and RNC2 respectively). Using equation (20), we can get a lower bound for the traffic generated on the Iur interface between the two RNC-s

$$\begin{aligned} &Iur^{lower}(RNC1, RNC2) = \\ &= \sum_{z_1 \in RNC1, z_2 \in RNC2} \; \sum_{\mathcal{Z}_{j,r} \ni z_1, z_2} \; \sum_{Z \subseteq \mathcal{Z}_{j,r}} \; \sum_{c=1}^{C} {}^{Z}Load_c(SHR_j^r) \times \\ &\times \big\{ I(z_1, z_2 \in Z,\ |Z| = 2) + I(z_1, z_2 \in Z,\ |Z| = 3) \big\} \end{aligned} \tag{21}$$

and an upper bound

$$Iur^{upper}(RNC1, RNC2) =$$

$$= \sum_{z_1 \in RNC1, z_2 \in RNC2} \sum_{\mathcal{Z}_{j,r} \ni z_1, z_2} \sum_{Z \subseteq \mathcal{Z}_{j,r}} \sum_{c=1}^{C} {}^{Z}Load_c(SHR_j^r) \times$$
$$\times \left\{ I(z_1, z_2 \in Z,\ |Z| = 2) + 2\, I(z_1, z_2 \in Z,\ |Z| = 3) \right\}, \tag{22}$$

where we summed up all those Z-set traffic loads of the soft handover regions, which are both in $RNC1$ and $RNC2$ (see the first two summations) for sets $Z \subseteq \mathcal{Z}_{j,r}$ that contain a cell of each RNC ($z_1 \in RNC1, z_2 \in RNC2$), so the proper 2 and 3-leg traffic loads (see the third summation and the indicator functions) for each call class (see the fourth summation). Note that we need the blocking probabilities here implicitly in the Z-set traffic loads through the Z-set call arrival rates.

Notice that we could determine the exact value for the traffic on the Iur interface, but here we omit it because it is contagious to formulate it (the 3-leg traffic gives one or two-legged traffic on the Iur interface depending on the route structure in the corresponding SHR).

2.5 Cell Parameters

Distribution of the Channel Occupancy Time. Consider the mean value of the cell residence time that is usually called *channel occupancy time* (it is the time during which an active call holds a channel in a cell).

The *mean channel occupancy time* of c class calls originating in cell z on route r is

$$h_c^{new}(z^{(r)}) = h_c - \frac{{h_c}^2}{T_0^{(z),r}}\left(1 - e^{-\frac{T_0^{(z),r}}{h_c}}\right), \tag{23}$$

similarily to equation (13) and the *mean channel occupancy time* of c class calls handed over to cell z is

$$h_c^{ho}(z^{(r)}) = h_c\left(1 - e^{-\frac{T_0^{(z),r}}{h_c}}\right), \tag{24}$$

similarily to equation (14), where $T_0^{(z),r}$ is the travelling time on route r through cell z, $T_0^{(z),r} \stackrel{def}{=} \sum_{i=1}^{I_{(z),r}} t_i^{(z),r} = \sum_{i=1}^{I_{(z),r}} \frac{L(l_i^{(z),r})}{v(l_i^{(z),r})}$ (see *Fig. 3*).

We can also derive the *distribution of the channel occupancy time* in the cell for the newly initiated calls of class c as follows:

Let $\xi_{(z)}^r$ be the random variable of time required for a trip from the call originating point to the boundary of cell z on route r. From assumption (A8) and denoting the random variable of the holding time of a class c call by τ_c, we can get the probability of $\{\xi_{(z)}^r < t\}$ given that $\{\tau_c > t\}$ and given that the class c call is generated on $l_i^{(z),r}$ as

$$p_c(\xi_{(z)}^r < t \mid \tau_c > t,\ l_i^{(z),r}) \stackrel{def}{=} \begin{cases} 0 & ,\ t \le T_i^{(z),r} \\ \frac{t - T_i^{(z),r}}{t_i^{(z),r}} & ,\ T_i^{(z),r} < t \le T_{i-1}^{(z),r} \\ 1 & ,\ T_{i-1}^{(z),r} < t, \end{cases}$$

where $T_i^{(z),r} \stackrel{def}{=} \sum_{h=i+1}^{I_{(z),r}} t_h^{(z),r}$ for each $i = 1, \ldots, I_{(z),r} - 1$ and $T_{I_{(z),r}}^{(z),r} \stackrel{def}{=} 0$.

On the other hand (similarily to the soft handover regions), the probability that a class c call is generated on $l_i^{(z),r}$ is equal to $p_c(l_i^{(z),r}) \stackrel{def}{=} \frac{t_i^{(z),r}}{T_0^{(z),r}}$, and thus unconditioning, we get $Pr\{\xi_{(z)}^r < t \mid \tau_c > t\} = \sum_{i=1}^{I_{(z),r}} p_c(\xi_{(z)}^r < t \mid \tau_c > t,\ l_i^{(z),r})\, p_c(l_i^{(z),r}) =$ $\frac{t}{T_0^{(z),r}} - \sum_{i=1}^{I_{(z),r}} \frac{I(T_i^{(z),r} < t \le T_{i-1}^{(z),r})\, T_i^{(z),r}}{T_0^{(z),r}} + \sum_{i=1}^{I_{(z),r}} \frac{I(T_{i-1}^{(z),r} < t)\, t_i^{(z),r}}{T_0^{(z),r}} = \frac{t}{T_0^{(z),r}}$, and introducing the density function of the above distribution, $g_c^{(z),r}(t) = \frac{1}{T_0^{(z),r}}$, we can get for the $\tau_c^{(z),r}$ random variable (the channel occupancy time in cell z on route r):

$$Pr\{\tau_c^{(z),r} > t\} = \int_0^{T_0^{(z),r} - t} g_c^{(z),r}(s)ds \int_t^{\infty} \frac{1}{h_c} e^{-\frac{h}{h_c}} dh = \frac{T_0^{(z),r} - t}{T_0^{(z),r}} e^{-\frac{t}{h_c}},$$ thus

the probability distribution of $\tau_c^{(z),r}$ is

$$Pr\{\tau_c^{(z),r} < t\} = 1 - \frac{T_0^{(z),r} - t}{T_0^{(z),r}} e^{-\frac{t}{h_c}}, \tag{25}$$

and the $f_c^{(z),r}$ probability density function of $\tau_c^{(z),r}$ is

$$f_c^{(z),r}(t) = \frac{T_0^{(z),r} - t + h_c}{T_0^{(z),r} h_c} e^{-\frac{t}{h_c}}. \tag{26}$$

The *mean channel occupancy time* for the new c class calls in cell z on route r is then $h_c^{new}(z^{(r)}) = \int_0^{T_0^{(z),r}} t\, f_c^{(z),r}(t)dt = h_c - \frac{{h_c}^2}{T_0^{(z),r}}(1 - e^{-\frac{T_0^{(z),r}}{h_c}})$, and this coincides with equation (23).

Blocking Probability and Offered Traffic Load. The *offered traffic load* for a cell is composed of all the newly initiated call intensities in the cell, the newly initiated and handover call intensities in the SHR preceding the cell on some route and succesfully reaching the boundary of the cell and finally the call arrival rates from outside the target area to the cell each multiplied by the corresponding mean channel occupancy times:

$$
\begin{aligned}
Load_c(z) = \sum_{Z_{j,r} \ni z} & h_c^{new}(z^{(r)})\, \lambda_c^{new}(SHR_j^r) + h_c^{ho}(z^{(r)}) \times \\
\times \{ & p_c^{new}(SHR_j^r | SHR_{j-1}^r)\big(\lambda_c^{new}(SHR_{j-1}^r) - {}^{\emptyset}\lambda_c^{new}(SHR_{j-1}^r)\big) + \\
& + p_c^{ho}(SHR_j^r | SHR_{j-1}^r)\big(\lambda_c^{ho}(SHR_{j-1}^r) - {}^{\emptyset}\lambda_c^{ho}(SHR_{j-1}^r)\big)\} + \\
& + \sum_{Z_{0,r} \ni z} h_c^{ho}(z^{(r)}) \lambda_c^0(p_1^r).
\end{aligned}
\tag{27}
$$

Having the offered traffic load, we can calculate the *blocking probabilities* of each call class c in each cell z by the *multirate Erlang B formula* (see [12]):

$$
B_c(z) = \frac{\sum\limits_{s=S_z-b_c+1}^{S_z} q(s)}{\sum\limits_{s=0}^{S_z} q(s)}, \tag{28}
$$

where b_c is the equivalent power of the calls of class c, S_z is the capacity of the cell z and the auxiliary function $q()$ is given by the following recursion:

$$
q(s) = \begin{cases} 1 & , \text{ for } s = 0 \\ \frac{1}{s} \sum\limits_{c=1}^{C} Load_c(z)\, b_c\, q(s - b_c) & , \text{ for } 0 < s \le S_z. \end{cases} \tag{29}
$$

Unfortunately, the offered traffic load and the blocking probability are not independent of each other, therefore we can not get them explicitly. We have a simultaneous system of equations for the parameters of interest that we are going to solve numerically using the following straight-forward *iterative algorithm:*

Step # 0: Calculate the *soft handover probabilities* for the new and the soft handover calls for each call class $c = 1, \ldots, C$ in each soft handover region SHR_j^r, $j = 1, \ldots, J_r$ on each route $r = 1, \ldots, R$ by equations (1), (2), (3), (4) and (5) and the *mean SHR residence time* by equations (13), (14), (15), (16) and (17) or by equations (18) and (19) knowing the soft handover probabilities already – all these will not change any more.

Step # 1: Calculate the *call intensities* for each call class $c = 1, \ldots, C$ in each soft handover region SHR_j^r, $j = 1, \ldots, J_r$ on all routes $r = 1, \ldots, R$ by equations (6), (8), (9), (10), (11) and (12) (the very first step is calculated assuming infinite cell capacities that is each blocking probability $B_c(z)$ equals 0 for all cells $z = 1, \ldots, N_{cell}$ and for each call class $c = 1, \ldots, C$).

Step # 2: Calculate the *Z-set traffic load* for each $Z \subseteq \mathcal{Z}_{j,r}$ for call class $c = 1, \ldots, C$ in each soft handover region SHR_j^r, $j = 1, \ldots, J_r$ on each route $r = 1, \ldots, R$ by equation (20) and the *offered traffic load* for each cell $z = 1, \ldots, N_{cell}$ by equation (27) respectively.

Step # 3: Calculate the *blocking probabilities* for each call class $c = 1, \ldots, C$ and for each cell $z = 1, \ldots, N_{cell}$ by using equations (29) and (28).

Repeat: Steps #1, #2 and #3 until a predefined stopping condition is not satisfied (some stopping conditions: given number of iterations; soft handover probabilities and/or offered loads and/or blocking probabilities do not change more than some predefined small positive real number(s)).

Remark: The simultaneous system of equations define an $f : \mathbb{R}^M \longrightarrow \mathbb{R}^M$ continuous function, where M is the number of parameters (as a function of the system parameters). We can consider function f as a continuous, bounded function from a bounded M-dimensional space (the probabilities are bounded and other parameters can be normalized for example by the sum of the corresponding parameters). Therefore, this function f has a fixed point $f(x) = x$ (this is exactly the claim of the *Brouwer's fixed point theorem*, see [13]) and this makes our iterative algorithm reasonable.

Soft Handover Intensities. We can derive a very important cell parameter, the *soft handover intensity* (mean number of handover requests in the cell (to the cell)) that consists of the non-blocked newly initiated calls in the SHR region preceding the cell on some route and the non-blocked soft handover calls in the same SHR that reach the boundary of the cell and the call arrival rates from outside the target area (if the cell is on the boundary of it):

$$
{}^{in}\lambda_c^{ho}(z) \stackrel{def}{=} \sum_{z_{j,r}\ni z\notin z_{j-1,r}} \{p_c^{new}(SHR_j^r|SHR_{j-1}^r)(\lambda_c^{new}(SHR_{j-1}^r) - {}^{\emptyset}\lambda_c^{new}(SHR_{j-1}^r)) + p_c^{ho}(SHR_j^r|SHR_{j-1}^r)(\lambda_c^{ho}(SHR_{j-1}^r) - {}^{\emptyset}\lambda_c^{ho}(SHR_{j-1}^r))\} + \sum_{z_{0,r}\ni z} \lambda_c^0(p_1^r). \tag{30}
$$

2.6 A Numerical Example

Let us consider the road network depicted in *Fig. 4*. The cell radius is $R_0 = 2$ (*units of length*), the cell center distance is $D = 3$ (*units of length*), the cell capacities are $S_z = 12$ (*units in channel*) for $z = 0, \dots, 9$. The O-D table used can be seen in *Table 1*. Streets $[c0, c1]$ and $[c1, c2]$ are avenues, street $[c0, c3]$ is main street and street $[c2, c3]$ is a minor street (see the load-speed profile, *Fig. 5*).

Three kind of call classes are considered (*Table 2*).

Table 2. Class parameters

	h_c (*hour*)	λ_c (*call/hour/user*)	b_c (*voice equivalent*)
voice ($c = 1$)	0.025	0.8	1
data ($c = 2$)	0.075	0.2	3
video ($c = 3$)	0.25	0.002	8

No call arrival rates were considered from outside the area ($\lambda_c^0(cj) = 0,\ j = 0, 1, 2, 3$).

We had two scenarios for calculating the traffic on the Iur interface (two pairs of RNC-s)
Scenario #1: $RNC1 = \{0, 1, 2, 3\}$, $RNC2 = \{4, 5, 6, 7, 8, 9\}$ (vertical cut),
Scenario #2: $RNC1 = \{0, 3, 4, 7\}$, $RNC2 = \{1, 2, 5, 6, 8, 9\}$ (horizontal cut).

The Iur traffic estimators for the two scenarios are:
Scenario #1: $6.557428 \le Iur(RNC1, RNC2) \le 8.065130$,
$(Iur^{lower}() \le Iur() \le Iur^{upper}())$
Scenario #2: $Iur(RNC1, RNC2) = 0.491769$.

Not suprisingly, the offered user-plane traffic on the Iur interface is much more bigger in the case of the "vertical cut" scenario (the vertical cut goes through the streets $[c0, c3]$ and $[c1, c2]$ with heavier traffic flowing on them).

In *Table 3*, we have summarized the offered traffic load, blocking probabilities and soft handover intensities for the cells.

Table 3. Both Scenario #1 and #2 parameters

cells	$L_2(z)$	$B_2(z)$	${}^{in}\lambda_2^{ho}(z)$
$z=0$	3.684217	0.540058	15.450449
$z=1$	0.485426	0.010870	2.065872
$z=2$	3.340962	0.483740	13.241960
$z=3$	3.397117	0.582286	10.021143
$z=4$	2.013934	0.334187	15.434090
$z=5$	2.021324	0.317490	12.582370
$z=6$	3.363575	0.559529	7.822015
$z=7$	3.561136	0.527074	15.594346
$z=8$	0.318122	0.003280	2.524403
$z=9$	3.808097	0.548885	13.647168

3 Conclusion

The traffic analysis method presented in [1] has been derived for a model considering soft handovers by introducing more detailed parameters (equations (6), (8), (9), (10) and (11)) for finite capacity cells.

The distribution and density of the channel occupancy time for a cell in equations (25) and (26) has been derived.

An iterative algorithm has also been detailed to solve our system of simultaneous non-linear equations (the offered traffic load, needed for calculating the blocking probabilities of the cells, have been expressed with the detailed parameters – equation (27)).

The traffic load for soft handover regions has been derived enabling the calculation of the overhead caused by the soft handover traffic in the transport network. The traffic load for soft handover regions were used to calculate the offered traffic on the Iur interface (equations (20), (21) and (22)).

We have shown on a simple numerical example that there can be significant difference in terms of inter RNC traffic among different RNC structures. Therefore, it is useful to solve a clustering problem for the grouping of cells to RNC-s optimizing (minimizing) the inter RNC traffic. For example, this way our results and our software could be a part of a network planning procedure.

A Appendix: Incremental Assignment Method

The incremental assignment mechanism is based on a multistep shortest route routing algorithm in which the *user optimal rule* is assumed. The user optimal rule means that it seems reasonable to assume that each subscriber always selects the shortest route between two centroids.

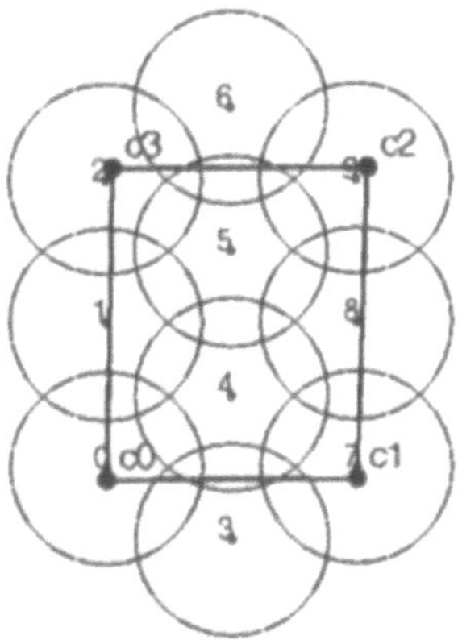

Fig. 4. A road network

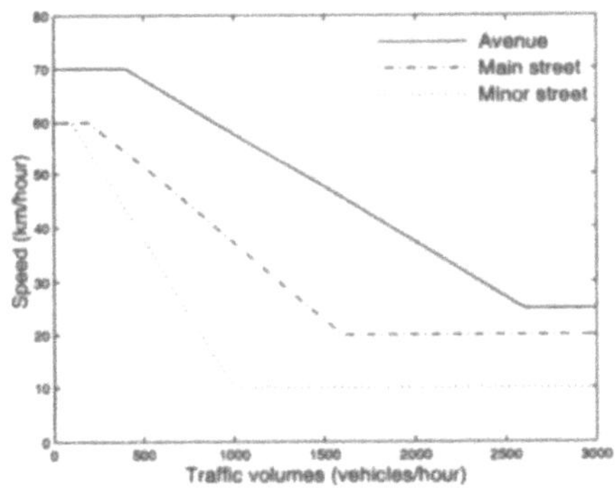

Fig. 5. Load-speed profiles

We define the transportation *usage cost of a street* as the travel time of vehicles moving along the street in our case as in [1]:

$$UsageCost(j)\ (hour) = \frac{L(j)\ (km)}{v(j)\ (km/hour)},$$

where $L(j)$ is the length of street j and $v(j)$ is the speed of vehicles in street j. $L(j)$ is obtained from the road systems model and $v(j)$ is given by the load-speed profile of street j, where the *load-speed street profile* characterizes the street and represents the relation between the traffic volume and the speed of vehicles in the street. Streets are classified into different types such as avenue, main street, minor street etc., according to the scale of the traffic they can carry. Each type of street has its own load-speed profile. *Fig. 5* shows an example of these profiles.

Let OD_{ij} be the traffic volume from centroid i to centroid j, which is given by the O-D table. The incremental assignment method approximately estimates the amount of traffic volume flowing along each possible alternative route from centroid i to centroid j. For the estimation, the method divides the traffic volume into m parts and in m number of steps it assigns $\frac{OD_{ij}}{m}$ amount of traffic volume to the shortest route between centroid i and centroid j (the well-known *Dijkstra-algotithm* is used for the calculation of the shortest route, see in [14] for example). In each step, the vehicular traffic gradually adapts to the street network environment (the traffic volume on each link changes, the usage cost of each street also changes), consequently the shortest route can change as well – the cost of a route is the sum of the usage costs of the streets of the trip between centroid i and centroid j. The larger the value of m, the more accurate approximation is achieved.

References

1. K. Nakano, K. Saita, M. Sengoku et al., Mobile Communications Traffic Analysis on a Road System Model, Performance and Management of Complex Communication Networks, International Federation for Information Processing (IFIP), Kluwer Academic Publishers, pp. 1-20, May, 1998
2. UMTS Phase 1., 3GPP standard 3G TS 22.100, ver. 3.2.0, Apr. 1999, http://www.3gpp.org
3. Tero Ojanperä and Ramjee Prasad, Wideband CDMA For Third Generation Mobile Communications, Artech House, 1998
4. Daniel Wong and Teng Joon Lim, Soft Handoffs in CDMA Mobile Systems, IEEE Personal Communications, 1997
5. J. G. Markoulidakis et al., Mobility Modeling in Third-Generation Mobile Telecommunications Systems, IEEE Personal Communications, pp. 41-56, August, 1997
6. Suwon Park et al., Modeling and Analysis of CDMA Soft Handoff, IEEE 46th Vehicular Technology Conference, New York, 1996
7. Jae Kyun Kwon and Dan Keun Sung, Soft Handoff Modeling in CDMA Cellular Systems, IEEE 47th Vehicular Technology Conference, New York, 1997
8. Moo-Ho Cho et al., The Handoff Rate of Two-Way Soft Handoff Scheme in DS-CDMA Cellular Systems, IEICE Transactions of Communications, Vol. E80-B, No.8 August, 1997
9. Mahmood M. Zonoozi and Prem Dassanayake, User Mobility Modeling and Characterization of Mobility Patterns, IEEE JSAC Vol. 15., No. 7., September, 1997
10. Edward Chlebus et al., Analysis of Channel Holding Time in Wireless Mobile Systems: Does the Probability Distribution of Cell Residence Time Matter?, 16th International Teletraffic Congress, Edinburgh, UK, June, 1999.
11. K. Tutschku et al., A framework for spatial traffic estimation and characterization in mobile communication network design, 16th International Teletraffic Congress, Edinburgh, UK, June, 1999.
12. A. Nilsson and M. J. Perry, Multirate Blocking Probabilities: Numerically Stable Computations, 15th International Teletraffic Congress, Washington DC, USA, June, 1997
13. L. E. J. Brouwer, Über Abbildung von Mannigfaltigkeiten, Mathematische Annalen 71, 1910
14. R. K. Ahuja, T. L. Magnanti and J. B. Orlin, Network Flows: Theory, Algorithms, and Applications, Prentice-Hall, 1993

Part VI

QoS Control

On the Tradeoff Between Effectiveness and Scalability of Measurement-based Admission Control

András Veres and Zoltán Richárd Turányi

Ericsson Traffic Analysis and Network Performance Laboratory, Budapest, Hungary, {Andras.Veres,Zoltan.Turanyi}@eth.ericsson.se

Abstract. One of the key problems to be solved in Mesurement-Based Admission Control is to efficiently utilise the information provided by the sources and measured by the network. There is a direct relationship between the amount of information available and the resulting effectiveness of the admission algorithm. In this paper we analyse the tradeoff between complexity and effectiveness using a practial class of AC methods in the context of Differentiated Services Internet.

1 Introduction

When a network aims to provide any guarantee on the quality of its service it must introduce some form of Admission Control (AC) to protect the network from overload. AC algorithms check incoming requests and decide if they can be served at the requested quality without degrading the quality of other connections. The decision is based on traffic information supplied by the sources, measured by switches or assumed by the AC algorithm.

Recently numerous papers discussed different AC approaches utilising a different mix of the above three information sources. In [1] the decision is based on a complex, deterministic descriptor, which, if tightly set, enables high utilisation without performing on-line measurements. Heuristic methods [2–5] make simple measurements and assume that some statistical properties of the traffic remain constant or predictable. [6] describes a theoretic approach where sources are assumed to be well described by MMPP processes. [7] gives effective bandwidth formulae for a class of non Markovian sources, where state transitions are modelled by a markov chain. [8] uses measurements of the average and variance of the rate and assumes Gaussian property of the aggregate. Sally Floyd's method [9] uses the Hoeffding bound to compute a theoretical upper bound on the overflow probability, expects only a peak rate from the sources, measures the aggregate mean rate and assumes the independence and weak stationarity of the sources. Brichet and Simonian [10] can achieve higher utilisation using token bucket parameters supplied by the sources. In [11] we proposed an AC algorithm that uses admitted peak rates and token buckets, and measurements of the variance and mean providing a tighter bound. Finally the work of S. Crosby et al. is mentioned [12], their Mosquito algorithm uses detailed per-flow measurements to approximate the SCGF of the flows and assumes weak dependence among consecutive time periods.

The efficiency of these AC methods can be evaluated by their ability to meet the primary and secondary goals of AC. The primary goal is to ensure that the admittance of a flow will not degrade the service level of already admitted traffic. The secondary goal is to admit as many traffic as possible while ensuring the primary goal.

Each information source has its own drawbacks and benefits meeting these goals:

- *Information provided by the sources.* This type of information can be the most reliable and also this can be enforced at the ingress point by traffic policing or shaping. However in many cases sources can not provide an exact and detailed characterisation of their traffic at the time of flow setup because of the highly varying nature of traffic generated by certain applications e.g., streaming video. As a result these descriptor sets usually can only be fitted very loosely thus limiting the efficiency of admission control algorithms that rely only on the supplied parameters. Furthermore we should not force applications to fit certain complex traffic models rather on the contrary: the traffic descriptors should be general and simple enough to fit all the needs of current and future applications.
- *Information based on assumptions.* Assumptions are very useful in approaching the AC problem with analytical tools. The drawback is that these assumptions can be very misleading and certain –traditionally used– assumptions in telecommunications such as Poission models have been proven to model IP traffic poorly [13]. Thus more complex models are suggested by most authors, but their applicability in AC have not been proven satisfactorily.
- *Information based on measurements.* Measurements are made to increase our knowledge about the traffic. The measured data is then used to fit the model we assumed the traffic fits best. On the other hand measurement has its limits as well and the infromation which can be gained using measurements is also limited due to practical or theoretical reasons. Measurements always have some variance (error) that will cause erroneous AC decisions. This variance can be decreased by measuring a larger number of flows or increasing the measurement period. This is not always possible, as the time interval that would be required for precise measuremnet may be larger than the length of the flows or the timescale during which the process can be considered stationary. There are also practical limitations of using measurements, as obtaining certain complex statistics or measuring a large number of parameters may require an uneconomical amount of resources.

In a previous paper [11] we argued that an efficient and scalable admission control algorithm should utilise these three information sources in the following way:

1. Information provided by the sources should be simple and easy to police.
2. The required assumptions should be simple and reasonable for example weak stationarity or independence.
3. Measurements should be done only on large traffic aggregates and not per-flow.

In this paper we analyse the tradeoff between the detailedness of measurements and the gain in AC effectiveness, i.e., the tightness of the effective bandwidth calculations based on the parameters. Our conclusion is that measuring the traffic in a small number of groups gives the same performance as per-flow measurements, while the implementation of the former does not exhibit the scalability problems of the latter. We present an optimal strategy to group the flows and also demonstrate our conclusions using measured real-life TCP/IP traffic.

2 AC model with flow grouping

We assume a class of AC methods, where at flow setup time the sources admit only their peak rates h_k, which is simple, general and easy to enforce. The router has a FIFO buffer on the outgoing interface with service rate C. For efficient admission control decision, the router makes measurements on the average bit-rate of the traffic. We can examine the tradeoff of efficiency vs. complexity by varying the detailedness of these measurements. The two ends of the spectrum under consideration are:

- Only the aggregate traffic on the link is measured. This is very easy to implement, just a simple counter is required on the interface.
- All packets are classified according to their flow (e.g., they are classified based on the source and destination addresses, port numbers, logical connection identifier) and the average rate of each flow is measured separately: m_k. This is much more complex to do as it not only requires to keep per flow states in each router, but also classification is needed for each packet to decide which flow k it belongs to.

The first end of the spectrum scales well but gives only an overall measurement althogh the flows may be very different in their traffic statistics and how they utilize their profiles. Using aggregate measurements, this information is lost, so a conservative effective bandwidth calculation has to be used. If per-flow measurements are available then we know more about the behaviour of the sources and can derive tighter bounds for the effective bandwidth.

In between the two extremes we can find methods in which we group a number of flows into a limited number of groups and we make measurements of the aggregate traffic of the groups only. We do not differentiate the flows further within a group. This way we have more information than just the aggregate and also classiffication may be easier to do for a limited number of groups. We predict that the more groups we make the tighter bound can be given. Furthermore, if the grouping is done in an optimal way even higher utilization can be achieved.

In this paper, the presented AC methods are based on the concept of *effective bandwidth* and zero buffer approximation. (A non zero buffer approximation can be found in [11].) The effective bandwidth of the aggregate traffic is BW if

$$\Pr\left(\sum X_k \geq BW\right) = \epsilon, \tag{1}$$

where ϵ is the saturation probability. A new flow is admitted if $BW \leq C$ where the new flow is included.

The rest of the paper is organized in the following way: in Section 3 we show how to construct an AC method with aggregate measurement, in Section 4 the method based on grouping is derived. Examples for the methods are given in Section 5. Section 6 and 7 discuss the way we group flows. In Section 8 a real life example is shown.

3 Effective bandwidths based on aggregate measurement

In [14] it is shown how the Chernoff bound leads to the Hoeffding bound using a suitable overestimation of the moment generating function. In this section we present another overestimation of the Chernoff bound which will make it easy to understand the extension we give later in the paper. Using the Chernoff bound the effective bandwidth of the aggregate traffic can be expressed as a function of s ($s > 0$) [14]

$$BW(s) = \frac{1}{s}\sum_{k=1}^{N} \ln\left(1 + \frac{e^{sh_k} - 1}{h_k} m_k\right) + \frac{\gamma}{s} \tag{2}$$

Where N is the number of flows, h_k and m_k are the peak and measured mean rates of flow k respectively. $\gamma = -\ln\epsilon$, where ϵ is the probability that the link capacity is exceeded. This inequality holds for any $s > 0$, so for a tight result (2) should be optimized for s.

If we know the individual values of m_k (per-flow measurements) then this expression can be directly used for AC. In this case the m_k of the new flow has to be estimated (e.g. by h_k or a token rate if provided). If we can measure only the aggregate then we need to modify this expression to contain only the aggregate mean. First we rewrite this expression as

$$BW(s) = \frac{1}{s}\ln\prod_{k=1}^{N}\left(1 + \frac{e^{sh_k} - 1}{h_k} m_k\right) + \frac{\gamma}{s} \tag{3}$$

Factorizing:

$$\prod_{k=1}^{N}\left(1 + \frac{e^{sh_k} - 1}{h_k} m_k\right) = \prod_{k=1}^{N}\left(\frac{e^{sh_k} - 1}{h_k}\right)\prod_{k=1}^{N}\left(m_k + \frac{h_k}{e^{sh_k} - 1}\right) \tag{4}$$

The geometric mean is smaller than the algebric

$$\sqrt[N]{\prod_{k=1}^{N}\left(m_k + \frac{h_k}{e^{sh_k} - 1}\right)} \le \frac{1}{N}\sum_{k=1}^{N}\left(m_k + \frac{h_k}{e^{sh_k} - 1}\right) \tag{5}$$

And the effective bandwidth expression is

$$\begin{aligned} BW(s) &\le BW_{\text{approx}}(s) \\ &= \frac{N}{s}\ln\left(\frac{M + \sum_{k=1}^{N}\frac{h_k}{e^{sh_k}-1}}{N}\right) - \frac{1}{s}\sum_{k=1}^{N}\ln\left(\frac{h_k}{e^{sh_k} - 1}\right) + \frac{\gamma}{s} \end{aligned} \tag{6}$$

Where $M = \sum m_k$ is the aggregate load measurement. The saturation probability is not exceeded for any value of $s > 0$, for optimal utilization we may find the optimal s where this expression is minimal

$$BW_{\text{approx}} = \min_s (BW_{\text{approx}}(s)) \tag{7}$$

This bound –which is an overestimation of the Chernoff bound– is always tighter than the Hoeffding bound, but cannot be expressed in a closed form. A good approximation for s can be found by using the first few elements of the Taylor series and then differentiating on s. (See [11].)

$$s_{opt} = \sqrt{\frac{\gamma}{\frac{1}{8}\sum_{j=1}^{N} h_j{}^2 - \frac{1}{2}\cdot\frac{1}{N}\left(M - \frac{1}{2}\cdot\sum_{j=1}^{N} h_j\right)^2}}$$

The closed form result using this approximation is still in most cases tighter than the Hoeffding bound, especially if the traffic is biased to high or low mean-to-peak ratio, which is the typical case for most current applications. (See again [11].)

Note that during the calculations a major step was (3) where using an approximation we replaced the expression with an other where only the *sum* of means is present. On a network link this means that only a simple counter is needed for the aggregate traffic measuring the transmitted amount of bytes during a given period.

4 Effective bandwidths based on measurements of flow groups

In this section an effective bandwidth expression is given which contains information on load measurements only on a per group basis, where the number of groups can range from 1 (in this case we have aggregate measurement only) to the total number of flows (in this case per flow measurements are done). First we give a solution which is true for arbitrary grouping. Then in Section 6 problem of grouping is addressed.

Let's group the N flows into G sets (groups), and denote these sets as $A_i, i = 1..G$ and let $n_i := |A_i|$ the number of flows in group i. Then using the algebric-geometric inequality among the groups we get

$$\begin{aligned}\prod_{k=1}^{N}\left(m_k + \frac{h_k}{e^{sh_k}-1}\right) &= \prod_{i=1}^{G}\left(\sqrt[n_i]{\prod_{k\in A_i}\left(m_k + \frac{h_k}{e^{sh_k}-1}\right)}\right)^{n_i} \\ &\le \prod_{i=1}^{G}\left(\frac{1}{n_i}\sum_{k\in A_i}\left(m_k + \frac{h_k}{e^{sh_k}-1}\right)\right)^{n_i}\end{aligned} \tag{8}$$

The full effective bandwidth expression is:

$$BW(s) \leq BW_{\text{groups}}(s) = \frac{1}{s}\sum_{i=1}^{G} n_i \ln\left(\frac{M_i + \sum_{k\in A_i} \frac{h_k}{e^{sh_k}-1}}{n_i}\right) - \frac{1}{s}\sum_{k=1}^{N} \ln\left(\frac{h_k}{e^{sh_k}-1}\right) + \frac{\gamma}{s} \tag{9}$$

Where $M_i = \sum_{k\in A_i} m_k$ is the average load measurement for group i.

A good approximation for s is derived in Appendix A. The result is

$$s_{opt} = \sqrt{\frac{\gamma}{\frac{1}{8}\sum_{i=1}^{N} h_i^2 - \frac{1}{2}\sum_{k=1}^{G} \frac{1}{n_k}\left(M_k - \frac{1}{2}\sum_{i\in A_k} h_i\right)^2}}$$

Clearly the tightness of this expression depends on the choices of:

- the number of groups (G)
- the distribution of flows among the groups (A_i)

Expression (4) can be optimized for both. For the second a simple "rule of thumb" exists: if we select "similar" flows into the same group then the geometric mean will be closer to the algebric in (4) resulting in tighter bandwidth estimation.

The term "similar" here denotes flows with similar $V(k) = m_k + h_k/(e^{sh_k} - 1)$ values. The problem is hard because s is still an unresolved parameter. But in any case flows with similar m_k and h_k are "similar" in this sense. As a practical example we can say that all flows belonging to similar application types (e.g. IP telephony, streaming video, file transfer) are "similar".

5 Is there any practical gain in grouping?

The gain is obvious if the grouping is done efficiently, and there are 'similar' types of flows. We did a test to see the magnitude of this gain. We assumed two typical applications and created two randomized sets of flows with similar mean and peak rates within a group. (See Figure 1a.) Type-1 flows have average peak rate 100kbit/s and average mean-to-peak ratio 0.5, while type-2 flows have average peak rate 1Mbit/s and average mean-to-peak ratio 0.1. The first type may model high quality streaming audio with low burstiness and lower peak rates, the second may correspond to MPEG compressed real-time video applications with high peak rates and relatively low mean-to-peak ratios.

On Figure 1b. the gain in the admitted number of flows can be observed if we group the flows into two groups and make load measurements per group instead of measuring the aggregate. The lower line is the admittance region if aggregate load measurement is used (Hoeffding bound based admission control), the upper line shows when we group the flows according to their types (group-1 contains type-1 and group-2 contains type-2 flows).

We can see that in case of distinct flow sets high gain can be achieved when we do correct grouping.

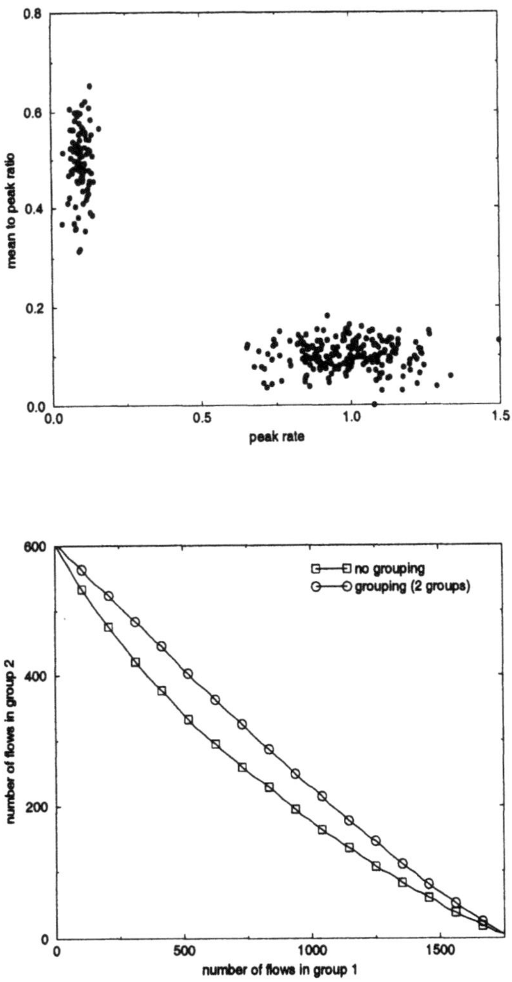

Fig. 1. a) Randomly generated test set of flows. b) Admittance region of flow type-1 versus flow type-2.

6 How to group flows?

A tight approximation of the per-flow formulae can be done if flows with similar $V(k)$ values are grouped together. On Figure 2 the $V(k)$ values of the flows from the previous example is plotted in ascending order.

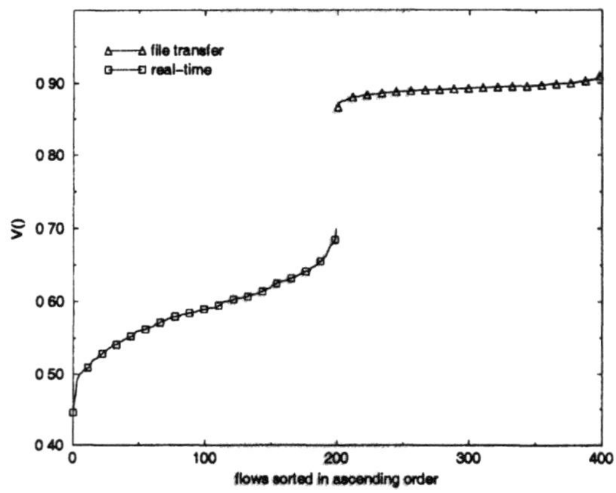

Fig. 2. V(k) values for flows (s=1.122).

After sorting - interestingly - all flows of type-1 can be found on the left side and all flows of type-2 on the right side, and there is a significant jump exatcly where the two groups can be separated. This suggests to define the groups according to where such jumps are visible on the $V(k)$ plot.

$V(k)$ is influenced by the transport protocol, the application and the behavior of the user as well. In real networks it can be expected that there will be typical combinations of the above three (e.g. telephony: RTP/UDP, ON/OFF type with peak rate around 10kbps and mean-to-peak ratio close to 0.5) resulting in clusters on the $(m_k, m_k/h_k)$ scatter plot.

In the following example there are more traffic types differing significantly in peak and mean rates. See Figure 3a. The corresponding sorted $V(k)$ plot is also diplayed on Figure 3b. This latter graph also shows the seasonal difference of the $V(k)$ curve which reveals three regions - suggesting the use of three groups.

7 How many groups are needed?

By increasing the number of groups we gain more knowledge and so we expect tighter bounds on the effective bandwidth. On the other hand if the number of

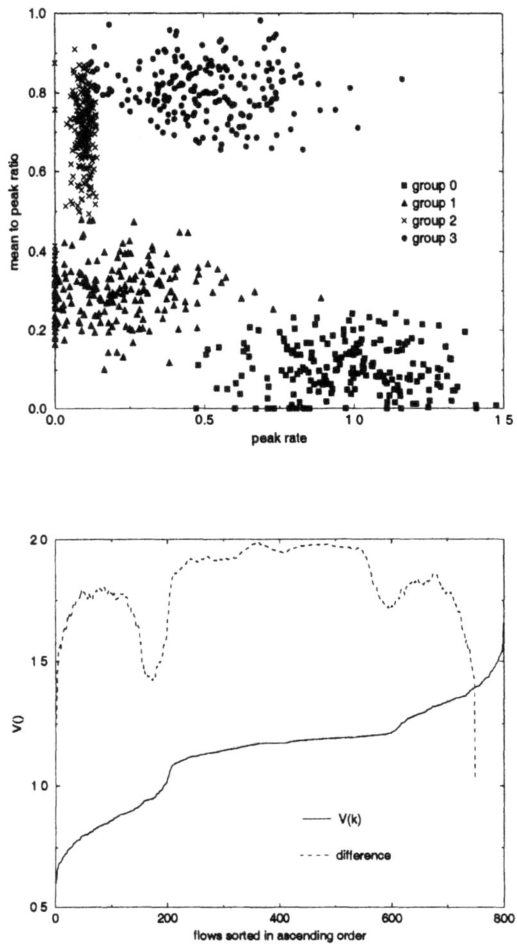

Fig. 3. a) Scatter plot of flows belonging to 4 groups. b) $V(k)$ plot of flows and its seasonal difference plot.

groups is large, more measurements are needed and classification may be more complex which may limit the scalability of the method.

To show this tradeoff a random set of flows are evenly placed on the (peak, mean-to-peak) space (see Figure 4a). Then the number of groups G is increased to see how much gain is achieved with different number of groups. For a certain G the flows are sorted by their $V(k)$ values and are evenly divided into G groups such that N/G flows fall into each group. Figure 4b shows the gain achieved by grouping, calculated as the ratio of $BW - M$ and $BW_{\text{groups}} - M$, where M is the total average rate of the traffic.

The figure suggests that it is enough to use only a very few number of groups, and any further increase in the number of groups (e.g. per-flow measurements) does not give significantly higher gain.

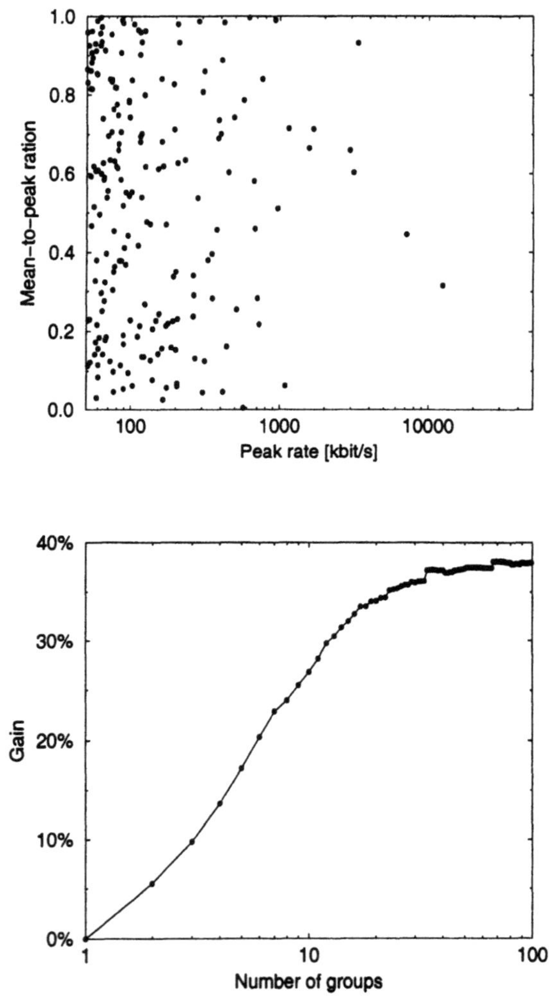

Fig. 4. a) Random flow set. b) The gain vs. the number of groups.

8 Analysis of a dial-in service.

Finally we used measurements of a real dial-in service. Flows were defined as all packets arriving to a specific IP address and with the same source TCP/UDP port number, as we tried to differentiate among different application types. We assumed that the peak rates of the flows equaled the modem connection speed which was in most cases 33Kbits/sec. It is likely that the peak rate values provided by the sources will not be more accurate in the near future (especially in the absence of appropriate real-time operating systems and protocols). The observed average rates are displayed on Figure 5. The TCP flows are spread out widely, but the UDP flows are more concentrated around 0.5 mean-to-peak (approximately 16kbps) which is caused by streaming video/audio applications (e.g. RealAudio). After grouping the flows according to their protocol, the admittance regions are plotted for different TCP-UDP traffic mixes on Figure 6.

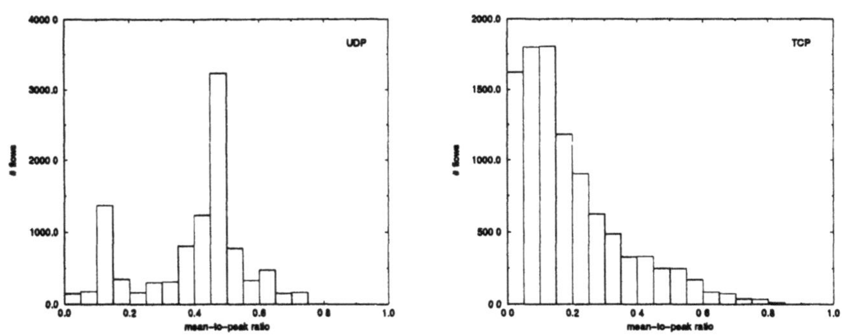

Fig. 5. Histogram of measured mean-to-peak ratios of modem flows containing a) UDP and b) TCP packets.

9 Conclusion

We introduced a class of admission control algorithms based on the effective bandwidth concept. We argued that between the two extremes suggested so far for MBAC (i.e., aggregate and per-flow measurements) there exists a range of algorithms based on flow grouping that differ in implementation complexity and in the extent they can exploit the statistical multiplexing gain. The paper contains closed form formulae for these AC algorithms. Through examples we demonstrated that by increasing the number of groups the aggregate effective bandwidth decreases but the benefit from this is significant only for very few groups which means that a simple implementation of only a few groups can utilise most of the gain. We also

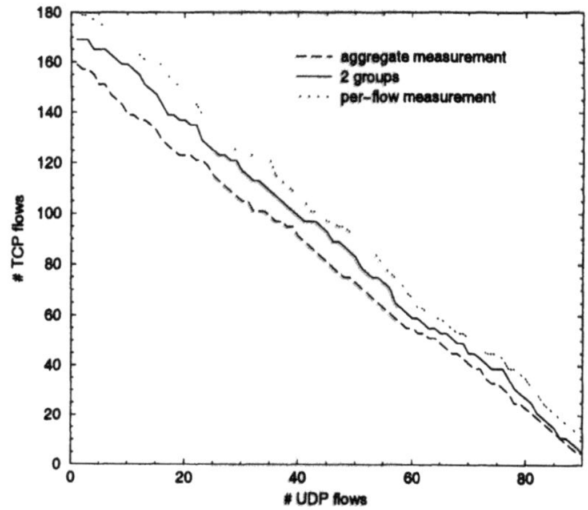

Fig. 6. Admittance region of UDP versus TCP flows, link capacity $C = 2Mbit/s$.

give a heuristics of how to group flows efficiently and finally we demonstrate the efficiency of the method on data collected in a real-life dial-in service.

A Finding an approximate value for s

The effective bandwidth formula for grouped flows is the following:

$$BW_{\text{groups}}(s) = \frac{1}{s}\sum_{i=1}^{G}\sum_{j\in A_i} \ln\left[\left(\frac{M_i}{n_i} + \frac{1}{n_i}\sum_{k\in A_i}\frac{h_k}{e^{sh_k}-1}\right)\frac{e^{sh_j}-1}{h_j}\right] + \frac{\gamma}{s}$$

Taking the series expansion of $h_k/(e^{sh_k}-1)$ we get:

$$\frac{h_k}{e^{sh_k}-1} = \frac{1}{s} - \frac{1}{2}h_k + \frac{1}{12}h_k^2 s - \frac{1}{720}h_k^4 s^3 + O(s^5)$$

Substituting the first 3 elements (higher terms would disappear during the next steps) we get

$$BW_{\text{groups}}(s)$$
$$= \frac{1}{s}\sum_{i=1}^{G}\sum_{j\in A_i} \ln\left[\left(\frac{M_i}{n_i} + \frac{1}{s} - \frac{1}{2n_i}H_i\frac{1}{12n_i}HH_i s\right) n_i\frac{e^{sh_j}-1}{h_j}\right] + \frac{\gamma}{s}$$

using the notation $H_i = \sum_{k\in A_i} h_k$ and $HH_i = \sum_{k\in A_i} h_k^2$. Taking the series of the logarithm:

$$\ln\left[\left(\frac{M_i}{n_i}+\frac{1}{s}-\frac{1}{2n_i}H_i+\frac{1}{12n_i}HH_is\right)n_i\frac{e^{sh_j}-1}{h_j}\right]$$
$$=\left(\frac{M_i}{n_i}-\frac{1}{2n_i}H_i+\frac{1}{2}h_j\right)s+\left(\frac{1}{12}\frac{HH_i}{n_i}+\frac{1}{24}h_j^2-\frac{1}{8}\frac{(2M_i-H_i)^2}{n_i^2}\right)s^2$$

Substituting back to BW_{groups}

$$BW_{\mathrm{groups}}(s)\approx\sum_{i=1}^{G}M_i+\frac{1}{8}HH_is-\frac{1}{8}\frac{(2M_i-H_i)^2}{n_i}s+\frac{\gamma}{s}$$

By deriving this solving for $s=0$ we get an approximation

$$s\approx\sqrt{\frac{8\gamma}{\sum_{i=1}^{G}HH_i-\frac{(2M_i-H_i)^2}{n_i}}}$$

and

$$BW_{\mathrm{groups}}(s)\approx\sum_{i=1}^{G}M_i+\sqrt{\frac{1}{2}\gamma\left(\sum_{i=1}^{G}HH_i-\frac{(2M_i-H_i)^2}{n_i}\right)}$$

which is always smaller than the Hoeffding bound and is equal when the group mean-to-peak ratio is 0.5 which is leads to the highest variance in case of on/off sources and is thus worst case in some sense.

$$BW_{\mathrm{Hoeffding}}(s)=\sum_{i=1}^{G}M_i+\sqrt{\frac{1}{2}\gamma\sum_{i=1}^{G}HH_i}$$

References

1. E. Knightly, H. Zhang, "Traffic Characterization and Switch Utilization using a Deterministic Bounding Interval Dependent Traffic Model", in *Proc. IEEE INFOCOM '95*, Boston, April 1995
2. C. Casetti, J. Kurose, D. Towsley, "A New Algorithm for Measurement-based Admission Control in Integrated Services Packet Networks", *Protocols for High Speed Networks '96*, INRIA, Sophia Antipolis, Oct. 1996
3. S. Jamin, S. Shenker, "Measurement-based Admission Control Algorithms for Controlled-load Service: A Structural Examination", *Internal report*, Apr. 1997

4. S. Jamin, P. Danzig, J. Shenker, L. Zhang, "A Measurement-Based Admission Control Algorithm for Integrated Service Packet Networks", *IEEE/ACM Transactions on Networking*, vol. 5. no. 1. Feb. 1997
5. S. Jamin, C. Jin, L. Breslau, "A Measurement Based Admission Control Algorithm for Controlled-Load Service with a Reference Implementation Framework",*Internet draft*, Nov. 1997
6. G. Kesidis, J. Walrand, C. Chang, "Effective Bandwidths for Multiclass Markov Fluids and Other ATM Sources", *IEEE Trans. Networking*, Vol. 1, No. 4, pp. 424-428, Aug. 1993.
7. K. Kontovasilis, N. Mitrou, "Effective Bandwidths for a Class of Non Markovian Fluid Sources", *Proc. ACM Sigcomm 98* pp. 263-274, Sept. 1998.
8. R. Guerin, H. Ahmadi, M. Naghshineh, "Equivalent capacity and its application to bandwidth allocation in high-spped networks", *IEEE Jurnal on Selected Areas in Communications*, 9(7), pp. 968-981, Sep. 1991
9. S. Floyd, "Comments on Measurement-based Admissions Control for Controlled-Load Service", *unpublished*, available at ftp://ftp.ee.lbl.gov/papers/admit.ps.Z
10. F. Brichet, A. Simonian, "Conservative Gaussian models applied to Measurement-based Admission Control", *submitted to IEEE INFOCOM '98*, San Francisco, March. 1998
11. Z. R. Turányi, A. Veres, "A Family of Measurement-Based Admission Control Algorithms", *Proc. IFIP Performance of Information and Communication Systems '98*
12. S.Crosby, I.Leslie, J.T.Lewis, R.Russell, F.Toomey and B.McGurk, "Statistical Properties of a Near-Optimal Measurement-based CAC algorithm", *Proceedings IEEE ATM '97, June 1997, Lisbon*
13. V. Paxson, S. Floyd, "Wide-area traffic: the failure of Poisson modeling", *IEEE/ACM Transactions on Networking 3:226-244*, 1995
14. R. J. Gibbens, F. P. Kelly, "Measurement-Based Connection Admission Control", *International Teletraffic Congress 15*, Jun. 1997
15. F. P. Kelly, "Notes on Effective Bandwidths", *In F. P. Kelly, S. Zachary and I. B. Ziedins, Stochastic Networks: Theory and Applications, Royal Statistical Society, Lecture Note Series 4, p 141-168, Oxford Univ. Press*

Overload Control Mechanisms for Web Servers

Ravi Iyer, Vijay Tewari, and Krishna Kant

Server Architecture Lab, Intel Corporation, Beaverton, OR {ravishankar.iyer | vijay.tewari | krishna.kant}@intel.com

Abstract. Web servers often experience overload situations due to the extremely bursty nature of Internet traffic, popular online events or malicious attacks. Such overload situations significantly affect performance and may result in lost revenue as reported by the recent denial of service attacks. Overload control schemes are well researched and understood in telecommunication systems. However, their use in web servers is currently very limited. Our focus in this paper is to propose effective overload control mechanisms for web servers. An important aspect in overload control is to minimize the work spent on a request which is eventually not serviced due to overload. This paper studies three simple schemes for controlling the load effectively. The first scheme selectively drops incoming requests as they arrive at the server using an intelligent network interface card (NIC). The second scheme provides feedback to a previous node (proxy server or ultimate client) to allow a gapping control that reduces offered load under overload. The third scheme is simply a combination of the two. The experimental results show that even these simple schemes are effective in improving the throughput of the web server by 40% and response time by 70% under heavy overloads, as compared with the case without any overload control.

1 Motivation

The exploding use of web-based user interfaces for conducting business on the Internet has brought to focus the problem of dealing with overloads to which the web servers (especially those that form the front end of an e-commerce site) are subjected to. This paper motivates the need for overload control and presents results of some preliminary experiments on overload control.

It is well-established by now that Internet traffic is very bursty over a large range of time-scales and shows asymptotic self-similarity and multi-fractal behavior at intermediate time scales. For example, our study of web-server request process in [2] shows that the busy-period traffic is asymptotically self-similar with a Hurst parameter of around 0.8. Several studies have also shown that WAN traffic is multifractal in nature [15,3]. Informally, self-similarity and multifractal behavior imply that there is considerable bunching of requests as they arrive at the server. Such an arrival process is known to lead to heavy-tailed queue-length distribution, which means that unless the web server is engineered for a rather low average operating load, it will experience huge swings in response times and may occasionally experience queue overflows (and the consequent "server-too-busy" errors sent to clients).

Our recent analysis of e-commerce sites suggests that e-commerce traffic usually cannot be assumed to be stationary for more than 10-15 minutes [4]. Nonstationarity further exacerbates loading problems on the web server and consequently degrades user experiences. In addition, e-commerce sites are increasingly affected by special events either directly (e.g., promotional sale held by the site itself) or indirectly (e.g., championship game broadcast by television along with advertisements that direct viewers to the site). Such events can easily subject the front ends (i.e., web servers) of the e-commerce sites to loads that are an order of magnitude higher than normal loads, and thus cannot be handled by conservative engineering practices. The massive overload of Victoria Secret's web site during the last Superbowl illustrates this point very well.

All these characteristics call for effective overload control of web servers. The recent denial of service (DoS) attacks on major web-sites has highlighted this need even further. The DoS attacks are usually carried out by a simple program, usually replicated on a large number of clients, that sends out a barrage of HTTP requests to the web-site and overloads it. Obviously, combating such DoS attacks requires an overload control scheme that can reject requests selectively from misbehaving clients. This would require rather sophisticated overload controls, which can be built on the foundation laid in this paper.

The outline for the rest of the paper is as follows. Section 2 discusses load management schemes both in the context of telecommunications systems (where they are most well developed) and as they are currently employed by the web servers. Section 3 describes the experimental setup, Section 4 presents the overload control methods that were tested, and section 5 presents the results. Finally, section 6 concludes the paper and discusses areas for further work.

2 Overview of Overload Control Mechanisms

Overload control is a very well-researched topic in telecommunications systems, and a carefully designed overload control scheme is a part of every level of the SS7 signalling stack [10–12,8]. In particular, telecommunications signalling nodes use a hierarchical structure to isolate each SS7 layer from congestion at other layers. Also, every signalling link, its associated processor, all network level (MTP3) processors, and all application level (ISUP/BISUP or TCAP) processors are protected by appropriate congestion control mechanisms. This section first discusses the general structure of these overload control schemes, and then points out special considerations for applying them to web-servers.

2.1 Overload control in Telecommunications Systems

The signalling network used in telecommunications systems (SS7) is a datagram network where most telephony related services do not establish any explicit virtual circuit for an end to end reliable communication. Only the link-layer uses transparent retransmissions to cope with (link level) errors; higher layers (including application) depend on other mechanisms such as timeouts, redundant messages, and explicit repetition of certain messages for lost/corrupted/duplicated/delayed messages. This approach is different from the web-server environment where the transport (TCP) layer is supposed to provide transparent protection against all losses/duplications of messages. Such a behavior considerably complicates feedback overload control as discussed later in the paper.

In the following, overload control schemes are described in general terms. Details of specific schemes for link, network and application level congestion control may be found in [10]. The overload control is typically effected by defining 3 thresholds for the monitored parameter (e.g., CPU queue length) (a) abatement, (b) onset, and (c) discard. When the onset threshold is crossed from below, a feedback message is sent out towards the traffic source to request cut-down in the traffic. The traffic throttling happens at some *control node*, which is usually just the previous node, but could in principle be any node on the path back to the source, including the source itself. If the discard threshold is crossed before the traffic has been adequately cut-down, the excess traffic is simply discarded. Once the onset threshold is crossed, feedback messages continue to be sent until the monitored parameter goes below the abatement threshold. The motivation of continuous feedback is both to guard against dropped feedback messages and also to effectively handle rapidly fluctuating loads. In order to limit the feedback related overhead, most schemes don't send feedback for every message. Instead, they may send feedback for every n messages or every τ seconds, where n or τ are parameters of the control mechanism.

One crucial aspect in overload control is the amount of resources expended on calls that are eventually dropped. Since setting up/tearing down a call involves a sequence of signalling messages, it helps to be more reluctant to drop messages well into the call setup process. Also, it undesirable to drop messages related to call teardown, since processing those messages would release resources. This brings in the concept of *congestion priorities* for various messages. That is, depending on their expected importance, messages are assigned *congestion priorities* for the purposes of overload control. Congestion priorities need not be identical to scheduling priorities at the server; for example, the server may process all accepted messages in a FCFS manner but give high congestion priorities to important messages so that they are unlikely to be dropped. Congestion priorities are implemented by using a separate sequence of abatement (A_i), onset (O_i) and discard (D_i) threshold for each priority level i. (Assuming that higher i represents higher congestion priority, one typically uses $A_{i+1} > D_i$.) The feedback message indicates the priority level at which onset threshold has been crossed. The highest onset level crossed essentially indicates the priority level such that all messages of same and lower priority are to be throttled.

Even when only one congestion priority is used, it may be useful to have a sequence of increasing onset levels, such that crossing level O_i indicates a "severity-level" of i to the control node. This can be used by the control node for deciding the how much traffic to throttle. The SS7 application-level congestion control mechanisms known as automatic congestion control (ACC) and automatic code gapping (ACG) use such an approach. The severity level is typically used in one of the following two ways for traffic throttling at the control node:

1. Percentage throttling: Traffic is throttled probabilistically (with higher drop probability at higher severity level).
2. Gapping control: The severity level is translated into a "gap" (i.e., minimum time between successive requests) and this gap is enforced by dropping all non-conforming requests.

Many considerations go into setting various congestion thresholds including feedback delays (which decides spacing between onset and discard thresholds), maximum acceptable response time (which decides location of onset threshold), probability of the queue running empty (which decides the abatement threshold), etc. These considerations apply in the web-server context as well, but a full discussion of these is beyond the scope of this paper.

2.2 Overload Control in Web servers

Before considering overload controls, let us first note a few important distinctions between web-servers and SS7 signalling nodes. Unlike the latter, current web servers have a monolithic architecture, with most of the processing performed on the main processor. A direct consequence of monolithic structure is large overhead associated with bringing a request up to the application layer. (The current architecture also results in unexpected bottlenecks which makes explicit overload control even more important [1].) In particular, an incoming packet will result in I/O interrupt handling, TCP processing, message assembly, possibly a copy from kernel to user space, and message analysis at the HTTP level, before the request can actually be dropped. This "wasted effort" would result in an unstable situation where the throughput continues to degrade as a function of overload amount. This is a critical consideration in any practical overload control scheme for web-servers. As the Web server architecture matures, it is expected that different software layers would run on different physical processors, which would make the task of isolating each processor easier. In fact, driven by the needs for QoS support, IPSec support [6], offloading of protocol processing from the main processor, etc., more distributed approaches are emerging rapidly. One such approach is to have several special purpose processors connected via a high-speed, low latency interconnect (e.g. InfiniBand^{TM}) for assisting the main processor (e.g., for TCP processing, security processing, etc.). Concurrently, the intelligence is percolating down into "protocol processors" and intelligent NICS that can do packet classification, packet forwarding, partial TCP processing (e.g., TCP checksums), etc. Assuming that these processors do not themselves get bottlenecked, much of the overhead of dropping packets is eliminated and good overload performance can be achieved. The experiments reported here used such a setup. It is important to note, however, that much of the benefit of overload control can also be achieved by implementing packet classification and dropping in the driver software of an ordinary NIC.

The overload control mechanisms in place in current web servers are rather rudimentary. Often, the only overload control technique is to return the HTTP *server too busy* message if the HTTP server queue exceeds some threshold. It is then up to the client to decide whether it wants to retry the request and with how much delay. A more sophisticated scheme is to do admission control at a node in front of the server farm (e.g., a load-balancer), so that web-servers don't have to deal with excess requests. Such an approach does not result in any wasted work on the web servers, however, it requires HTTP processing capability in the front end, which is used more and more as the overload increases. Thus, the detrimental effect of overload (i.e., wasted work) occurs at the front end, which is highly undesirable. More sophisticated load balancers also try to do balancing between multiple clusters of web-servers, perhaps located in geographically diverse areas [14,13]. The geographical diversity allows the use of spare capacity in another region if the servers in one region are overwhelmed. It is also possible to exploit geographically distributed servers in minimizing the effects of network congestion and delays. Although sophisticated load-balancing can help considerably in alleviating overloads caused by the highly variable nature of the web traffic, it must be noted that *load-balancing cannot replace proper overload control.* Thus, a good overload control is essential for commercial grade servers even with load-balancing. Moreover, for small (e.g., single node) web-servers where load-balancing does not apply, or when not all servers can handle all web-pages, overload control becomes even more essential.

In this paper, a direct implementation of overload control on the web-server is explored without necessarily assuming the presence of a "front-end" node that does admission control or load balancing. The motivation is to decentralise the rather heavy duty tasks of packet classification (needed for QoS based overload control), maintaining information on TCP sessions in progress, maintaining status about (possibly a large number of) traffic sources, etc. With the availability of intelligent NICs, packet classification and dropping can be done on each server without incurring significant "wasted work". Also, instead of directly sending "server-too-busy" message from the server application level to the client application, lower level feedback messages are sent which can be used either for throttling the traffic or for simply indicating "server too busy" on the client end. Whenever feasible, the feedback is sent to a server on the path back to the client (e.g., the front-end load balancer, a proxy server, firewall, re-director node, etc.) so that traffic throttling can be effected easily. Dropping traffic as close to the source as possible helps cut down unnecessary traffic through the network.

As noted in the last subsection, a major difference between web-servers and SS7 nodes in terms of overload control issues is the connection oriented (TCP) environment. In an SS7 node, it is usually okay to simply drop any messages that cannot be handled since this would eventually result in a dropped call and subsequent retry by the user. In contrast, dropping a packet over TCP layer results in repeated attempts by the other side to retransmit the packet. The TCP-initiated retries are more troublesome because of much smaller gaps between retries compared to retries by a human and the overhead of maintaining the connection related information. A feedback mechanism that can avoid automatic retries in this case would be very helpful, but would surely require some changes to the default TCP behavior.

Although the use of intelligent NICs in a monolithic server allows dropping of packets without incurring much overhead on them, it introduces a mismatch between the relevant data units that the application layer may wish to drop (i.e., an entire message) and the data units that the NIC deals with (i.e., packets). Although most requests to a web-server fit in one ethernet packet, a general solution demands that all packets corresponding to a request be identified and dropped. When there is a one to one mapping between TCP connections and application layer requests, this mismatch is easy to handle. In particular, by dropping the TCP connection request (i.e., the SYN message), the entire data exchange is avoided. However, a TCP connection may be used for multiple application level data transfers (using keep-alive feature of HTTP/1.0 or by default in HTTP/1.1). In such a case, overload control can be considered at two levels of granularity: (a) TCP connection level, and (b) HTTP transaction level. If the time for which the TCP connection remains open is well-bounded, a connection level control is not only adequate but also highly desirable. However, if a single TCP connection is left open for an entire user session, a connection level control is clearly flawed. The increasing use of secure HTTP protocol for safeguarding sensitive transactions at e-commerce sites illustrates both of these extremes. Secure HTTP transactions typically use secure socket layer (SSL) which involves 3-4 rounds of message exchanges between the client and the server for mutual authentication and key exchange, followed by one or more data transfers [7]. In some environments, such as on-line retailing, a secure channel is needed only for purchase related transactions and therefore the duration of a secure HTTP session is well-bounded. A connection level control would work the best in this case. However, in certain other applications such as on-line banking, the entire user session is secured by going through the handshake process only in the beginning and then keeping the TCP connection open. In such cases, it may be unreasonable to keep additional users out until those already in are done.

In order to do transaction level control, the intelligent NIC needs to examine the IP header of the packets and recognize packets belonging to the same message. Thus dropping all packets of a request is straightforward. However as stated earlier this interferes with the normal TCP functioning and the only way to handle this situation is for the intelligent NIC to be able to generate special type of acknowledgements that are treated by TCP as true acknowledgements, but delivered to the application layer as indications of packet drops. In situations where certain classes of requests involve only a few transactions whereas other involve many, it may be desirable to implement both connection and transaction level controls simultaneously. If no such a priori classification exists and the number of transactions per connection varies widely, transaction level controls would have to be implemented.

The discussion above indicates that overload control for exchanges in a connection oriented environment is difficult and would require changes to the transport layer so that proper feedback can be generated by the intelligent NIC and acted upon properly on the receive end. The feedback itself can be provided using a UDP channel and is easy to handle. Given a wide deployment of TCP, any changes to it need to be evaluated carefully. Also, in more general internet servers that use both TCP and UDP for data exchange (e.g., streaming media servers), the impact of overload control (or more appropriately admission control) must be examined on both types of transport. A related issue is of how feedback should be used by the ultimate clients (or browsers), where a direct traffic throttling makes no sense. The browsers could either transparently convert the feedback into a "server-too-busy" message to the user, or attempt to filter attempts by the user in repeatedly hitting the same site under overload conditions.

3 Experimental Methodology and Setup

This section describes the experimental setup that was used for performing overload control experiments, including such essential components as packet classification, overload detection, feedback mechanism, and traffic throttling policies used. It may be noted that the main purpose of this paper is to demonstrate the advantages of overload control in web-servers and to highlight the need for appropriate O/S and protocol support, rather than to explore the policy or parameter space for optimal overload performance. The latter issues need to be examined for each application environment and are beyond the scope of this paper. Also, the experiments address only the simple environment where each connection carries just one request, and thus there is no need to distinguish between connections and the request within a connection.

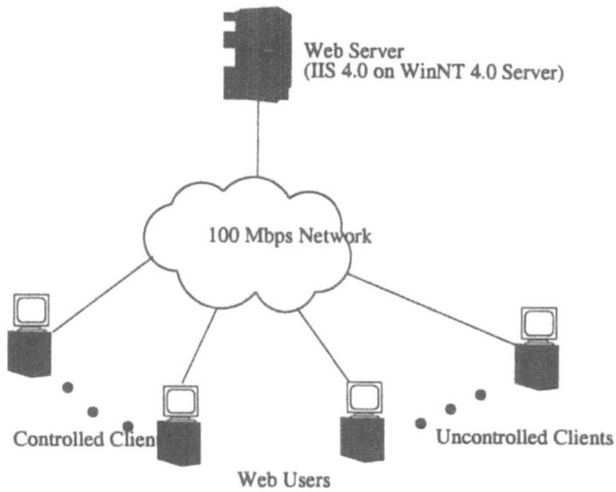

Fig. 1. Experimental Setup

3.1 System Configuration

Our experimental setup consists of a few clients and a web server as illustrated in Figure 1. The web server was running Microsoft Internet Information Service 4.0 (Web server) on Windows NT 4.0 and equipped with 4 NICs (Network Interface Cards) designed to operate on 100 Mbps ethernet. In order to allow targeting a desired amount of load on each NIC, each of them was configured on a separate subnet. The clients were partitioned into two categories called *controlled* and *uncontrolled*. The controlled clients made requests to the web server and responded to information about overload on the server. In contrast, the uncontrolled clients did not respond to changing load on the web server, and can be considered to be given preferential treatment. This treatment was introduced to understand the impact of overload control mechanisms in the presence of QoS constraints. The uncontrolled clients can be considered to be users that have paid a higher subscription rate to the service provider, revenue-generating traffic (e.g., purchases transactions in e-commerce), or others.

One of the four server NICs were targeted by uncontrolled clients and did not need any packet classification capability. All others required such a capability and intelligent NIC's from Netboost^{TM} were used. Apart from the functionality of ordinary NICs, these NICs consist of two major components called *classification engine* and *policy engine*. The classification engine understands an imperative language through which one could specify what fields in the packet headers (IP header, TCP

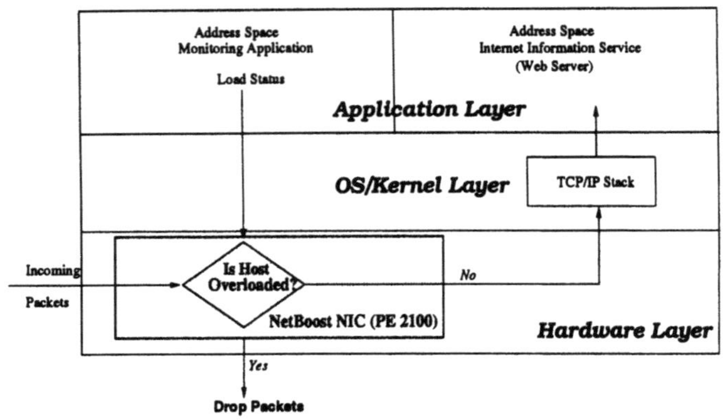

Fig. 2. Dropping Incoming Requests using the intelligent NIC

header, or HTTP header) or even arbitrary strings in the message body (e.g., URL) are to be examined for packet classification. The policy engine deals with treatment of the packets (dropping, forwarding, sending them up the host stack, etc.). For our purposes, incoming packets were classified into two categories: (a) connection request packets destined to port 80 (HTTP) and (b) all other packets. Connection request packets were sent to the policy engine, while all the other packets were directly passed to the host networking stack. In order to drop connection packets under overload, functionality was developed within the policy engine using intelligent NIC's APIs. The host application monitored the load on the system (via NT's Perfmon utility) and activated selective dropping of packets within the policy engine when the system got overloaded. This mechanism avoids the overhead of protocol stack processing for dropped packets. The flow of packets as implemented using the intelligent NIC is illustrated in Figure 2.

3.2 Traffic Generation

All clients generated HTTP GET requests resulting in the execution of a single active server page (ASP) script on the server that generated the web page to be sent out dynamically. The traffic was generated by a locally developed tool that generates aggregate traffic seen by the web-server without necessarily having to individually emulate each user. The generator has two parts: request parameter generator and actual request generator. The request parameter generator generates the time, size, target web page and other characteristics of the requests and could even be run offline to generate a trace. The request generator then formulates actual HTTP requests that accurately reflect these parameters. In order to achieve correct timing, successive requests are generated by a round-robin scheme between all the O/S threads on a client. When multiple client machines are used for load generation, a round-robin scheme is used between the clients as well. The traffic generator is capable of generating traffic with complex behavior (e.g., asymptotic self-similarity and multifractality at intermediate time scales for the arrival process, flexible description of skewness in web-page accesses, etc.), but most of those features were not used in the experiments reported here.

The uncontrolled client made 30 requests per second for all runs. In order to study the impact of the arrival process on the performance, two extreme cases were considered: (a) deterministic inter-arrival times, and (b) $M/G/\infty$ traffic [3], which is known to be extremely bursty and asymptotically self-similar. Overload situations were simulated by providing the web server with additional load from the controlled clients. Each controlled client generated an average of 20 requests per second. Each experimental run was for about 5 minutes and the controlled clients were started up after the load from the uncontrolled clients had stabilized at the server.

3.3 Data Collection

During each experimental run, the clients collected data about each request sent. The parameters monitored were the latency for the response and the type of the response. In addition, aggregate parameters for each test were collected which gave information about the total number of OK (HTTP 200) responses received and the total number of Server Error (HTTP 500) responses received. The aggregate data was collected in the time window that the overload occurred. Separate data was collected to determine the length of the tail until the load returned to normal.

During the experiments, data collection on the web server was done using the performance monitoring utility of the host O/S (PERFMON). This utility provides access to various performance counters pertaining to several system objects such as the processor, active server page (ASP) statistics, TCP and UDP related counters, etc. In particular, processor utilization, the ASP queue length, connection requests per second and connection requests rejected and UDP packets sent during the experiment were used to analyze the performance of the proposed mechanisms. One limitation of this utility is that it can update performance parameters at most every second. For a large server, 1 second updates may be too slow. For certain metrics, such as the number of queued requests, this limitation can be easily overcome by explicitly monitoring the desired parameter.

4 Proposed Mechanisms

4.1 Overload Detection

Although a high processor utilization is a good indicator of server load, it is not very useful for overload control since it does not indicate the severity of the overload beyond 100%. The real measure of interest is, of course, the unfinished work, often also referred to as virtual waiting time [5]. However, because a direct estimate of unfinished work is either infeasible or expensive, simple approximations (e.g., number of unserved requests) are typically used. In our experiments, every request targets an ASP file (actually the same ASP file), therefore, a good metric is the ASP queue length. It is important to note here that if the requests invoke a number of ASP scripts with widely different characteristics, the ASP queue length alone may not be a good measure of the unfinished work. The host operating system provides access to the registry from where data regarding queue length for the ASP requests can be obtained by an application. The ASP object from the Windows registry facilitates monitoring of various parameters related to ASP requests. Using the API related to this object the queue length of outstanding ASP requests was monitored via an application thread on the web server.

The following subsections describe three simple overload control schemes that were used to demonstrate potential performance gains due to overload control. Optimization of the schemes or their parameter values is beyond the scope of this paper.

4.2 Dropping Incoming Requests

This is the "baseline" scheme that simply drops new connection requests under overload. As stated earlier, most web-servers currently use a similar scheme (except that they incur a higher overhead by generating the "server too busy" response). Thus, to be useful, more sophisticated schemes have to provide better overload performance than this simple scheme.

In this scheme, two thresholds are defined for the ASP queue length: abatement and discard. Whenever discard threshold is crossed, the host software makes a down call into the policy engine on the intelligent NIC so that all new connection requests arriving from the controlled clients are dropped. This dropping continues until the ASP queue goes below the abatement threshold. (Once again, the crossing of abatement threshold requires a down call to the intelligent NIC so that its policy engine can start redirecting traffic to the local TCP stack).

One important consideration in such a scheme is how the requests are dropped. As discussed earlier, in current web-server architectures, there is no real dropping of the request; instead, either the web-server or a front-end load balancer responds to the HTTP request by a "server-too-busy" response. When done by the web-server itself, this approach obviously results in a lot of wasted work and thus will not perform well under heavy overload. In our scheme, the connection request is actually dropped by the NIC without even the TCP layer knowing about it. Consequently, the TCP on the other end considers this as a "lost packet" situation and reattempts the connection with increasing gaps (1.5 seconds initially, and doubling every time thereafter for a certain number of attempts). This puts additional burden on the clients and the network, which is undesirable.

4.3 Traffic Throttling

In this scheme, two thresholds are defined for the ASP queue length: abatement and onset. Whenever onset threshold is crossed, a feedback message is sent by the application towards the proxy server. The feedback mechanism is implemented by using UDP because of its lightweight nature. The server does not discard any packets in this case.

Upon receiving the feedback indicating overload, the proxy server needs to reduce the rate at which it sends requests to this web server. It can accomplish this by forcing a minimum time gap between successive requests served from its queue.

This introduces the risk of queue buildup at the proxy, a possible performance bottleneck. Since most proxy servers have the ability to send back custom messages to the web clients, a queue length threshold can be introduced to protect the proxy server from overload. When the proxy server queue length exceeds beyond this chosen threshold value, it can send a "Server Too Busy" (STB) message directly to the user, thus saving the additional overhead of establishing a connection with the web server and receiving this response from the server. This also saves the server from performing wasted work (processing STBs).

4.4 A combined mechanism for overload control

A third mechanism involving the use of both the above mentioned methods is described here. In this method, three thresholds are defined to classify the load on the server (in increasing order): an abatement threshold, an onset threshold and a discard threshold. The proxy server starts gapping the requests when the onset threshold is crossed. If the load does not increase further, this mechanism acts exactly like the throttling mechanism presented earlier. However, if the load increases further and the discard threshold is crossed, subsequent connection requests are dropped. When the server load falls below the onset threshold, dropping is disabled and all subsequent connections are accepted. When the load eventually drops below the abatement threshold, a feedback message is sent to the clients to resume full traffic.

5 Results and Analysis

5.1 Performance without Overload Control

We start by analyzing the impact of varied levels of load on throughput, utilization and response time. Figures 3 and 4 present the gathered results from this experiment. The offered load (requests to active server pages) is varied from approximately 30 requests per second to roughly 100 requests per second. If the connection is accepted by the server, the request is typically processed by the web service (IIS 4.0) in two ways depending on the number of requests already queued at the server. If the number of requests are below a certain threshold, IIS processes the request and sends back an OK response code to the client. If the number of queued requests has crossed the threshold, then the web service rejects the request and sends a "Server Too Busy" (STB) response to the client. This threshold, called "RequestQueueMax", is a O/S registry parameter that can be configured based on the observed performance of the server.

Figure 3 presents the number of OK and STB responses as the offered load increases from 32 requests per second (leftmost) to 100 requests per second (rightmost). From the figure, it is found that the web server is capable of sustaining approximately 45 requests/s without generating STBs at approximately 100% processor utilization. As the offered load increases to 71 requests/s (160% of optimal load), almost half the requests experience STB responses, thus lowering the OK throughput to 39 requests/s (86% of server capacity). Furthermore, when the offered load increases to 100 requests/s (225% of server capacity), the OK throughput reduces to 32 requests/s (71% of server capacity).

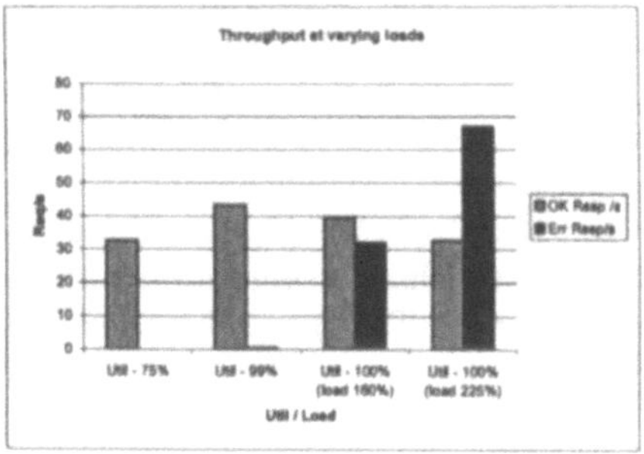

Fig. 3. Throughput vs. offered load (no overload control)

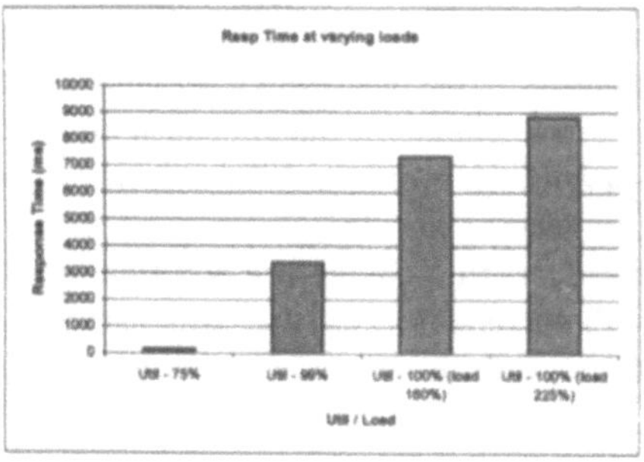

Fig. 4. Response time vs. offered load (no overload control)

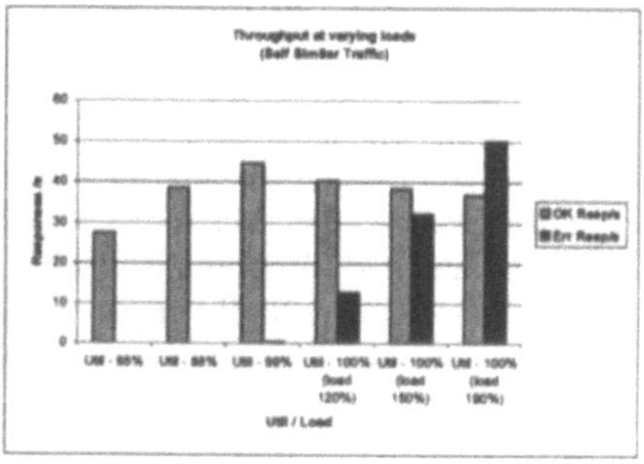

Fig. 5. Throughput vs. offerred load (Self Similar Traffic, no overload control)

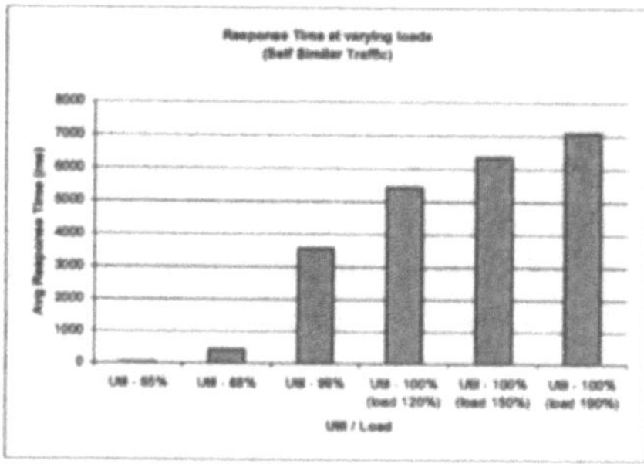

Fig. 6. Response time vs. offerred load (Self Similar Traffic, no overload control)

Table 1. CPU utilization and ASP queue length without overload control

Offered Load (requests/s)	Processor Utilization (%)	Avg Queue Length (Reqs)
32.58	73.67	0.043
43.97	99.95	105.02
71.42	99.87	249.52
99.55	99.82	252.68

Another important metric to measure performance is the response time as seen by the client. Figure 4 presents the impact of increasing load on the average response time. It is clear from the figure that when the system is running at 75% processor utilization (32 requests/s load), the average response time seen by the client is very small (about 170 ms) and a negligible average queue length. However, when the system is almost 100% utilized (at roughly 44 requests/s), the response time increases to over 3.3 seconds. The increase in response time is attributed the time spent in the request queue (average queue length of 105 requests) waiting to be serviced. Furthermore, as the load increases beyond 45 requests/s, the average response time increases to 7.3 seconds corresponding to an average queue length of 249 requests (at 160% load) and to 8.8 seconds corresponding to an average queue length of 252 requests (at 225% load). Table 1 shows the processor utilization and queue length at various loading levels.

The results above used deterministic interarrival times. Figures 5 and 6 show the results for the $M/G/\infty$ self-similar request arrival process. Here the average offered load is varied from approximately 27 requests per second to roughly 85 requests per second. It is seen that the web server is still capable of sustaining approximately 45 reqs/sec without generating STBs at approximately 100% processor utilization. As the offered load increases to 70 reqs/sec (150% of optimal load), almost half the requests experience STB responses, consequently lowering the OK throughput to 38 reqs/sec (85% of server capacity). Furthermore, when the offered load increases to 85 reqs/sec (190% of server capacity), the OK throughput reduces to 36 reqs/sec (80% of server capacity). As to the response time, Figure 5 shows that when the web server is loaded to 88% processor utilization (corresponding to a load of 38 req/sec) the average response time is close to 500 ms. At 150% overload however, the response time increases to 6.2 seconds even though the throughput is 38 requests/sec.

It can be seen from this discussion that the overload behavior does not differ much even though the nature of the traffic changes drastically (from deterministic to self-similar). This is not surprising because if the ASP queue is running close to being full, the nature of the traffic hardly matters.

5.2 Dropping Incoming Requests

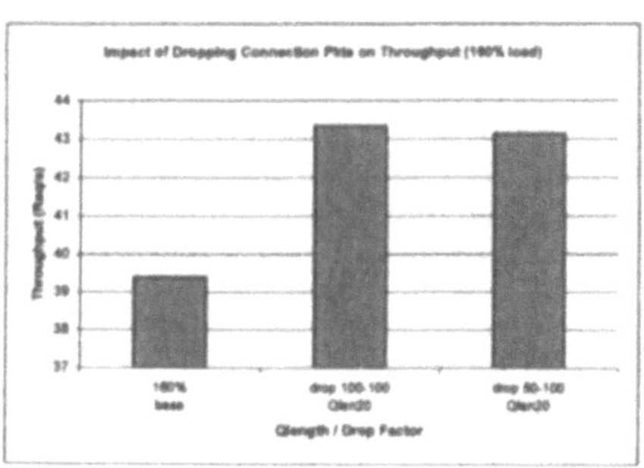

Fig. 7. Throughput for 160% overload (dropping only)

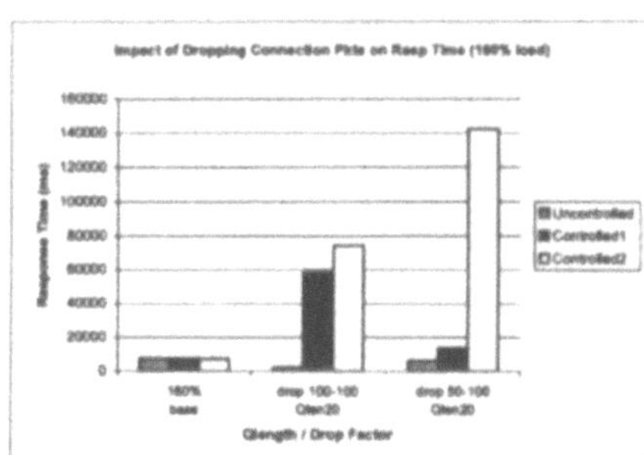

Fig. 8. Response time for 160% overload (dropping only)

This section studies the performance improvement obtained by simply dropping excess new connection requests at server. For this, two controlled clients were used so that it is possible to choose different dropping percentages for them. Figures 7 and 8 present the performance of this scheme under 160% overload. Three different cases are shown here: (a) base case (i.e., no overload control), (b) control triggered at onset threshold of 20 and a 100% drop, for both controlled clients, and (c) control triggered at onset threshold of 20 and (50%,100%) dropping at the two controlled clients. Note here that requests from the uncontrolled client are not dropped and thus it can be considered a preferred client. The results indicate that this mechanism raises the throughput close to the server capacity (an improvement of 9%). Also, the change in response time is different for controlled vs uncontrolled clients. The average response time for controlled clients goes up significantly since dropping the connections results in several TCP retransmissions until the connection is accepted or the number of retries exceeds a predefined maximum. The control increases the response time of controlled clients by a factor of almost 20 whereas the response time of the uncontrolled client decreases substantially.

For brevity, detailed results for the self-similar traffic are omitted here. The overall behavior is similar to that for the deterministic traffic except that the highly bursty nature of the traffic precludes achieving the optimum throughput of 44 requests/sec. Instead, the best achieved throughput is only about 41 requests/sec. As in the deterministic case, the response time of the controlled clients increases substantially, but that of the uncontrolled client drops by about 30%.

5.3 Impact of Traffic Throttling

This section studies the performance improvement obtained by detecting the overload and sending the feedback. Ideally, the feedback should be sent to a control node (e.g., proxy server) as described in Section 4.3. However, this scenario was simulated by sending a UDP message directly to the controlled clients. Upon receiving the message, the controlled clients increase the time gap between their consecutive requests. In the future, we intend to incorporate this mechanism into a proxy server (e.g., Squid). Here we analyze the performance of this mechanism and its relation to the following parameters: level of overload, queue length thresholds and time gap.

Note that in these experiments, an explicit feedback is sent out to the clients whenever the ASP queue size falls below the abatement threshold. This is different from the more traditional approach where the control node chooses a time-interval for traffic throttling. An explicit feedback certainly gives more accurate information and thus will work better; however, it suffers from the weakness that if the feedback message is lost or excessively delayed, an overcontrol will result. It is possible to rectify this by using probe messages, but this aspect has not been explored at this stage.

Figure 9 and 10 show the effect of traffic throttling on transaction throughput and response times when the offered load is 225% of the server capacity. The base case (leftmost bars), representing 225% overload with no control, shows a throughput of roughly 33 requests/s and a response time of approximately 9 seconds. In this situation, the average queue length was approximately 252 (close to the maximum value). Thus, the overload control was studied with onset thresholds of both 200

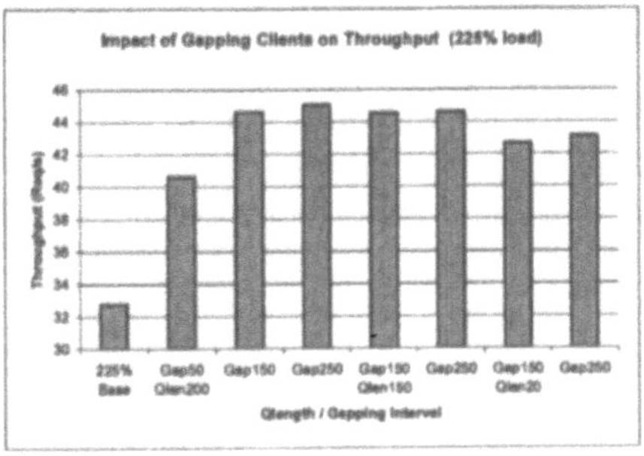

Fig. 9. Throughput for 225% overload (throttling only)

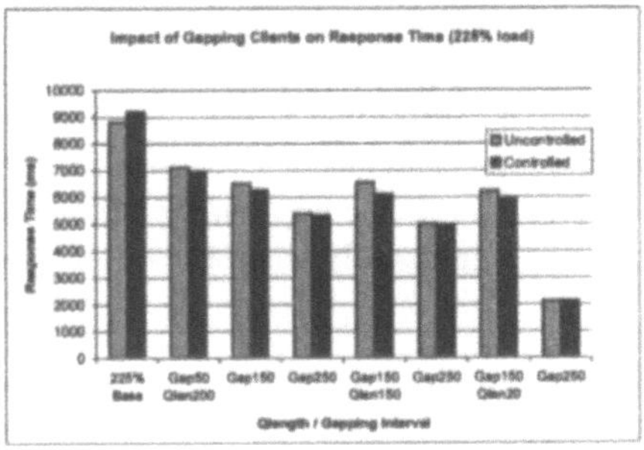

Fig. 10. Response time for 225% overload (throttling only)

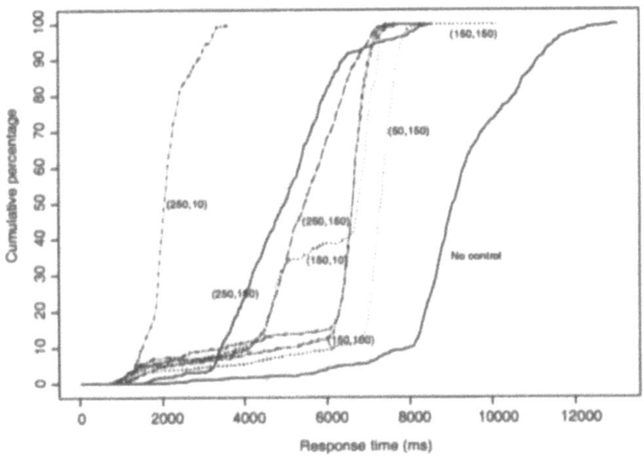

Fig. 11. Uncontrolled client resp. time dist. for 225% overload (throttling only)

and 150. The gapping interval used by the clients was also varied (50 ms, 150 ms & 250 ms respectively). From Figures 9 and 10, it may be seen that the performance gains are significant up to a time gap of 150 ms. Since the three controlled clients each generate a uniform load of 20 requests/s under no control, increasing the time gap by 150 ms reduced the traffic during overload to roughly 5 requests/s (a total of 15 requests/s from the three uncontrolled clients). Since the uncontrolled client generated roughly 30 requests/s, the overall load to the server was about 45 requests/s. Since this is the capacity of the server, the queue length remains constant and the performance improvement achieved by throttling traffic at a gap 150 ms is close to optimal. As the time gap was increased further to 250 ms, the gains in performance were minimal. We hypothesize that using a lower onset threshold to activate the overload control mechanism can further reduce the average queue length and correspondingly reduce the average response time. The data shown in the figure not only supports this hypothesis but also shows that the overall throughput does not change significantly when the onset threshold is reduced.

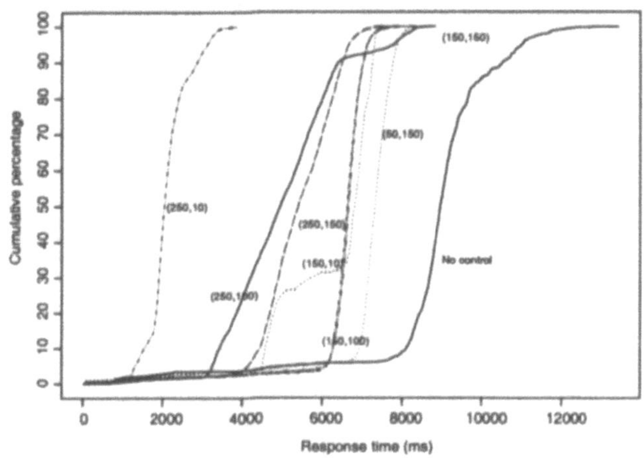

Fig. 12. Controlled client resp. time dist. for 225% overload (throttling only)

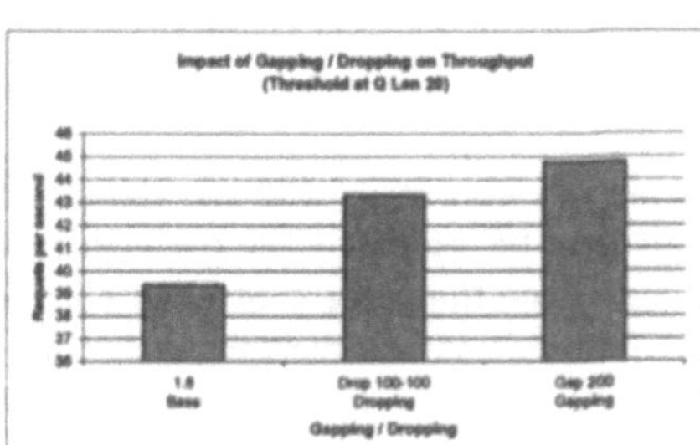

Fig. 13. Throughput for 160% overload (dropping vs. throttling)

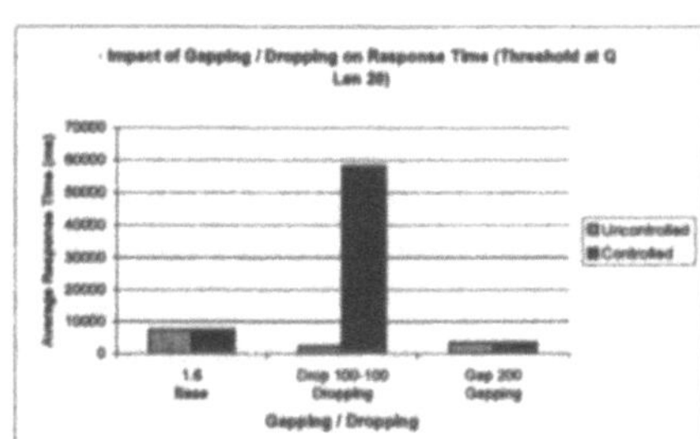

Fig. 14. Response time for 160% overload (dropping vs. throttling)

The impact of the queue length threshold and time gap values on throughput and average response time was illustrated in the above. However, the service provider may also be interested in the level of service provided to the incoming web traffic during overload. One way of quantifying the level of service is by looking at the overall distribution of latencies observed for all request-response pairs during the overload. Figures 11 and 12 plot the cumulative distribution of requests (y-axis) versus the response time (x-axis). In other words, a given point (X, Y) denotes that Y% of the requests experienced a service time of X milliseconds or less. In the figure, there are eight different curves corresponding to the 225% overload and the throttling mechanism run with different values for (time gap, queue length) as indicated by the labels. The case of no overload contro is depicted by *No control* and shows the highest average response time due to the largest queue buildup (about 260 reqs). As the threshold queue-length is reduced , the curves shift to the left. Similarly, as the time gap is increased under a fixed queue length threshold, the response time decreases. Also, for a large time gap of 250 ms, the curves increase very gradually from left to right. This can be attributed to the high fluctuations in the queue length caused by periods of underload and overload.

The results above are for deterministic arrival process. With self-similar process, the throttling scheme is able to acheive a throughput of 43.5 requests/sec, which is very close to the optimal value of 44 requests/sec. The response time also decreases substantially as in case of the deterministic traffic.

Figures 13 and 14 compare the performance of traffic throttling and dropping mechanisms for deterministic traffic under 160% overload. At this loading level, throttling achieves a throughput improvement of roughly 12% and response time reduction by a factor of 3.6. As expected, throttling incoming traffic is more effective than dropping connection requests since the former can better control the number of requests sent to the overloaded web server. In terms of response time, the average response time including controlled and uncontrolled clients is lower for the throttling mechanism. However, the average response time for the uncontrolled clients is much lower with dropping mechanism since dropping affects the controlled clients severely by causing several retransmissions.

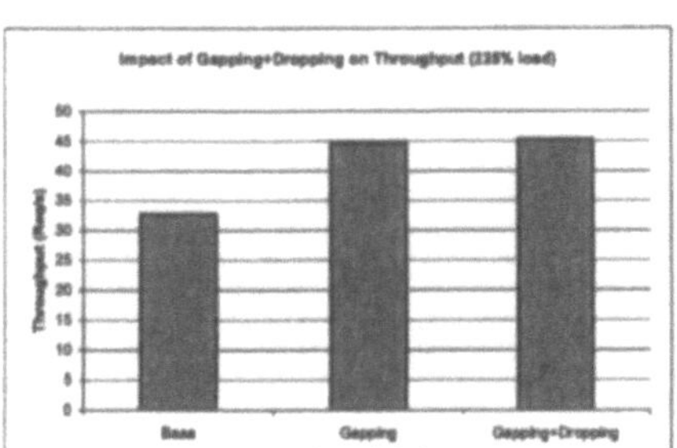

Fig. 15. Throughput for 225% overload (dropping and throttling)

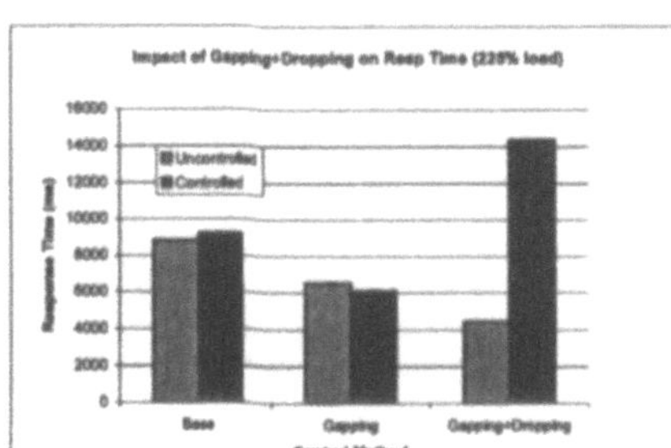

Fig. 16. Response time for 225% overload (dropping and throttling)

5.4 Impact of combining the two schemes

In this case, the onset threshold is chosen as 150 and discard threshold as 200. Figures 15 and 16 show the performance improvement of this scheme under 225% overloaded and deterministic arrivals. While the improvement in throughput is not significant when compared to the throttling mechanism, the average queue length is lower by 40% resulting in a lower response time for the uncontrolled clients as shown in Figure 16. This is an important result since it shows that revenue-generating traffic can be supported even when the server is heavily overloaded. The combined scheme performs better than the throttling only scheme for self-similar traffic as well.

5.5 Multiple Congestion Thresholds

Section 2.1 mentioned the use of congestion priorities in implementing multiple grades of service. In order to demonstrate this, the controlled clients were divided into two classes such that one class had higher congestion thresholds than the other (and hence given a preferential treatment under overload conditions). The uncontrolled clients were still retained and obviously received the best treatment during overload. In these experiments, self-similar input traffic was used and overload control was achieved by traffic throttling. The experiments were run with traffic loads of 120% and 150%.

Figures 17, and 18 presents the data for 150% overload. The client with the lowest threshold for overload control has a reduction in the overall throughput by about 22% while the other clients see an improvement in the throughput. It can be seen that the overall throughput is not affected much by the control, however, the impact of different congestion thresholds is clear. Similarly, the response time for the client with no control (corresponding to the client being given the highest QoS) shows an improvement of 55% as compared to an improvement of 35% for the client with the lowest threshold for overload control (corresponding to the lowest quality of service). Thus, multiple thresholds can be used to tune the performance according to the QoS requirements for various classes of traffic.

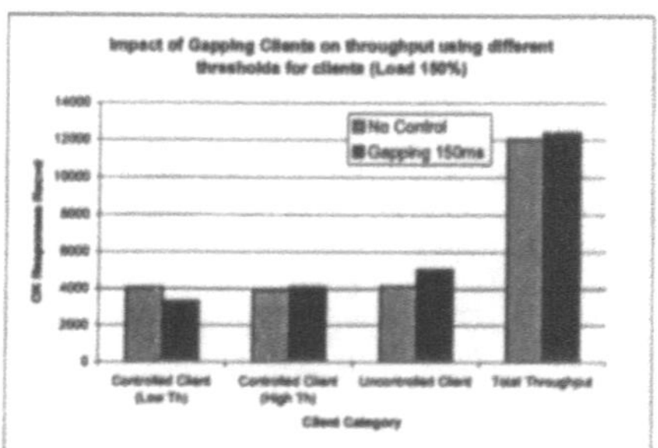

Fig. 17. Throughput for 150% overload (throttling w/ multiple thresholds)

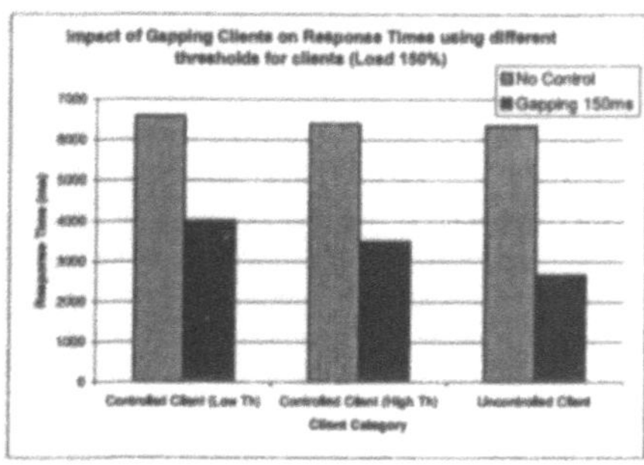

Fig. 18. Response time for 150% overload (throttling w/ multiple thresholds)

6 Conclusions and Future Work

This paper proposed several simple overload control mechanisms to improve the performance of heavily stressed web servers. One such mechanism is to use an intelligent NIC at the web server to selectively drop connection request packets when overload occurs. Another mechanism is to send load feedback backwards towards the traffic source (another native server, a proxy server, or the ultimate client) to enable traffic throttling at or closer to the source. Through extensive experimental runs, it was shown that the performance improvements using these mechanisms can improve the throughput by as high as 40% and reduce the response time by 70% when the web server is 225% overloaded. The two mechanisms can also be combined and help reduce the response time further while maintaining the improvement in throughput. These results demonstrate how intelligent NICs can be exploited for an effective overload control without requiring a separate dispatcher node for the web-server.

In order to judge the true worth of an overload scheme, it is necessary to repeat experiments for a wide variety of scenarios in terms of arrival process characteristics, level of overload, duration of overload, and type of overload. Also the parameter space for the overload control scheme (e.g., monitored congestion index, location of various thresholds, etc.) needs to be explored in order to find suitable operating regions. This paper has not concentrated on these issue since the purpose was simply to demonstrate the usefulness of more sophisticated overload control schemes than simply returning "server-too-busy" message to the client.

As mentioned in section 2.1, application level overload feedback messages (e.g., ACC or ACG) used in telecommunications systems go only back to the previous application node. This is done because retracing the path backwards and directing feedback to them is expensive (in fact, not even possible according to the current standards). Reference [9] proposes a scheme for a more general and unified congestion control mechanism. A similar problem occurs in case of web-servers as well. If a proxy server receives overload feedback, a considerable extra work would be needed to direct it back to the originating client (e.g., additional UDP channel back to each client and maintenance of the context so that the feedback can be pushed backwards). It remains to see if the overhead of such a scheme is worthwhile. It appears that effectively combating DoS attacks would require such a capability.

Transaction level overload control was mentioned in section 2. Efficient implementation of such controls would require some protocol support so that it is possible to drop individual requests and not have them retried. In certain situations, successive requests on a TCP connection are strongly dependent on previous ones — a prime example of this is the SSL handshake required by a secure HTTP transaction where the SSL handshake itself requires 3-4 request-response pairs. In such a situation, either the entire handshake should be allowed to go through or it should not be allowed to start at all. However, once the secure connection is established, a transaction level control might be used for individual requests. An appropriate use of congestion priorities can achieve such a behavior; however, the difficulty arises in recognizing congestion priorities at the NIC level. Again, a protocol support (e.g., congestion priorities encoded in HTTP headers) would help greatly in implementing such schemes.

Identifying precisely what protocol changes are required at TCP and HTTP level to support feedback overload control is a major issue for further work on the topic. It is important to actually make these changes, do an extensive testing under a variety of loading conditions, and use that as a basis for proposing changes to existing standards. As mentioned earlier, the browsers also should be able to deal effectively with feedback messages. Demonstrating the usefulness of such changes to browsers is also a topic for further work in the area.

References

1. G. Banga, P. Druschel, and J.C. Mogul, "Better Operating System features for faster network servers," Proc. of workshop on internet server performance (WISP), June 1998.
2. K. Kant and Y. Won, "Server capacity planning for web traffic workload", IEEE transactions on data and knowledge Engineering, Oct 1999, pp731-747.
3. K. Kant, "On Aggregate Traffic Generation with Multifractal Properties", Proceedings of GLOBECOM 2000, Rio de Janeiro, Brazil.
4. K. Kant and M. Venakatachalam, "Modeling traffic non-stationarity in e-commerce servers, working paper, Sept 2000, Available at `kkant.ccwebhost.com`.
5. K. Kant, *Introduction to Computer System Performance Modeling*, McGraw Hill, 1992.
6. S. Kent and R. Atkinson, "Security Architecture for the Internet Protocol," RFC 2401, Nov 1998.
7. K. Kant, R. Iyer, and P. Mohapatra, "Architectural Impact of Secure Socket Layer on Internet Servers", to appear in International conference on computer design (ICCD), Sept 2000.
8. K. Kant, "Performance of Internal Overload Controls in Large Switches", 28th Annual Simulation Conference, Phoenix, AZ, April 1995, pp 228-237.
9. K. Kant, "A Unified Global Congestion Control Scheme for Broadband Signaling Networks", unpublished report. Available at `kkant.ccwebhost.com`.
10. D.R. Manfield, G. Millsteed, and M. Zukerman, "Congestion Controls in SS7 Signalling Networks", IEEE Communications Magazine, June 1993, pp 50-57.
11. B. Northcote and M. Rumsewicz, "An investigation of CCS Tandem Overload Control Issues," Proceedings of GLOBECOM, Nov 1995, Singapore, pp718.
12. M.P. Rumsewicz and D.E. Smith, "A Comparison of SS7 Congestion Control Options During Mass Call-in Situations", IEEE Trans. on Networking, Vol 3, No 1, Feb 1995.
13. M. Rumsewicz, M. Castro, and M.T. Le, "Eddie Admission Control Scheme: Algorithm description, prototypt design details and capacity benchmarking", available at `www.eddieware.org`.
14. M. Rumsewicz, "Load balancing and control algorithm for a single site Web server", available at `www.eddieware.org`.
15. A. Feldmann, A.C. Gilbert and W. Willinger, "Data Networks as Cascades: Investigating the multifractal nature of Internet WAN traffic", Proc. 1998 ACM SIGCOMM, pp42-55.

Performance Comparison of Different Class-and-Drop Treatment of Data and Acknowledgements in DiffServ IP Networks

Stefan Köhler and Uwe Schäfer

University of Würzburg
Department of Computer Science
Lehrstuhl für Informatik III
Am Hubland, D-97074 Würzburg, Germany
koehler@informatik.uni-wuerzburg.de

Abstract. In differentiated services IP networks the sender of a TCP connection determines the class of data packets he emits. The receiver chooses the class of the acknowledgements sent back, independently of the received class. In this work, we examine the impact of different drop precedence within a class and the assignment of different classes for data and acknowledgements. The results show that the throughput of a TCP connection depends not only on the data class, but also on the right choice for the acknowledgements. Some combinations of classes for data and ACKs could even lead to an "unfair" use of bandwidth. On the other hand, for a high throughput the selection of the drop precedence is in most cases only important for data packets.

1 Introduction

Internet Service Providers are looking for strategies to offer "differentiated services" to satisfy customer demand for quality of service (QoS) and of its potential to increase revenues. How to support these differentiated services is still a subject of research. One of the models, the DS service architecture [6], tries to implement different levels of services for individual or aggregated flows. The current framework allows the sender to mark its packets by setting the appropriate bits in an IP header field [2]. The network provider then checks the packets at the network border for conformance to service contracts. If the packet conforms to the contract it is marked as IN (in profile) otherwise the packet is marked as OUT (out of profile). The treatment of the IN and OUT packets in the core network depends on the per-hop behavior of the traffic class. Per-hop behavior (PHB) is defined as the externally observable forwarding behavior applied at a DS-compliant node to a DS behavior aggregate [5]. The DS framework is independent of the routing decision and thus does not define any end-to-end service. It achieves scalability by implementing complex classification and conditioning functions only at network boundary nodes. The service is defined through the different treatment (PHB) of the marked packets in the routers.

There are several ongoing discussions in the Internet Engineering Task Force (IETF) DiffServ working group about the interaction between UDP and TCP traffic. In contrast to UDP, TCP uses congestion control based on a window mechanism to reduce its transmission rate. This leads in situations with congestion to unfairness between UDP and TCP. There are already several papers [12,13] discussing the effects between UDP and TCP and trying to clarify whether it is necessary to use three instead of two levels of drop precedence. As far as we know, no investigation on the effect of different classes for TCP data and acknowledgments in a DS network has been done. Thus, we concentrate on the behavior of TCP connections in

- different proposed traffic classes and
- a single traffic class with different drop precedences.

The aim of this paper is to evaluate the behavior of TCP under these conditions and to clarify the impact on the goodput of a TCP connection.

The paper is organized as follows. Section 2 describes different PHBs, the traffic conditioners which are located at the boundary nodes to mark the packets appropriately and the active queue-management to implement the different PHBs. The topology, configuration and parameters used in the simulations are presented in Section 3. Section 4 discusses the results. Finally, we give a conclusion and an outlook in Section 5.

2 Components of a Differentiated Service Network

We consider in this paper the following defined [3,4] forwarding per-hop behaviors:

2.1 Expedited Forwarding PHB

The intention of the expedited forwarding (EF)-PHB is to build a low loss, low latency, low jitter, assured bandwidth, end-to-end service through DS domains. Such a service appears to the endpoints like a point-to-point connection or a "virtual leased line" [4] and has also been described as Premium Service [11]. In RFC 2598 no explicit treatment of the marked packets is defined, but packets marked for EF-PHB may be remarked at a DS domain boundary only to other codepoints that satisfy the EF-PHB. Packets marked for EF-PHB should not be denoted or promoted to another PHB by a DS domain. Consequently, packets which exceed the agreed rate are dropped.

2.2 Assured Forwarding PHB

The Assured Forwarding (AF)-PHB [3] specifies four traffic classes with three drop precedence levels (colors) in order to provide differentiated services to the customers in IP networks. The level of forwarding assurance of an IP packet in the AF class depends on

1. the forwarding resources that have been allocated to the AF class,
2. the current load within the AF class and,
3. in case of congestion within the class, the drop precedence of the packet.

A DS node does not reorder IP packets of the same microflow, no matter if they are in or out of the profile, as long as they belong to the same AF class. This is motivated by the fact that reordering of TCP packets might cause severe performance problems for a TCP connection. A DS node should implement at least one of the four AF classes, but it is not required to implement all of them. More details on the behavior of the AF-PHB can be found in [3]. Unmarked traffic is treated as Best Effort.

2.3 Traffic Conditioners

Traffic conditioners are used to shape respectively meter the traffic entering or leaving a DS domain. Beside the "normal" Token Bucket algorithm which guarantees that the burstiness of a flow is bounded in such a way that the flow never exceeds the rate $b + r \cdot t$, where b is the bucket size, t the time and r the token arrival rate, we implemented the following proposal by Heinanen and Guerin [1].

Two Rate Three Color Marking (trTCM) with Three Drop Precedence

The Three Color Marker (TCM) uses two token buckets (P and C) to meter an IP packet stream and marks its packets either green (DP0), yellow (DP1), or red (DP2). The two token buckets have different depths called Peak Burst Size (PBS) and Committed Burst Size (CBS). The first bucket is filled with the Peak Information Rate (PIR) and the second with the Committed Information Rate (CIR). The conditioner operates in one of two modes, Color Aware and Color Blind mode, where the colors stand for particular codepoints in the IP header. In case of the AF-PHB, the color can be coded as the drop precedence of the packet. In both modes the token buckets P and C are initialized to PBS respectively CBS at time 0. Thereafter, the token count T_p is incremented PIR times per second by one to an upper bound of PBS and the token count T_c is incremented CIR times per second by one up to CBS. If a packet of size B bytes arrives at time t in Color Blind mode, the following happens:

- the packet is red, if $T_p(t) - B < 0$,
- otherwise the packet is yellow and T_p is decremented by B, if $T_c(t) - B < 0$,
- otherwise the packet is green and both T_p and T_c are decremented by B.

If the traffic conditioner is in the Color Aware mode, it reacts as follows:

- the packet is red, if it has been precolored as red or if $T_p(t) - B < 0$,
- otherwise the packet is yellow and T_p is decremented by B, if the packet has been precolored as yellow or if $T_c(t) - B < 0$,
- otherwise the packet is green and both T_p and T_c are decremented by B.

2.4 Active Queue-Management

Beside traffic conditioners which are necessary to control the incoming traffic of a DS domain, a further mechanism is needed to determine the PHBs of the nodes.

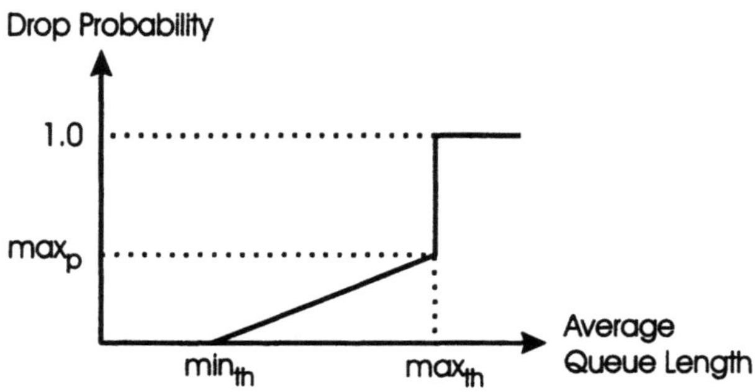

Fig. 1. Increasing drop probability in RED.

Random Early Drop (RED) Queue

In RED Queues four parameters need to be configured. Two parameters define the thresholds, min_{th} and max_{th}, where random packet drops occur driven by the average queue length (cf. Figure 1). The value max_p determines the drop probability at max_{th} and a weight w_q is used to calculate the average queue length (AQL). An arriving packet at time t is accepted with probability $1 - P(AQL(t))$, where $P(AQL(t))$ is the drop probability. The average queue size is calculated using a low-pass filter with an exponential weighted moving average [9]. The intention behind RED is to control the average queue length through max_{th}, min_{th}, and max_p, as well as the degree of burstiness reflected through w_q.

n-RED

An extension of RED is n-RED in order to support the drop precedences of several classes, where the parameter values differ between the classes. We use the overlap

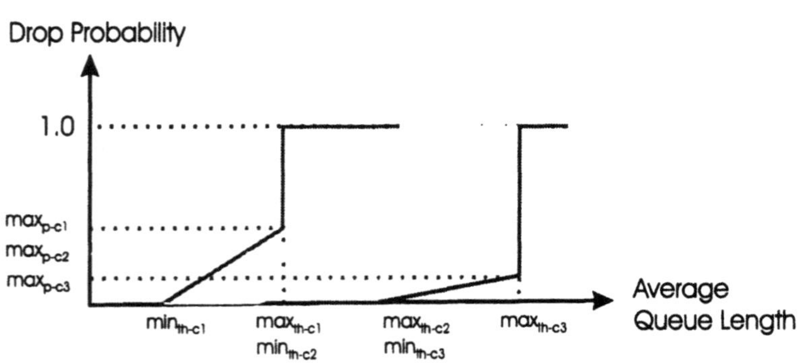

Fig. 2. Drop probability for different traffic classes in n-RED ($n = 3$).

n-RED model as shown in Figure 2. The parameter n defines the level of drop precedence. The average queue length for packets at level n is the sum of packets in class 0 to n. The average queue length determines the drop probability as explained in the previous section. If the different drop precedences are adopted to a color as shown in Figure 2, the n-RED model can be used to define an AF-PHB. There are further possibilities to extend the RED model, for instance, the RIO (RED with IN and Out) mechanism calculates an average queue length for IN packets (AQL_{IN}), and another one for all packets in the queue (AQL_{ALL}). Both packet classes use the same threshold parameters, whereas the drop probability for IN packets depends on AQL_{IN} and the drop probability for OUT packets on AQL_{ALL}.

The mechanisms we use to build EF- and AF-PHB are only proposals and not defined in an RFC. There is ongoing discussion in the DS working group about optimal mechanisms to obtain service discrimination.

3 Example Network and Parameters

For our simulations we used the network simulator **ns** version 2.1b5 [8] developed at UC Berkeley, LBL, USC/ISI and Xerox PARC. The code has been modified to implement the traffic conditioners, multi-color RED and RIO queues.

3.1 Network Topology

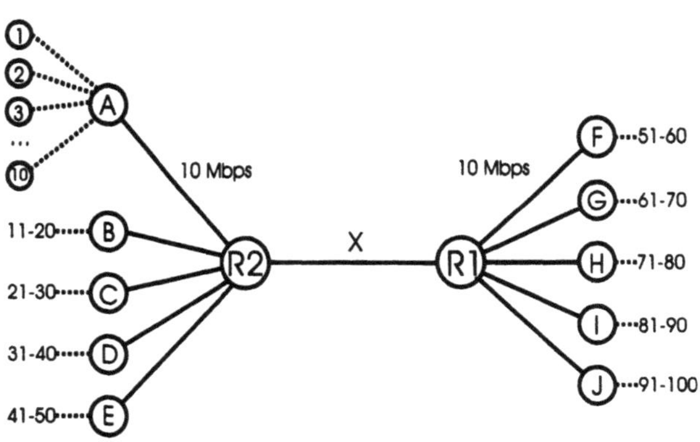

Fig. 3. Network topology used for simulation.

For simplicity and comparability with other simulations we chose a similar topology to [7] or [12]. As shown in Figure 3, sources and destinations are connected via a bottleneck link between router R1 and router R2. Five nodes with traffic conditioners to meter the incoming traffic are placed on both sides. Each node is connected over a 10 Mbps link to the router. Each node contains ten individual TCP sources. All sources are ftp-sources and use NewReno to transfer ftp-packets. Two sources send according to the simulation scenario Premium respectively Green traffic (20%), three send Assured respectively Yellow (30%) and five sources use Best Effort respectively Red (50%) to transfer the packets. Thus, there are 50 sources on each side connected to a sink on the opposite side and vice versa. The link between router R1 and router R2 has a bandwidth of $X = 50$ Mbps[1] . To simulate a bottleneck we vary the bandwidth X from 100 Mbps to 35.7 Mbps (50% to 140% maximum offered traffic). We observe the ftp connection between source 1 and source 51. The remaining 99 traffic sources create background traffic during the whole simulation. The background sources begin their transmission randomly in the first 500ms. After 10 seconds the background traffics get stabilized and the examined source starts to transmit. The observed source is interrupted after 60 seconds of transmission for 20 seconds. This procedure is repeated a total of 10 times. The background TCP connections send ACKs of the same class (drop precedence) as the data. Only the examined connection uses different classes respectively drop precedences for data and ACKs.

Two scenarios are investigated:

- **Two-Bit Differentiated Services Architecture**
  The "classical" model from [11] is implemented. It contains three classes: Premium (EF-PHB), Assured (AF-PHB), and Best Effort. The traffic within a class has identical drop precedence. This scenario shows the impact of different classes for data and ACKs in a TCP connection. Two Token Buckets are placed at each node but not in the routers. The first Token Bucket controls the Premium Rate. If the Premium traffic exceeds its contracted rate the packets get lost. Premium traffic which follows the contract is placed in a preferentially treated FIFO queue. The second Token Bucket observes the Assured Rate. If Assured traffic exceeds its rate, the packets are only remapped to Best Effort. Beside these traffic conditioners, RIO queues are placed in each node and in the routers to treat Assured and Best Effort traffic.
- **Two Rate Three Color Marking with Three Drop Precedence**
  To concentrate on the effects of different drop precedence in an AF class, the traffic in the simulations belongs to the same class. The only difference between the TCP connections is the different drop precedence. A trTCM traffic conditioner is placed at each node to control the Color and a n-RED queue. The routers use only n-RED queues without traffic conditioners.

3.2 Parameters for Simulation Study

For all simulations we used a certain set of reference parameters:

Delay

To cover a wide range of different scenarios, the RTT is varied from 20ms to 200ms (see Table 1). For every RTT value a scenario is simulated where all the connections have an equal RTT and a scenario where different RTTs are mixed to avoid possible oscillations of the TCP connections. A mixed scenario consists of traffic sources with 3 different RTTs. As shown in row two in Table 1 the link of the first traffic source to the router has a delay of 3ms, the second a delay of 23ms, the third a delay of 3ms and so on. Both scenarios show nearly equivalent results (see Appendix).

[1] This corresponds to 5 nodes connected with 10Mbps.

Table 1. Simulated delays, RTT is the round trip time of the observed connection.

node to router [ms]	R1 to R2 [ms]	router to destination [ms]	RTT [ms]
3	4	3	20
3-23-3-23-48	4	3-23-3-23-48	20
10	10	10	60
15	20	15	100
23-48-3-23-48	4	23–48-3-23-48	100
30	40	30	200
48-23-3-23-48	4	48-23-3-23-48	200

Maximum offered traffic

It is difficult to define something similar to load for a TCP connection, because TCP adapts its window size to the available resources. In the simulation scenario each node is connected over a 10 Mbps link to the router. Five nodes are connected to each router. Thus, if the link from R1 to R2 has a bandwidth of 50 Mbps in every direction, the system is well defined and has a maximum offered traffic of 100%. To create a bottleneck, the bandwidth of the router link is varied in the following way:

Table 2. Maximum offered traffic at the the bottleneck link

maximum offered traffic [%]	50	100	110	120	130	140
bandwidth [Mbps]	100	50	45.5	42	38.5	36

Traffic Conditioners and Queue Management

The values for the traffic conditioners are motivated by the fact that 20% of the traffic should be Premium respectively AF_{11} and 30% of the traffic should be Assured respectively AF_{12}.

Table 3. Two-Bit DS token bucket parameters

Premium CIR	250 kByte/s = 2 Mbps
Premium CBS	50 kByte
Assured CIR	375 kByte/s = 3 Mbps
Assured CBS	70 kByte

Table 4. trTCM token bucket parameters

CIR	250 kByte/s = 2 Mbps
CBS	50 kByte
PIR	375 kByte/s = 3 Mbps
PBR	70 kByte

Theoretical studies still have to be done to formulate general rules which determine the right parameters for traffic conditioners and queues in a DS network. To be comparable, we oriented our choice on previous simulations [7,10,12,13] and performed some studies to find appropriate parameters.

Table 5. RIO queue parameters

min_{out}	35
max_{out}	50
p_{out}	0.1
min_{In}	55
max_{In}	65
p_{In}	0.05
w_q	0.002 (=500 packets)
queue limit	90

Table 6. trTCM parameters

min_{red}	35
max_{red}	50
p_{red}	0.3
min_{yellow}	45
max_{yellow}	60
p_{yellow}	0.2
min_{green}	60
max_{green}	70
q_{green}	0.1
w_q	0.002 (=500 packets)
queue limit	90

The maximum transfer unit (MTU) of the ftp connections is 1000 Bytes for all sources. In the following section, we concentrate on the most significant results. More results can be found in the Appendix.

4 Simulation Results

In this section we present numerical results for the TCP goodput based on the parameters given in the previous section. Each point in the figures represents the average amount of data that was transferred within 60s. The errorbars are the 95% confidence intervals. We group the points according to their data class and connect the points with different acknowledgement classes.

In the rest of the paper we use yx as abbreviation for a connection with class y for the data and class x for its ACKs (e.g. PB for Premium data and Best-Effort ACKs). The notation Premium, Assured, Best Effort express Px, Ax, and Bx.

4.1 Two-Bit Differentiated Services

We implement the Two-Bit DS model to estimate the impact of different classes for data and ACKs in a TCP connection. In the following two sections we present the results for different delays and maximum offered traffic.

Influence of Different Delays

Figure 4(a) shows the results for different maximum offered traffic at a fixed round trip time of 20ms. In this scenario the influence of the ACK class is very clear. As expected the rate for PP connections is independent of the load and – compared to Assured data connections – very low since in Two-Bit DS the dropper discards all Premium packets that are out of profile. The boundary for the throughput depends of course on the chosen parameters for the Premium class. PA and PB connections have an even lower goodput since beside the control of the data rate the ACKs have a higher delay and a probability to get lost in the RIO queue. Due to their small size, acknowledgements are not dropped or remapped at the traffic conditioner. The goodput of PB connections is about 60% in comparison to PP connections. Thus,

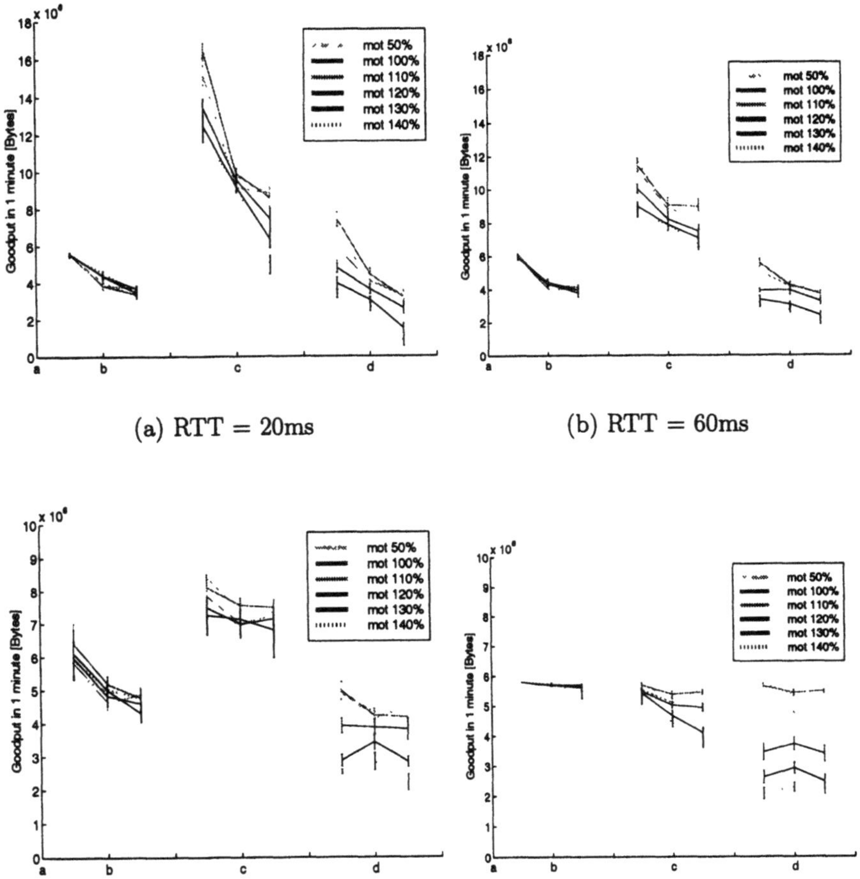

(a) RTT = 20ms

(b) RTT = 60ms

(c) RTT = 100ms

(d) RTT = 200ms

Fig. 4. Comparison of goodput for different class combinations with varying maximum offered traffic (mot = maximum offered traffic)

even for connections in the Premium class with small RTTs the appropriate choice of ACKs is important.

The effect that the goodput does not only depend on the class with which the data is sent, but also on the class of the ACKs, is even more evident for connections that use Assured data. The throughput of AP connections in Figure 4(a) is up to 140% higher than that of AB connections. However, for Assured and for Best Effort connections the variation of the goodput at different loads is significantly higher than for Premium connections. This results from the absence of droppers and the use of RIO queues. These mechanisms cause also that the Assured connections always have a higher goodput than Best Effort connections.

Figures 4(a) to 4(d) show the evolution of the goodput, when the RTTs of the connections increase. With rising round trip time the effect of the Premium dropper lessens as can be see in Figure 4(c). This results in a higher variation of the PP goodput. In general, for a higher round trip time the difference between the classes decreases.

Due to a higher round trip time, the token buckets influence the throughput not any longer and the Premium connections get a higher goodput than the other classes, see Figure 4(d). This is based on the preferential treatment of Premium packets.

In the Figures 4(a) to 4(d) all connections in the respective simulations have an identical round trip time. The results are equivalent for an environment in which the connections have varying round trip times (see Appendix).

Influence of the Bottleneck Link

The figures are based on the same simulation presented in the previous section. The goodput is depicted with varying RTTs (including the mixed scenarios) over the different class combinations for a specified maximum offered traffic.

Figure 5(a) shows the goodput in dependence of the round trip times and a maximum offered traffic of 100%. The figure indicates that with rising RTT the goodput of the different classes approaches an equal level. For low round trip times the Assured connections raise their throughput at the expense of the Best Effort connections. As seen in the figures the goodput is nearly doubled for Assured than for Best Effort connections. Further investigations are needed to clarify the amount of unfairness in these situations.

Whereas at low load the class of the ACKs is the main factor for goodput, at higher loads the ACKs get less important for the goodput. This evolution is presented in the Figures 5(a) to 5(d). In Figure 5(d) more Best-Effort data packets are thrown away due to the high level of congestion and therefore these connections get a lower throughput. This effect could also be observed for low RTT connections. In Figure 5(a) and 5(d) the goodput of an AB connection with a RTT of 20ms is lower than the goodput for the same connection and a RTT of 60ms, whereas the AP connections take profit out of the situation.

All simulations show that Premium and Assured connections are extensively protected by the used mechanisms and that sending Assured data with Premium acknowledgements results in the highest throughput, especially at low round trip times. This behavior should be taken into consideration for pricing models in DS networks. With Premium data packets the throughput can only reach the contracted rate.

4.2 Three Color Marking with Three Drop Precedence

To investigate the influence of different drop precedence in an AF class the whole traffic in the simulations belongs to the same class. Only three different levels of drop precedence are distinguished, labeled AF_{11}, AF_{12}, and AF_{13}, where the first number in the index describes the class and the second the drop precedence. Thus, AF_{11} has the lowest drop probability, followed by AF_{12}. Instead of the labels AF_{11}, AF_{12}, and AF_{13} the shorter labels D_1, D_2 and D_3 are used in the figures.

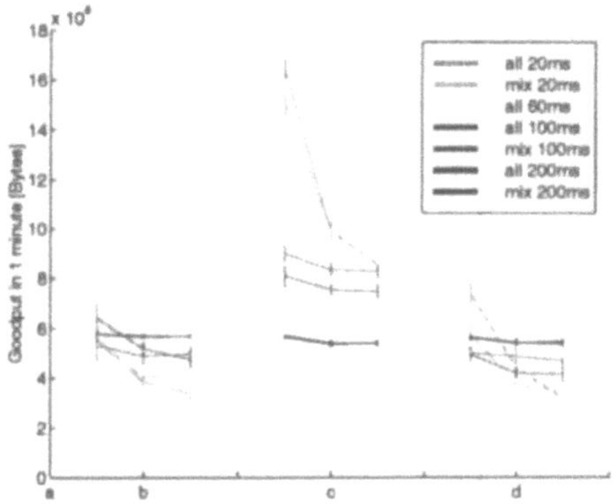

(a) 100% maximum offered traffic

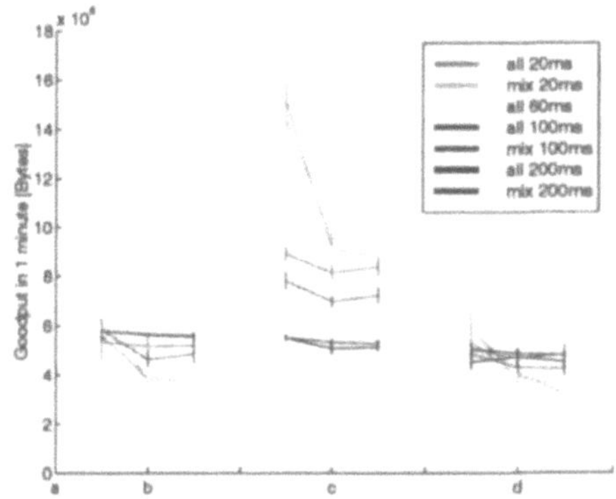

(b) 110% maximum offered traffic

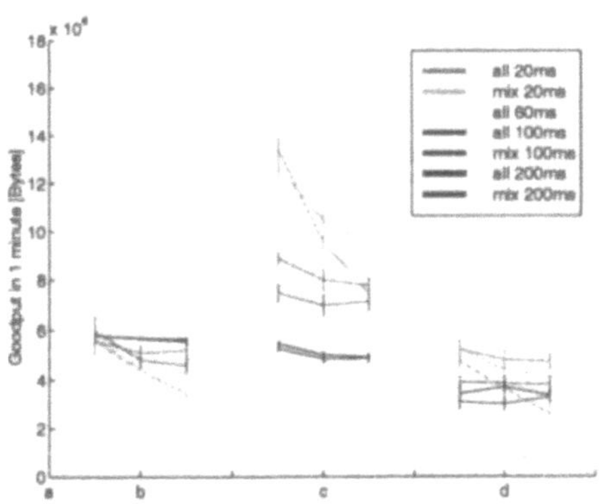

(c) 120% maximum offered traffic

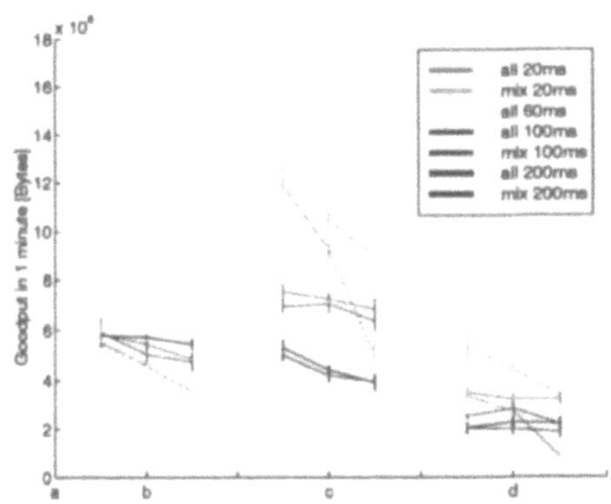

(d) 140% maximum offered traffic

Fig. 5. Comparison of goodput for different class combinations with varying RTTs

Influence of Different Delays

The goodput is illustrated with varying maximum offered traffic over the different drop precedence combination for a specified RTT in Fig. 6.

In Figure 6(a) the goodput for each data drop precedence remains within a relatively small range of values. In comparison to the Two-Bit model, the throughput depends only on the drop precedence for the data except for very high loads and not on the drop precedence for ACKs. In this situation the n-RED queue protects AF_{11} and AF_{12} at the expense of AF_{13} packets.

As seen in the Figures 6(a) to 6(d), the difference in the goodput between the single drop precedences decreases with increasing round trip times. It is the same behavior we have seen in the Two-Bit DS model. In Figure 6(d) the goodput is independent of the different drop probabilities for low loads. In case of an increasing

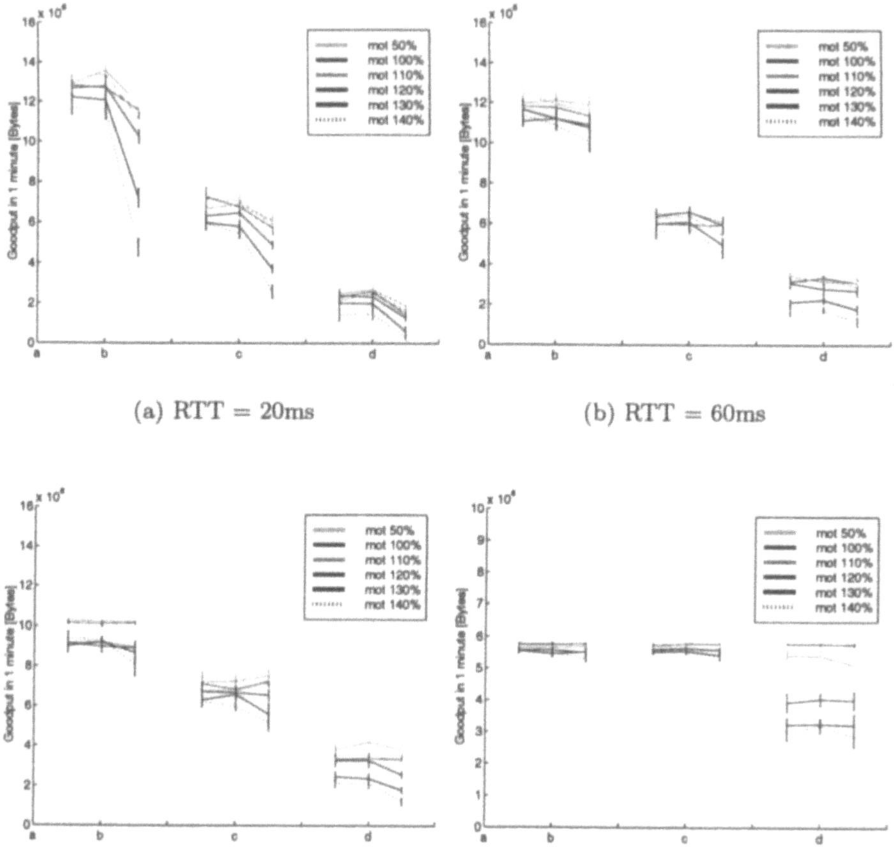

(a) RTT = 20ms

(b) RTT = 60ms

(c) RTT = 100ms

(d) RTT = 200ms

Fig. 6. Comparison of goodput for different combinations of drop precedences with varying maximum offered traffic (mot)

load the classes with lower drop probability are protected at the expense of the classes with higher drop probability.

Influence of the Bottleneck Link

The following figures are based on the same simulations presented in the previous section. The goodput is depicted in Figure 7 with different RTTs (including mixed scenarios) over the combination of drop precedence for a specified maximum offered traffic.

In comparison to Two-Bit DS, the confidence intervals of all measured samples are smaller. This is caused by the use of the same queue for all packets. In Two-

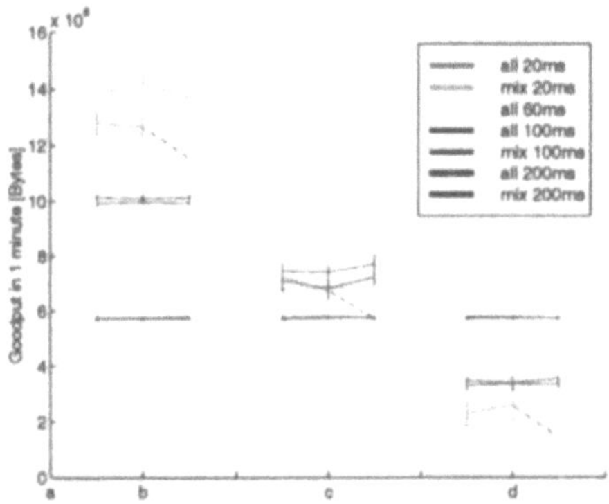

(a) 100% maximum offered traffic

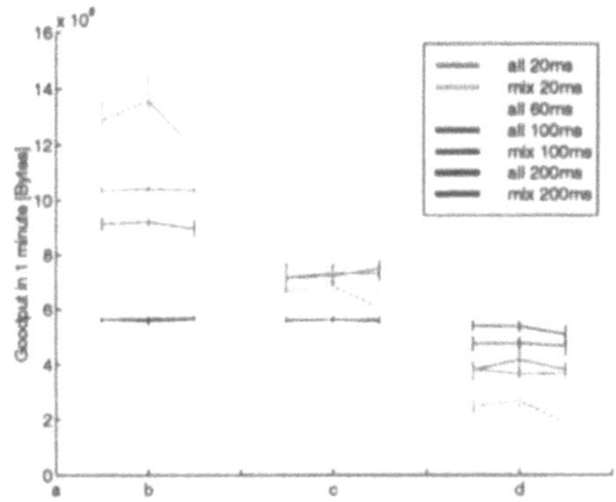

(b) 110% maximum offered traffic

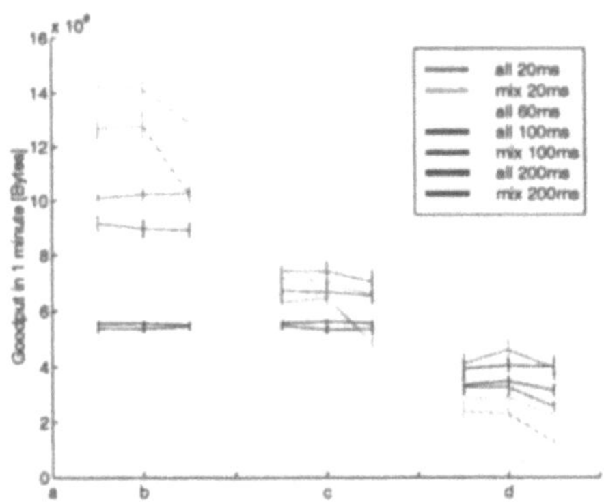

(c) 120% maximum offered traffic

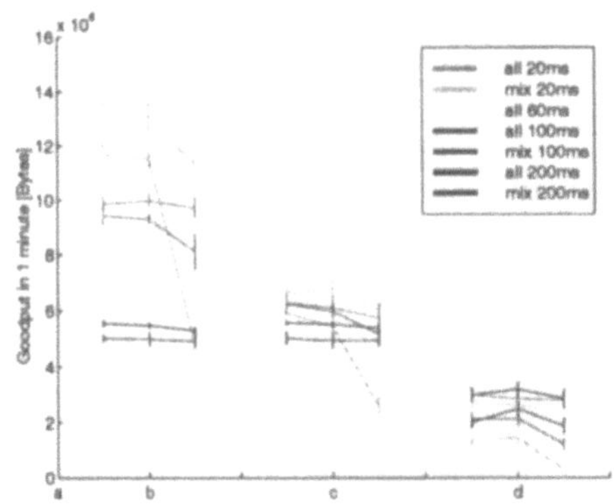

(d) 140% maximum offered traffic

Fig. 7. Comparison of goodput for different combinations of drop precedence with varying RTTs

Bit DS two queues are used, one priority queue for Premium packets and one for Assured and Best Effort packets.

In contrast to Two-Bit DS the acknowledgement class has not a huge influence. The performance of a TCP connection mainly depends on the data class. This results from the fact that the loss of an ACK can be compensated by the TCP protocol, whereas this is not the case when a data packet is lost. Only in regions where the RTT is small and the network is overloaded the class of the ACKs influences the performance, cf. Fig. 7(a). Under these special conditions the goodput could be lowered more than 50%, see Figure 7(d), but as previously mentioned, in most scenarios only the drop precedence of the data is the important parameter for the goodput.

5 Conclusions and Outlook

In this paper we investigated the impact of different class and drop treatment for data and ACKs in a DS environment. The results of our simulations lead to the following conclusions:

- The throughput of a TCP connection in a DS network does not only depend on the sender but also on the receiver. The appropriate choice of the ACK class has a big influence on the throughput.
- In most cases, the use of different drop precedences within a class for ACKs has no influence on the performance worth mentioning. In general, the drop probability of the data determines the goodput. The drop precedence of the acknowledgements is only important for high loads and low round trip times.
- The simulations show that it is possible to get a better throughput with Premium respectively EF-PHB ACKs. In some cases an even unfair gain is made.
- The performance gain for some combinations of data and ACKs should be considered in pricing models.
- Our studies facilitate the use of the EF class for real time traffic because the user gets a fixed, load-independent share of the bandwidth.

Some questions arise from our investigation. They are left to further studies and discussions:

- It should be specified which traffic should be transported in the EF class. In our opinion TCP can not take advantage of this traffic class except when the network is overloaded.
- EF-PHB traffic can not exceed a specified rate. To get a higher throughput the TCP connection may be be split into an EF and AF class part. In this case there is the possibility of reordering TCP packets which affects the throughput. Further studies are needed to estimate if it is worth to use tow data classes for TCP or not.
- A more complex topology and realistic traffic types, especially adaptive, non-adaptive and short-lived (e.g. WWW) traffic flows should be used to investigate the influence of different classes and drop precedences.
- The influence of the routing decision has to be considered (e.g. shortest path for premium traffic).

A Additional Simulation Results

In this section we present additional results of simulations where the background traffic has different RTTs. The parameters are described in Section 3. As previously mentioned, the results are similar to the results in Section 4 for a non-mixed environment.

Two-Bit DS – Influence of different delays

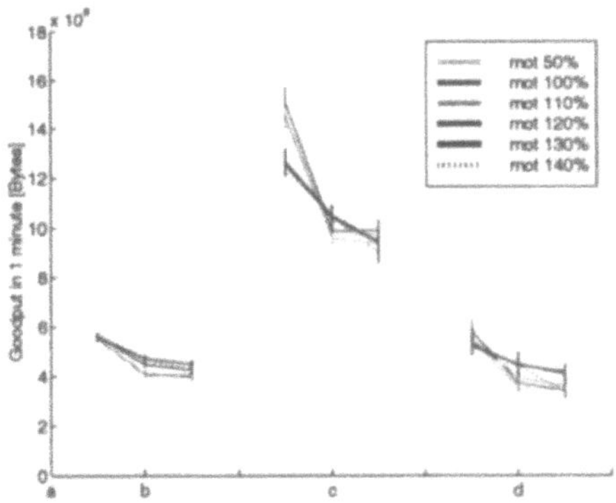

(a) RTT=20ms (connections with different RTTs)

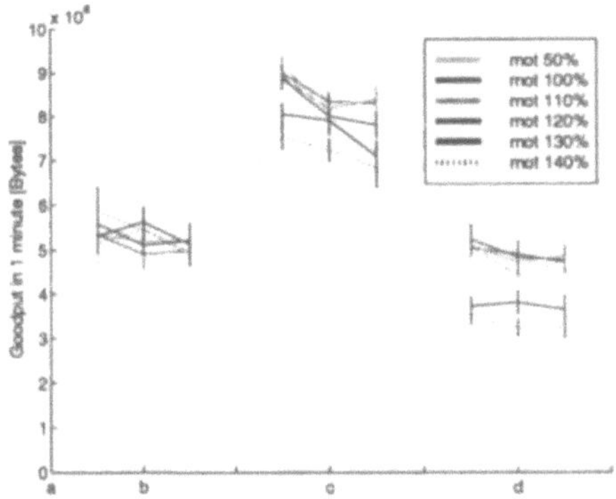

(b) RTT = 100ms (connections with different RTTs)

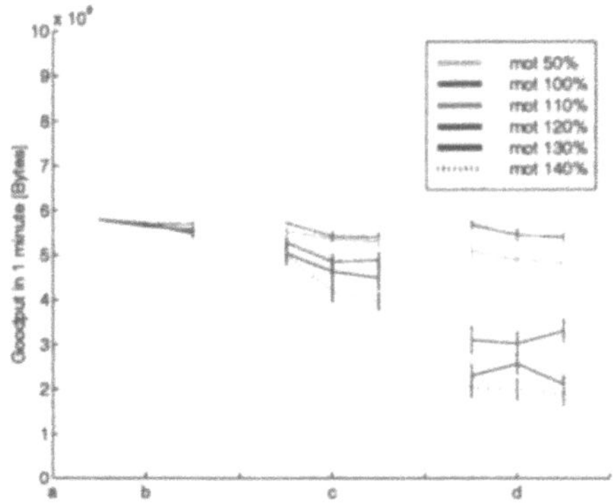

(c) RTT = 200ms (connections with different RTTs)

Fig. 8. Comparison of goodput for different class combinations with varying maximum offered traffic (mot)

Three Color Marking with Three Drop Precedence – Influence of Different Delays

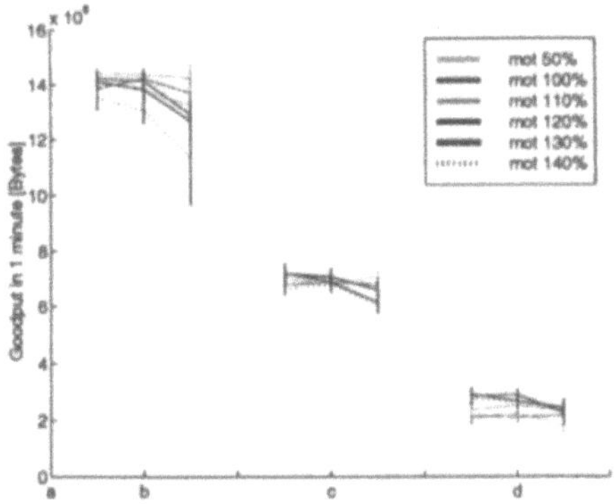

(a) RTT = 20ms (connections with different RTTs)

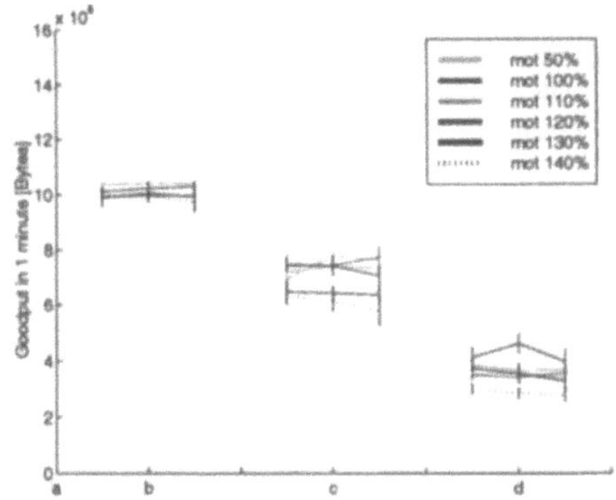

(b) RTT = 100ms (connections with different RTTs)

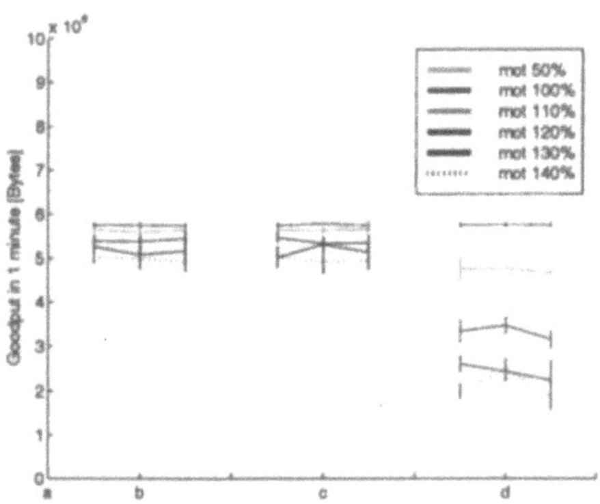

(c) RTT = 200ms (connections with different RTTs)

Fig. 9. Comparison of goodput for different combinations of drop precedences with varying maximum offered traffic

Acknowledgements

The authors would like to thank Norbert Vicari and Kenji Leibnitz for the productive discussions of the presented results. The financial support of the Deutsche Telekom AG (Technologiezentrum Darmstadt) is appreciated.

References

1. J. Heinanen, R. Guerin (1999) A Two Rate Three Color Marker. RFC 2698
2. K. Nichols, S. Blake, F. Baker, D. Black (1998) Definition of the Differentiated Services Field (DS Field) in the IPv4 and IPv6 Headers. RFC 2474
3. J. Heinanen, F. Baker, W. Weiss, J. Wroclawski (1999) Assured Forwarding PHB Group. RFC 2597
4. V. Jacobson, K. Nichols, K. Poduri (1999) An Expedited Forwarding PHB. RFC 2598
5. M. Carson, W. Weiss, S. Blake, Z. Wang, D. Black, E. Davies (1998) An Architecture for Differentiated Services. RFC 2475
6. E. Davies, et al. (1999) A Framework for Differentiated Services. Internet Draft draft-ietf-diffserv-framework-02.txt
7. J. Ibanez, K. Nichols (1998) Preliminary Simulation Evaluation of an Assured Service. Internet Draft draft-ibanez-diffserv-assured-eval-00.txt
8. NS simulator, Version 2.1b5. Available at http://www-mash.cs.berkeley.edu/ns
9. S. Floyd, V. Jacobson (1993) Random Early Detection gateways for Congestion Avoidance. IEEE/ACM Transactions on Networking, Vol 1, Num 4, pp. 397-413
10. D. Clark, W. Fang (1998) Explicit Allocation of Best Effort Packet Delivery Service. IEEE/ACM Transactions on Networking, Vol 6, Num 4, pp. 362-373
11. K. Nichols, V. Jacobson, L. Zhang (1997) A Two-bit Differentiated Services Architecture for the Internet. RFC 2638
12. M. Goyal, A. Durresi, P. Misra, C. Liu , R. Jain (1999) Effect of Number of Drop Precedences in Assured Forwarding. Proceedings of GLOBECOM99, Brazil
13. N. Seddigh, B. Nandy, P. Pieda (1999) Study of TCP and UDP Interaction of the AF PHB. Internet Draft draft-nsbnpp-diffserv-udptcpaf-01.txt

Providing QoS Guarantee for Individual Video Stream via Stochastic Admission Control

John C. S. Lui* and X. Q. Wang

Department of Computer Science & Engineering
The Chinese University of Hong Kong

Abstract. In this paper, we consider a broadband video/audio delivery service. One important feature of this type of service is that it is necessary to provide high quality-of-service (QoS) guarantee to individual user. In this work, we consider a theoretical framework for performing admission control in a Video-on-Demand (VoD) system. The approach presents in this paper is a two-steps admission control algorithm. The QoS requirements are the average bandwidth and the average packet dropping rate. These two QoS requirements are either specified by users or can be a class specific QoS requirements. We illustrate how one can statistically guarantee the bandwidth requirement of the disk storage system using the Chernoff's theorem. We also illustrate how one can use the strong conservation law to determine the admissible region for a given set of packet dropping rates for different classes of requests. Once we determine the admissible region, we illustrate how to derive the transmission schedule which has a packet dropping rate vector that is less than or equal to the required packet dropping rate vector. We use example to illustrate how one can use our proposed method so as to achieve effective resource utilization by admitting more clients and thereby making the VoD service more cost-effective.

1 Introduction

Recent advances in distributed multimedia systems and networking technologies create the possibility of delivering many interesting broadband services. One of the most important broadband service is the real-time video/audio delivery such as Video-on-Demand (VoD) service. Designing a cost-effective VoD system is very challenging because the system needs to deliver the data to the admitted users with a very stringent time constraint. For example, if a user demands to view a video at a particular frame rate (i.e. 25 or 30 frames per second), then the system needs to guarantee this user the sufficient amount of I/O bandwidth, buffer space, as well

* This research is supported in part by the RGC Earmarked Grant and the CUHK Direct Grant.

as network transmission bandwidth so that this user is able to view the requested video at the requested frame rate, without or with small amount of jitter if possible. Since video data are variable-bit-rate (VBR) in nature, to guarantee the frame rate and at the same time, admit as many users as possible, the VoD system needs to perform some form of I/O bandwidth and network bandwidth allocation so as to deliver the data to the user in a timely manner.

One way to accomplish this goal is to perform some form of admission control. In general, when a request arrives at a VoD system, the admission control algorithm needs to decide whether this request can be accepted or not based on the following criteria:

- Whether the VoD storage system has enough resources (e.g., disk I/O bandwidth, buffer space, etc.) to satisfy the QoS requirement of this request.
- Whether the network has enough resources (e.g., transmission bandwidth) to satisfy the QoS requirement of this request.

If the system can satisfy the QoS requirement of this request and at the same time, QoS requirements of other existing requests will not be violated, this newly arrive request can be admitted, otherwise, it will be rejected.

Let us begin with the brief literature survey in this area. There is a lot of research work on the admission control algorithms for VoD servers, for example, [1–5]. For VoD services, video data have to traverse through the communication network, therefore, it is also very important to perform the necessary admission control and traffic scheduling on the communication network. There are many papers which address the admission control of communication networks, e.g. [6], however, there is no work reported on the relationship and the interaction of the I/O retrieval and the packet transmission process. Since a VoD service has rather stringent QoS requirements, the admission control algorithm should be carefully designed and should be easily implemented. The approach presents in this paper is a two-steps admission control algorithm. The QoS requirements are the average bandwidth and packet dropping rate (e.g., the expectation of the maximum number of packets which can be dropped during a transmission time frame). These two QoS requirements are either specified by users or can be a class specific QoS requirement. We illustrate how one can statistically guarantee the bandwidth requirement of the disk storage system using the Chernoff's theorem. We also illustrate how one can use the strong conservation law to determine the admissible region for a given set of packet dropping rates for different classes of requests. Once we determine the admissible region, we illustrate how to derive the transmission schedule which has a packet dropping rate vector that is less than or equal to the required packet dropping rate vector.

This paper is organized as follows. The system architecture is presented in Section 2. In Section 3, we present the admission control in the VoD storage sub-system. In Section. 4, we present the admission control as well as the video transmission scheduling algorithm in the network sub-system. Experiment results are presented in Section 5 and conclusion is given in Section 6.

2 System Architecture

A typical VoD system consists of a storage sub-system, processing sub-system and a network sub-system. Users view the videos on their display stations at the other

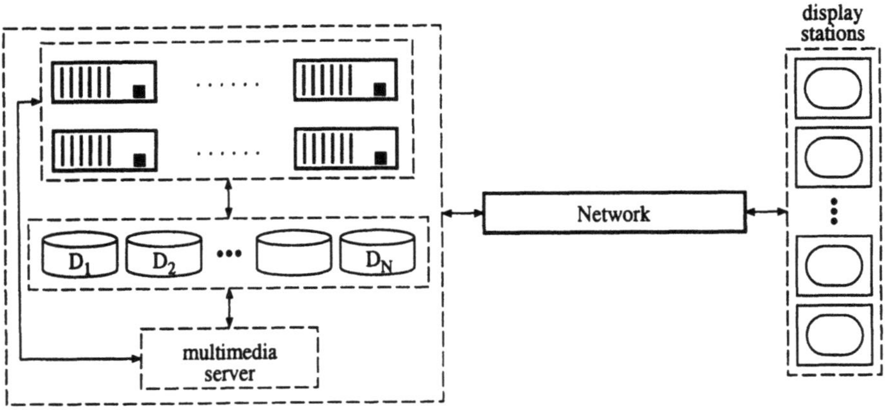

Fig. 1. A typical VoD system

end of network. Figure 1 illustrates a typical VoD system we consider in this paper. The VoD system has a farm of disks for storing popular videos. It also has sufficient amount of memory for buffering the video data retrieved from the storage sub-system before they are transmitted in the network sub-system.

In general, a VoD server needs to perform storage management for all video files. There are many different data placement techniques reported in the literatures [7], for example, for a large video file (e.g. movie), we may consider to stripe the data evenly across all disks in the system. For many small and popular video files (e.g. news clips or commercials), we may have to classify these short clips into different video objects and assign each video object to a disk so as to achieve load balancing property in the storage system. In this paper, we consider the data placement scheme that uses multiple disks to store video files. Each movie is striped across all disks in the VOD server. For example, data block 1 of a movie is stored in disk 1, data block 2 of this movie is stored in disk 2,..., data block N of this movie is stored in disk N, data block $N+1$ of this movie is stored in disk 1,.. etc. See Fig. 2 for an illustration. Since these disks can be accessed in parallel and independently, therefore, the system can support more users who want to view a movie simultaneously, possibly starting from the different parts of the same movie.

In order to satisfy the timing requirement of a video application, the server uses some form of cycle-based scheduling algorithm[9–11] to retrieve data blocks from the storage sub-system. This type of algorithm can be described as follows. The data retrieval process is performed on a cycle basis. Let T be the duration (in seconds) of a disk retrieval cycle. Within each cycle, the server needs to retrieve video data to serve n video requests. The server will sweep the disk[1] and retrieve the data blocks based on the relative distance of the request's target track number and the current read/write head's position. For each request, the disk has to seek to the appropriate track (therefore incurring seek latency), rotate to the appropriate sector (therefore incurring rotational latency) and read out data blocks of the request

[1] Sweeping is done such that in cycle i, the disk arm will move from the inner most track to the outer most track and in cycle $i+1$, the disk arm will move from the outer most track to the inner most track.

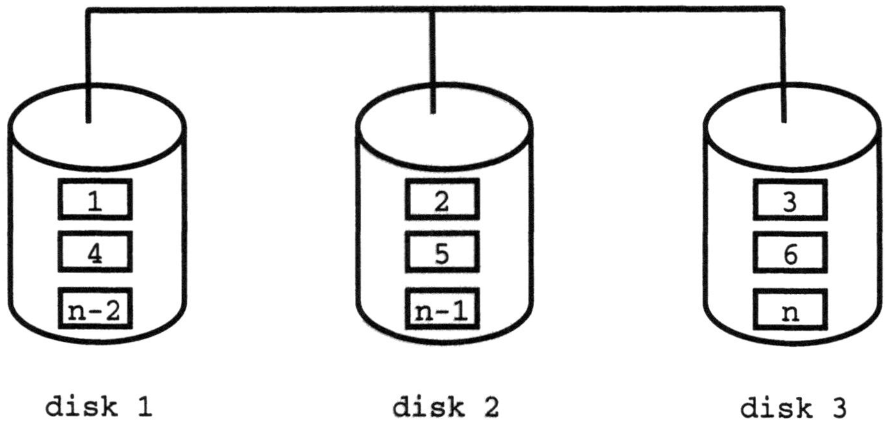

Fig. 2. Data placement scheme with $N = 3$.

(therefore incurring transfer latency). Under the cycle-based scheduling algorithm, the transmission of data retrieved from the storage sub-system in the ith cycle does not start until the beginning of the $(i+1)^{th}$ cycle. The motivation of using the cycle-based scheduling algorithm is that we can improve the efficiency of the I/O bandwidth by reducing the total amount of seek time within a cycle. Note that there is a side effect of the cycle-based scheduling, that is, the server needs additional buffer space to hold the retrieved data until the start of the next service cycle. At that time, all retrieved data is ready for transmission. In [9,12,10,11], authors illustrate the tradeoff between the improved I/O bandwidth utilization due to seek optimization and the need for additional buffer space to store the retrieved data. In general, higher number of admitted requests per cycle can reduce the overall seek overhead but also results in larger buffer space requirement at the server. Figure 3 illustrates the cycle-based scheduling algorithm in the storage sub-system.

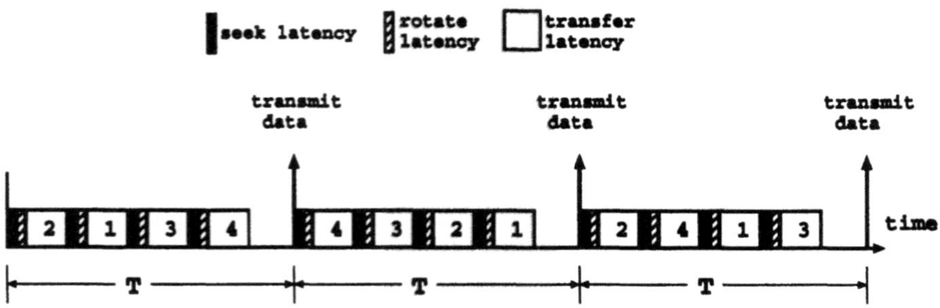

Fig. 3. Cycle-based scheduling algorithm for serving 4 video streams in the VoD storage sub-system

In the figure, the video server retrieves data for 4 video streams. The requirement is that all retrieved data must be serviced (e.g. retrieve from the storage sub-system) by the end of every retrieval cycle of length T. That is, the retrieved data will be ready for network transmission at the end of every retrieval cycle. Notice that due to the nature of the retrieval algorithm, the *order* of data retrieval for various video streams are different from cycle to cycle.

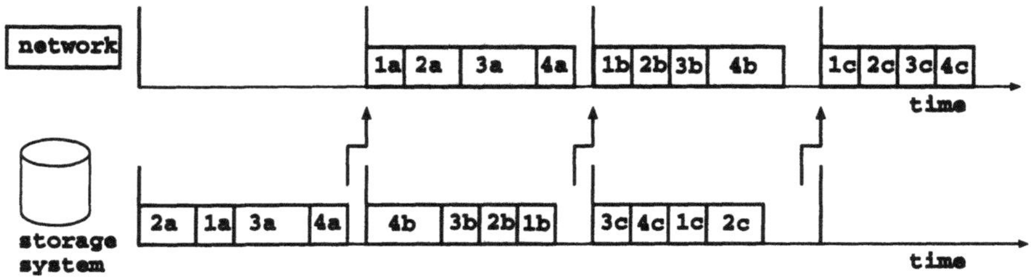

Fig. 4. Pipelining effect between storage retrieval and the network transmission

Without loss of generality, we assume the network sub-system can transmit a fixed amount of data in every period T. The data retrieved from the storage sub-system will be "packetized" (e.g. format into the network transmission unit) and transmitted across the network. Figure 4 illustrates the *pipelining effect* of the data retrieval from the storage sub-system to the network sub-system. Assume that we need to support four video streams, during the first transfer cycle, the storage sub-system retrieves one data block (e.g. 1a,2a,3a and 4a) for each of these four video streams. These data blocks will be transmitted across the network sub-system in the *next* network transmission period. It is important to note that the ordering of data packet transmission can be different from the ordering of data retrieval from the storage sub-system. This implies that the storage sub-system and the network sub-system can operate *independently* on selecting which data to retrieve or to transmit (or even to drop packet, if possible).

The QoS requirements we consider are bandwidth requirements and packet dropping rate requirements which are specified by users.

Let $\tilde{r_i}$ be the random variable denoting the bandwidth requirement of the ith video request and $r_i = E[\tilde{r_i}]$ be its average bandwidth requirement. One way to guarantee the quality of service for the ith request is to make sure that both the storage sub-system *and* the network sub-system can sustain $\tilde{r_i}$, that is, the storage sub-system can retrieve $\tilde{r_i}T$ amount of data per cycle and the network sub-system can transmit $\tilde{r_i}T$ amount of data per period. To achieve this, one way is to provide *worst-case* resource allocation, that is, let $r_i^* = \max\{\tilde{r_i}\}$, then we need to allocate enough disk I/O bandwidth to retrieve r_i^*T amount of data per cycle and allocate enough transmission bandwidth to transmit r_i^*T amount of data per transmission time frame. This type of *pessimistic* admission control usually implies that resources of the VoD system are usually under-utilized. The goal of our admission control algorithm is to guarantee QoS requirements of each individual class connection and at the same time, maximize the number of concurrent video requests that the system can support.

In the following sections, we will present our admission control algorithm. Let us first define some notations which will be used throughout this paper.

n	=	number of request currently admitted in the VoD.
N	=	number of disks in the storage sub-system.
T	=	the length of a time frame in the transmission network. This is also the length of a transfer cycle in each disk.
T_i	=	the random variable denoting the non-idling time in a transfer cycle of disk i, $i = 1, 2, \ldots, N$.
d_i	=	the packet dropping rate requirement specified by user i.
r_i	=	average bandwidth requirement for request i.
p_{ij}	=	the average probability of a block of video required by viewer j is in disk i.
p	=	the maximum overflow probability in disk transfer cycle. In order to make higher utilization of disk transfer bandwidth, we allow the disk transfer cycle to overflow with probability less than or equal to this probability.
$\tau_{seek}^{max}(n)$	=	a random variable denoting the worst case total seek time for serving n requests.
$\tau_{rot-j}^{max}(n)$	=	a random variable denoting the rotational latency for serving the jth request.

3 The Admission Control for Storage Sub-system

For the n requests currently admitted, the bandwidth requirements are $\mathbf{r} = (r_1, \ldots, r_n)$, the dropping rate requirements are $\mathbf{d} = (d_1, \ldots, d_n)$. The average number of videos requesting data from disk i is

$$n_i = \lceil p_{i1} + \cdots + p_{in} \rceil.$$

Therefore, for disk $i \in \{1, 2, \ldots, N\}$, we have:

$$T_i = \tau_{seek}^{max}(n_i) + \sum_{j=1}^{n_i} \tau_{rot-j} + \sum_{j=1}^{n_i} \tau_{trf-j}$$

where $\tau_{seek}^{max}(n_i)$ representing the worst case seek time for service n_i requests, i.e., when these n_i requests are evenly spaced out on the disk surface [7].

Let there be a new request (*e.g.*, $the(n+1)^{th} request$) arriving with bandwidth requirement r_{n+1} and dropping rate requirement d_{n+1}. Assume that the data block requested by the $(n+1)^{th}$ request is in disk i with probability $p_{i(n+1)}, i = 1, \ldots, N$. If it is admitted, T_i becomes:

$$T_i = \tau_{seek}^{max}(n_i') + \sum_{j=1}^{n_i'} \tau_{rot-j} + \sum_{j=1}^{n_i'} \tau_{trf-j}$$

where $n_i' = \lceil p_{i1} + \cdots + p_{i(n+1)} \rceil$. Let $F_{D_i}^*(s)$ be the Laplace transform for the random variable T_i and let $F_{rot-j}^*(s)$ and $F_{tfr-j}^*(s)$ be the Laplace transforms for the random variables τ_{rot-j} and τ_{trf-j}, respectively. Since $\tau_{rot-j}, \tau_{trf-j}, j = 1, \ldots, n$ are independent, using the convolution property of Laplace transform, we get:

$$F_{D_i}^*(s) = e^{[-s\tau_{seek}^{max}(n_i')]} \prod_{j=1}^{n_i'} F_{rot-j}^*(s) \prod_{j=1}^{n_i'} F_{trf-j}^*(s)$$

Let $M_i(s)$ be the moment generating function for the random variable T_i. Since $M_i(s)$ is equal to $F_{D_i}^*(-s)$, applying Chernoff's theorem to bound the tail of the random variable T_i, we have the following [1]:

$$Prob[T_i > T] \leq \inf_{\theta \geq 0} \{ \frac{M_i(\theta)}{e^{\theta T}} \}. \qquad (1)$$

Using standard numerical solution techniques, we can obtain the optimal θ^* which gives the tightest upper bound. If there exists a T_i such that $\frac{M_i(\theta^*)}{e^{\theta^* T}} > p$, the disk transfer cannot satisfy the bandwidth requirement of request $(n+1)$, since it means that at least one disk transfer cycle will violate the overflow probability requirement if $(n+1)^{th}$ request is admitted. On the other hand, if all $\frac{M_i(\theta^*)}{e^{\theta^* T}} \leq p$, it means the disk transfer can satisfy the bandwidth requirements of all $n+1$ requests.

If the bandwidth requirements of users' can be satisfied, then we need check whether the packet dropping rate requirements of the users' can be satisfied or not. We explain how one can check these requirements in the next section.

4 The Admission Control for Network Sub-system

Before we present the admission control algorithm for network sub-system, we need give some preliminaries which we use in deriving the the admission control algorithm for network sub-system.

4.1 Preliminaries

Let $E = \{1, \ldots, n\}$ be a finite set; $\mathbf{x} = (x_i)_{i=1}^{n}$ is a n-dimension vector. We first need to define the meaning of polymatroid [13]:

Definition 1 *The following polytope*

$$\mathcal{P}(f) = \{x \geq 0 : \sum_{i \in A} x_i \leq f(A), A \subseteq E\} \tag{2}$$

is termed a polymatroid if the function $f : 2^E \rightarrow \Re_+$ *satisfies the following properties: (i) (normalized)* $f(\emptyset) = 0$*; (ii) (increasing) if* $A \subseteq B \subseteq E$*, then* $f(A) \leq f(B)$*; (iii) (submodular) if* $A, B \subseteq E$*, then* $f(A) + f(B) \geq f(A \cup B) + f(A \cap B)$*. Specially, if*

$$\mathcal{B}(f) = \{x \geq 0 : \sum_{i \in A} x_i \leq f(A), A \subset E; \sum_{i \in E} x_i = f(E)\},$$

then $\mathcal{B}(f)$ *is the base of polymatroid* $\mathcal{P}(f)$.

The following definition defines the vertex of the base of polymatroid [14].

Definition 2 *Let* π *denote a permutation of* $\{1, 2, \ldots, n\}$ *and* π_j *is the jth component of* π*,* x^π *defined below are "vertices" of the base of the polymatroid defined in Definition 1.*

$$\begin{aligned}
x^\pi_{\pi_1} &= f(\{\pi_1\}) \\
x^\pi_{\pi_2} &= f(\{\pi_1, \pi_2\}) - f(\{\pi_1\}) \\
&\vdots \\
x^\pi_{\pi_n} &= f(\{\pi_1, \pi_2, \ldots, \pi_n\}) - f(\{\pi_1, \pi_2, \ldots, \pi_{n-1}\})
\end{aligned}$$

Definition 3 *[15] A scheduling control policy is said admissible if it is non-idling and non-anticipative, that is, if no server is allowed to be idle when there are jobs waiting to be served, and the control is only allowed to make use of past history and the current state of the system. Any admissible control cannot affect the arrival processes or the service requirements of the job.*

Next, we will give the definition of *strong conservation laws* which was reported in [15]. Let $E = \{1, 2, \dots, n\}$ denote the set of all job types. For any $A \subseteq E$, let $|A|$ denote the cardinality of A. Let $\mathcal{U}$ denote the set of all admissible policies. Let $\mathbf{x}^u$ denote the performance measure under an admissible policy $u \in \mathcal{U}$. Let $\pi = (\pi_1, \dots, \pi_n)$ denote a permutation of the integers $\{1, \dots, n\}$, which represents an admissible priority rule, that is, type π_1 jobs have the highest priority, and type π_n jobs have the lowest priority.

Definition 4 *[15] The performance vector x is said to satisfy strong conservation laws, if there exists a set function b (or respectively, f): $2^E \to \Re_+$, satisfying*

$$b(A) = \sum_{\pi_i \in A} x_{\pi_i}, \ \forall \pi : \{\pi_1, \dots, \pi_{|A|}\} = A, \ \forall A \subseteq E; \tag{3}$$

or respectively,

$$f(A) = \sum_{\pi_i \in A} x_{\pi_i}, \ \forall \pi : \{\pi_1, \dots, \pi_{|A|}\} = A, \ \forall A \subseteq E; \tag{4}$$

(when $A = \emptyset$, by definition, $b(\emptyset) = f(\emptyset) = 0$); such that for all $u \in \mathcal{U}$ the following is satisfied:

$$\sum_{i \in A} x_i^u \geq b(A), \ \forall A \subset E; \ \sum_{i \in E} x_i^u = b(E); \tag{5}$$

or respectively,

$$\sum_{i \in A} x_i^u \leq f(A), \ \forall A \subset E; \ \sum_{i \in E} x_i^u = f(E). \tag{6}$$

If the performance measure in a particular question is minimized (or maximized) by the admissible priority rules, then the function b (or f) applies in this question [16].

This definition states two requirements that a performance vector must satisfy in order to satisfy strong conservation laws [15]:

1. The summation of all components of the performance vector in question is invariant under any admissible control policy. This requirement is reflected in following equations

$$\sum_{i \in E} x_i^u = b(E);$$

or

$$\sum_{i \in E} x_i^u = f(E).$$

2. The summation of components of the performance vector in question who represents job types in A is minimized (or maximized) by any absolute priority rule giving the job types in A over the other job types. This requirement is reflected in following equations and inequalities:

$$b(A) = \sum_{\pi_i \in A} x_{\pi_i}, \ \forall \pi : \{\pi_1, \ldots, \pi_{|A|}\} = A, \ \forall A \subseteq E;$$

$$\sum_{i \in A} x_i^u \geq b(A), \ \forall A \subset E;$$

or

$$f(A) = \sum_{\pi_i \in A} x_{\pi_i}, \ \forall \pi : \{\pi_1, \ldots, \pi_{|A|}\} = A, \ \forall A \subseteq E;$$

$$\sum_{i \in A} x_i^u \leq f(A), \ \forall A \subset E;$$

The following theorem gives the relationship between the strong conservation laws and the base of a polymatroid. Unless otherwise stated, we use $\mathcal{B}(b)$ to denote the polytope $\{\mathbf{x} \geq 0 : \sum_{i \in A} x_i \geq b(A), A \subset E; \sum_{i \in E} x_i = b(E)\}$, which is also the base of polymatroid by setting $b(A) := b(E) - f(E - A)$.

Theorem 1 *[15] Assume the performance vector x satisfies the strong conservation laws (3) and (5) [(4) and (6)], then:*

a) the convex polytope $\mathcal{B}(b)$ [$\mathcal{B}(f)$] is the performance space;

b) $\mathcal{B}(b)$ [$\mathcal{B}(f)$] is the base of a polymatroid; and

c) the vertices of $\mathcal{B}(b)$ [$\mathcal{B}(f)$] are the performance vectors of the absolute priority rules.

Since $\mathcal{B}(b)$ [$\mathcal{B}(f)$] is a convex polytope, any vector in $\mathcal{B}(b)$ [$\mathcal{B}(f)$] can be expressed as a convex combination of its vertices. This implies that if a performance vector satisfies the strong conservation laws, we can easily derive the space of this performance vector. Also, given a vector, we can easily find out whether there exists an admissible priority rule under which the performance measure can achieve this vector. In the following, we illustrate how we can use the concept of polymatroid to determine whether we can satisfy a given dropping vector requirement.

4.2 The Admission Control Algorithm for Network Sub-system

In a transmission time frame, we use an admissible policy defined in last subsection because allowing the server to be idle when there are packets to be transmitted in current time frame will not benefit the system. Also, in practice, if we use a non-work-conserving scheduling policy, we have to consider carefully the stability condition of the system. Here we regard packets which should be transmitted in current time frame but cannot be transmitted as violating the deadline and must be dropped. Since the probability of overflow in a disk transfer cycle is very small (e.g., this is upper bounded by the parameter p), so we ignore the bits dropped in disk transfer cycle. We assume that the distribution of the size of video stream which need to be transmitted in a transmission time frame is the same as the disk read-size in the corresponding disk transfer cycle.

Assume the packet size is same for all the video streams (this is the case in ATM networks where each packet(cell) has the fixed length), it is easily seen that:

1. The overall dropping rate, over all video streams in E is invariant under any admissible policy;
2. The dropping rate over any given subset $A, A \subset E$ is minimized by offering absolute priority to video streams in the subset A over the video streams in $(E - A)$.

Base on this observation, we know that the dropping rate vector $\mathbf{d} = (d_1, \ldots, d_{n+1})$ satisfies the strong conservation laws.

From the preliminaries, we know the space of dropping rate $\mathbf{d}$ is a base of polymatroid with each vertex corresponding to a specific absolute priority scheduling policy. Since this base of polymatroid is a convex polytope, any point of this base of polymatroid can be expressed as a convex combination of its vertices. So if the required dropping rate vector is in this polytope, we know we can find a convex combination of absolute priority rules to achieve it.

Let $b(A)$ denote the lower bound of dropping rate over given subset A, $A \subset E$. If the dropping rate vector $\mathbf{d} = (d_1, \ldots, d_{n+1})$ satisfies[2]:

$$\sum_{i \in A} d_i \geq b(A),\ A \subset E$$

and

$$\sum_{i \in E} d_i > b(E)$$

we can always find a point in this base of polymatroid (i.e. the space of $\mathbf{d}$) which is better than given $\mathbf{d}$ and can be realized by a convex combination of absolute priority rules.

5 Experiment

In this section, we present some experimental results obtained by using our proposed admission control algorithm. We also compare these results with the results

[2] In here, $E = \{1, \ldots, n+1\}$

obtained by using average bandwidth allocation strategy. Since using peak bandwidth allocation (or worst case resource allocation) will result in worst bandwidth utilization than that of using average bandwidth allocation, therefore, we will not use peak bandwidth allocation policy in comparing with our proposed method.

In our experiment, there are two disks in the VoD storage system. The related characteristics of each disk are listed in Table 1.

Table 1. The parameters of the disk used in the experiment

Number of cylinders	5288
Transfer rate	80 Mbps
Maximum rotational latency	8.33 milliseconds
Seek time function (secs)	$seek(d) = \begin{cases} 0.6 * 10^{-3} + 0.3 * 10^{-3} * \sqrt{d} & \text{if } d < 400 \\ 5.75 * 10^{-3} + 0.002 * 10^{-3} * d & \text{if } d \geq 400 \end{cases}$

In Table 1, d is the seek distance for a requested video block in a disk transfer cycle. We use the worst case seek distance as the value of d, which is the seek distance for a requested video block when all the requested video blocks in a disk are evenly spaced out on the surface of this disk. So

$$d = \lceil \frac{\text{Number of cylinders in the disk}}{\text{Number of requested video blocks in the disk}} \rceil$$

and $seek(d)$ is the seek time for finding a video block in a disk. Also, we assume the rotational latency for the requested video block is uniformly distributed in the range $[0, 8.33]$ millisecond. The compressed video files are stored in the disks using the data placement scheme we talked about in Sect. 2. The size of one data block is exponentially distributed with mean 1.5 Mb for each MPEG-1 video block and 5 Mb for each MPEG-2 video block.

The VoD system can provide service in three classes. Class $\mathcal{A}$ service is provided to users whose requested videos are compressed by MPEG-1 and whose average bandwidth requirements are 1.5 Mbps and dropping rate requirements are 2% (that is, the average number of bits dropped in a transmission time frame should be less than $1.5 * 2\% = 0.03$ Mb). Class $\mathcal{B}$ service is provided to users whose requested videos are compressed by MPEG-1 and whose average bandwidth requirements are 1.5 Mbps and dropping rate requirements are 6%. Class $\mathcal{C}$ service is provided to users whose requested videos are compressed by MPEG-2 and whose average bandwidth requirements are 5 Mbps and dropping rate requirements are 10% [17]. We call a user's request "class i request" if the user will get class i service when his request is admitted, $i \in E = \{\mathcal{A}, \mathcal{B}, \mathcal{C}\}$. The length of a transfer cycle in the disk is 1 second. During each transfer cycle, the read-size of video is exponentially distributed with mean 1.5 Mb for each MPEG-1 video and 5 Mb for each MPEG-2 video. The given overflow probability of disk transfer cycle is $1.0 * 10^{-4}$. The transmission network we employed is ATM network with bandwidth 155 Mbps. The length of a transmission time frame is 1 second.

From the previous sections, we know that a new request (or new requests)can be admitted if and only if the system has enough resource to satisfy the QoS requirements of the new request(s) and at the same time, the QoS requirements of other existing viewers will not be violated.

We assume that at the beginning of the experiment, there are 51 video streams in the VoD system. Each class has 17 video streams (one can easily verify that the VoD system can support these video streams, both in storage sub-system and the network sub-system). Now there are one class $\mathcal{A}$ request, two class $\mathcal{B}$ requests and one class $\mathcal{C}$ request arrive at the same time. To see if these new requests can be admitted, first we need check whether the bandwidth requirements of the users' can be satisfied. Assuming the admitted requests are well-balanced, if the new requests are admitted in storage sub-system, applying the formula in Table 1, the worst case seek time for a video is 0.0047 secs, then applying (1), we got

$$P[T \geq 1] \leq 6.6107 * 10^{-5}$$

so the probability of overflow of the disk transfer cycle is less that given $p = 1.0 * 10^{-4}$. The bandwidth requirements of all users' can be satisfied.

Next, we need check whether the dropping rate requirements of all users' can be satisfied. We treat the video streams belonging to the same service class as one "big" video stream and compute the lower bound of dropping rate $b(A)$, where $A \subseteq E, E = \{\mathcal{A}, \mathcal{B}, \mathcal{C}\}$ represents the set of three "big" video streams. The results are listed in Table 2, also listed are dropping rate requirements $d(A), A \subseteq E$. By comparing the dropping rate requirements $d(A)$ with $b(A)$, we got:

$$d(A) \geq b(A) \quad \text{where } A \subseteq E,\ E = \{\mathcal{A}, \mathcal{B}, \mathcal{C}\}$$

So these new requests can be admitted. The admissible region includes all the points whose coordinates (x_1, x_2, x_3) satisfy:

$$\begin{aligned}
&x_i \geq 0, \qquad i = 1,2,3 \\
&x_1 \geq b(\mathcal{A}) \\
&x_2 \geq b(\mathcal{B}) \\
&x_3 \geq b(\mathcal{C}) \\
&x_1 + x_2 \geq b(\mathcal{AB}) \\
&x_1 + x_3 \geq b(\mathcal{AC}) \\
&x_2 + x_3 \geq b(\mathcal{BC}) \\
&x_1 + x_2 + x_3 \geq b(\mathcal{ABC})
\end{aligned}$$

The admissible region is illustrate in Fig. 5 where $HIJKO$ is a base of polymatroid, $H = (4.6783, 0.6895, 0.0442), I = (4.6783, 0, 0.7337)$, $J = (0.5996, 4.7682, 0.0442), K = (0, 4.7682, 0.6438), O = (0, 0, 5.4121)$.

From the previous section, we know we can find an admissible scheduling policy to achieve each point in the base of polymatroid $HIJKO$. For each of other points in admissible region, we can find an admissible scheduling policy to achieve a "better" point than the point itself.

Because $d(E) = 10.764 > 5.4121 = b(E)$, we know we can find a point which is in the base of polymatroid $HIJKO$ and which is better than the given dropping rate vector. Here, the given dropping rate vector is $(0.54, 1.71, 9)$ (see Table 2), by deducting 1 from d_2, deducting 4.8379 from d_3, we got the "better" point $(0.54, 0.71, 4.1621)$, where $0.54 + 0.71 + 4.1621 = 5.4121 = b(E)$. After solve the

Table 2. $d(A)$ and $b(A)$ in the experiment

Item	Value	Item	Value
$d\{\mathcal{A}\}$	0.54	$b\{\mathcal{A}\}$	1.2397e-25
$d\{\mathcal{B}\}$	1.71	$b\{\mathcal{B}\}$	7.2971e-25
$d\{\mathcal{C}\}$	9	$b\{\mathcal{C}\}$	0.0442
$d\{\mathcal{AB}\}$	2.25	$b\{\mathcal{AB}\}$	3.7213e-14
$d\{\mathcal{AC}\}$	9.54	$b\{\mathcal{AC}\}$	0.6438
$d\{\mathcal{BC}\}$	10.71	$b\{\mathcal{BC}\}$	0.7337
$d\{\mathcal{ABC}\}$	10.764	$b\{\mathcal{ABC}\}$	5.4121

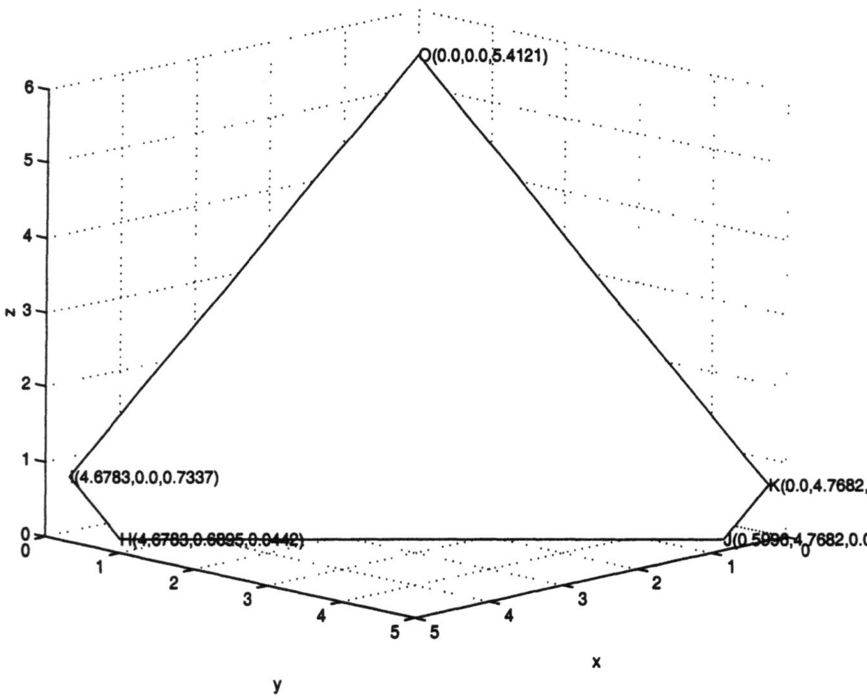

Fig. 5. The base of polymatroid $HIJKO$

following linear equation systems,

$$a_1 \begin{pmatrix} 4.6783 \\ 0.6895 \\ 0.0442 \end{pmatrix} + a_2 \begin{pmatrix} 4.6783 \\ 0.0 \\ 0.7337 \end{pmatrix} + a_3 \begin{pmatrix} 0.5996 \\ 4.7682 \\ 0.0442 \end{pmatrix} +$$

$$+ a_4 \begin{pmatrix} 0.0 \\ 4.7682 \\ 0.6438 \end{pmatrix} + a_5 \begin{pmatrix} 0.0 \\ 0.0 \\ 5.4121 \end{pmatrix} = \begin{pmatrix} 0.54 \\ 0.71 \\ 4.1621 \end{pmatrix}$$

$$\sum_{i=1}^{5} a_i = 1$$

$$a_i \geq 0, \ i = 1, 2, \ldots, 5$$

we got the scheduling policy $(0, 0.0963, 0.1489, 0, 0.7548, 0)$, that is, at the beginning of a time frame, with probability 0.0963, the scheduling policy is (2,3,1); with probability 0.1489, the scheduling policy is (3,1,2); with probability 0.7548, the scheduling policy is (2,1,3). Within each class, the scheduling policy is allocating the bandwidth proportional to the number of bits each video stream has at the beginning of the time frame. We can keep adding requests into the VOD system. Based on our experiment, we obtain the result that, using our proposed scheduling policy, the VOD system can support 20 video streams for class $\mathcal{A}, \mathcal{B}$ and $\mathcal{C}$, respectively.

Now we compare the results obtained by using our proposed algorithm with the result obtained by average bandwidth allocation strategy. Here "average bandwidth allocation strategy" refers to the strategy that allocates transmission bandwidth to the video streams in network sub-system according to the average bandwidth requirements of the users'. This strategy also considers the VBR nature of the video streams, so it can achieve higher bandwidth utilization than peak bandwidth allocation strategy. In our experiment, employing average bandwidth allocation strategy means allocating 1.5 Mbps bandwidth to each video streams in class $\mathcal{A}$ and class $\mathcal{B}$ and 5 Mbps bandwidth to each video streams in class $\mathcal{C}$. If there are 20 video streams for class $\mathcal{A}, \mathcal{B}$ and $\mathcal{C}$, then the overall bandwidth requirement is 160 Mbps which exceeds the bandwidth of the transmission network which is 155 Mbps. So by employing average bandwidth allocation strategy, we can't admitted 20 video streams for class $\mathcal{A}, \mathcal{B}, \mathcal{C}$, respectively. And we cannot determine whether the dropping rate requirement of each user's can be satisfied. This is in contrast with our proposed algorithm which can accommodate 20 video streams for class $\mathcal{A}, \mathcal{B}$ and $\mathcal{C}$, respectively, and at the same time, we can find the scheduling policy to achieve the dropping rate requirements easily.

6 Conclusion

In this work, we considered a theoretical framework of performing admission control in a VoD system. Previous work on the admission control in VoD is usually performed on a aggregated traffic basis. Our admission control algorithm can guarantee the QoS requirements of individual classes of connections and at the same time, achieve high bandwidth utilization. In the storage sub-system, some form of cycle-based scheduling algorithm is used and we determine the conditions in which the storage sub-system can satisfy the bandwidth requirements of the users'. For the network sub-system, we derive the admissible region of the packet dropping rate vector and the scheduling policy to achieve the dropping rate vector within the admissible region. Experiments show that our proposed algorithm can achieve high bandwidth utilization, making VoD service cost-effectively.

References

1. L. Golubchik, John C.S. Lui, Edmundo de Souza E. Silva, Richard Gail (1999). Evaluation of Tradeoffs in Resource Management Techniques for Multimedia Storage Servers. IEEE International Conference on Multimedia System, pp.292-296.
2. M. Krunz, S.K. Tripathi, (1997) Impact of Video Scheduling on Bandwidth Allocation for Multiplexed MPEG Streams. Multimedia Systems, **5**, 347-357
3. T.J. To and B. Hamidzadeh, (1999) Dynamic Real-time Scheduling strategies for Interactive Continuous Media Servers. Multimedia Systems, **7**, 91-106
4. K. Lee, H.Y. Yeom, (1999) An Effective Admission Control Mechanism for Variable-bit-rate Video Streams. Multimedia Systems, **7**, 305-311
5. X. Jiang, P. Mohapatra, (1999) Efficient Admission Control Algorithms for Multimedia Servers. Multimedia Systems,**7**, 294-304
6. M.J. Capone, I. Stavrakakis, (1996) Achievable QoS and Scheduling Policies for Integrated Services Wireless Networks. Performance Evaluation. **27 & 28**, 347-365
7. S. Ghandeharizadeh, R.R. Muntz, (1998) Design and Implementation of Scalable Continuous Media Servers. Special Issue of Parallel Computing Journal on Parallel Data Servers and Applications.
8. D.J. Gemmell, H.M. Vin, D.D. Kandlur et al. (1995) Multimedia Storage Servers: A Tutorial. IEEE Computer. **28**, 40-49
9. M. Chen, D. Kandlur and P. Yu, (1993) Optimization of the Grouped Sweeping Scheduling with Heterogeneous Multimedia Streams. ACM Multimedia. 235-242
10. F. A. Tobagi, J. Pang, R. Baird et al. (1993) Streaming RAID - A Disk Array Management System For Video Files. ACM Multimedia Conference. 393-399
11. P. S. Yu, M.-S. Chen, D. D. Kandlur, (1992) Design and Analysis of a Grouped Sweeping Scheme for Multimedia Storage Management. Third International Workshop on Network and Operating System Support for Digital Audio and Video. 44-55
12. S. Berson, L. Golubchik, R. R. Muntz, (1995) Fault Tolerant Design of Multimedia Servers. Proc. of the ACM SIGMOD Conf. on Management of Data. May, 364-375

13. D. Welsh, (1976) Matroid Theory. Academic Press, London
14. J. Edmonds, (1970) Submodular Functions, Matroids and Certain Polyhedra. Prof. Int. Conf. on Combinatorics.
15. J.G. Shanthikumar, D.D. Yao, (1992) Multiclass Queueing Systems: Polymatroidal Structure and Optimal Scheduling Control. Operations Research, **40, Suppl. No. 2**, S293-299
16. D.D. Yao, L. Zhang, (1997) Stochastic Scheduling via Polymatroid Optimization. Lectures in Applied Mathematics. **33**, 333-364
17. D. Wijesekera, J. Srivastava, A. Nerode et al. (1999) Experimental Evaluation of Loss Perception in Continuous Media. Multimedia Systems. **7**, 486-499

Part VII

Invited Tutorial

Stochastic Petri Nets and Their Applications *

Kishor S. Trivedi[1], Hairong Sun[2], Yonghuan Cao[1], and Yue Ma[2]

[1] Center for Advanced Computing and Communications
Department of Electrical and Computer Engineering
Duke University, Durham, NC 27708
Email: {kst, ycao}@ee.duke.edu
[2] Corporate Software Technology Center
Motorola, Inc. Elk Grove Village, IL 60007

Abstract. Although continuous-time Markov chains have been widely used to analyze the performance of communication networks, constructing and solving a continuous-time Markov chain is a tedious and error-prone procedure, especially when the systems are complex. A relief from the burden is provided by stochastic Petri nets and the corresponding software packages, which provide automated generation and solution of continuous-time Markov chains. This paper gives an overview of stochastic Petri nets. Two examples in communication networks (i.e., one for ATM networks and the other for wireless networks) are presented and studied to illustrate how to use stochastic Petri nets for performance, availability and performability analysis of communication networks.

1 Introduction

First introduced by A. A. Markov in 1907, Markov chains have been in use in performance analysis since around 1950. A Markov chain consists of a set of states and a set of labeled transitions between the states. A state of the Markov chain can model various conditions of interest in the system being studied. For example, these could be the number of packets in the buffer, the number of active users in the network, etc. After a sojourn in a state, the Markov chain will make a transition to another state. Such transitions are described either with *transition probability matrix* $\mathbf{P}$ (in case of discrete-time Markov chain, DTMC) or *infinitesimal generator matrix* $\mathbf{Q}$ (in case of continuous-time Markov chain, CTMC). Steady-state behavior of Markov chains, if it exists, can be studied by using a system of linear equations with one equation for each state, i.e., $\boldsymbol{v}\boldsymbol{P} = \boldsymbol{v}$ and $\boldsymbol{v}\mathbf{1} = 1$ for DTMC or

* This research was supported in part by the National Science Foundation under Grant No. EEC9418765 and by the Air Force Office of Scientific Research under MURI Grant No. F49620-00-1-0327.

$\boldsymbol{\pi Q} = 0$ and $\boldsymbol{\pi 1} = 1$ for CTMC, where $\boldsymbol{v}$ and $\boldsymbol{\pi}$ are the steady-state probability vectors for DTMC and CTMC, respectively. Transient behavior of a CTMC gives rise to a system of first-order, linear, ordinary differential equations[19,14], *i.e.*, $\frac{d\boldsymbol{\pi}(t)}{dt} = \boldsymbol{\pi Q}$.

The most commonly used method for the performance evaluation of communication networks is abstracting the physical system into a CTMC or DTMC, and then setting up ordinary differential equations (for transient solution) or linear equations (for steady-state solution) manually, and finally writing a program for the numerical solution to the above equations. It is a rather tedious and error-prone procedure, especially when the number of states becomes very large. There are some efforts on developing software packages to solve CTMC or DTMC automatically, while still requiring manual construction of the CTMC. Note that the Markov model of a system is sometime far removed from the shape and general feel of the system being modeled. System designer may have difficulty in directly translating their problem into a Markov chain. Since late 1980s, some scientists have been developed a new modelling formalism and software packages for the automated generation and solution of Markovian stochastic systems. The efforts led to the emergence of a formalism called stochastic Petri nets (SPN) which is more concise in its specification and whose form is closer to a designer's intuition about what a model should look like. Some software packages such as SPNP[11,13], DSPNexpress[5,6], TimeNet[26], and GreatSPN[8], are available, which can translate an SPN model into a CTMC and then solve it automatically. The users' effort is now just in building an SPN model from the real system.

In this paper, we will show how to build an SPN model to study the performance issue in communication networks. In section 2, the concept of SPN, GSPN and SRN are described. As an example, we construct a stochastic Petri net model for an Ethernet/ATM bridge in section 3. In section 4, a performability model for the RF channel recovery in wireless communication systems is built. In section 5, we briefly discuss the recent advances in SPN. The conclusions are drawn in section 6.

2 Stochastic Petri Nets

2.1 Petri Nets

Petri nets were originally introduced by C. A. Petri in 1962. Formally, a Petri net (PN) is a 5-tuple $PN = (P, T, F, W, M)$, where

- $P = \{p_1, p_2, ..., p_m\}$ is a finite set of *places* (drawn as circles);
- $T = \{t_1, t_2, ..., t_n\}$ is a finite set of *transitions* (drawn as bars);
- $F \subseteq (P \times T) \cup (T \times P)$ is a set of arcs connecting P and T;
- $W : F \rightarrow \{1, 2, 3, ...\}$ is weight (multiplicity) function;
- $M : P \rightarrow \{0, 1, 2, ...\}$ is the *marking* which denote the number of tokens (drawn as black dots) in places P and the initial marking is denoted as M_0.

Graphically, Petri net is a directed graph with two disjoint types of nodes: *places* and *transitions*. A directed arc connecting a place (transition) to a transition (place) is called an input (output) *arc* of the transition. A positive integer called multiplicity can be associated with each arc. Places connected to a transition by input arcs are called the input places of this transition, and those connected by means of output arcs are called the output places. Each place may contain zero or more tokens. A transition is *enabled* if each of its input places has at least as many tokens as the multiplicity of the corresponding input arc. A transition can *fire* when it is enabled, and upon firing, a number of tokens equal to the multiplicity of the input arc is removed from each of the input places, and a number of tokens equal to the multiplicity of the output arc is deposited in each of its output places. Therefore, the firing of a transition may transform a PN from one marking into another. With respect to a given initial marking M_0, the *reachability set* is defined as the set of all markings reachable through any possible firing sequences of transitions, starting from the initial marking [14,30].

Fig. 1 is a simple example of a PN [14]. In Fig. 1 a, only transition t_1 is enabled because place P_1 contains two tokens, t_2 is disabled because P_2 is empty. If the firing takes place, one token is removed from input place P_1 and one token is deposited in both output places P_2 and P_3 (see Fig. 1 b). In Fig. 1 b, both transitions t_1 and t_2 are enabled. If t_1 fires, the PN will reaches Fig. 1 d, while Fig. 1 c will be reached if t_2 fires. The reachability set in this example is given by $\{(2, 0, 0, 0), (1, 1, 1, 0), (0, 2, 2, 0), (0, 0, 1, 1)\}$.

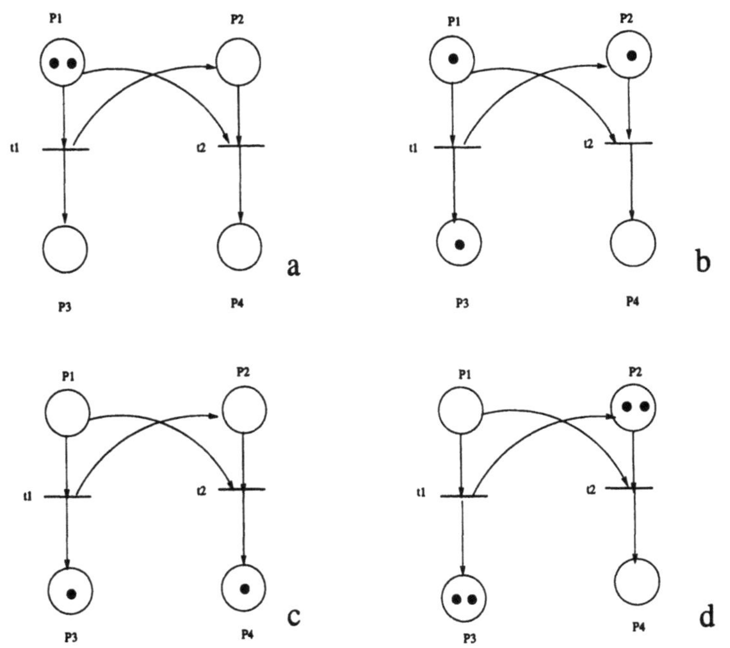

Fig. 1. An Example of Petri Net

PN can be used to capture the behavior of many real-world situations including sequencing, synchronization, concurrency, and conflict. In computer networks, it can be used to describe and verify the communication protocols. However, the concept of time is not explicitly given in the original definition of Petri nets, while for the performance evaluation of dynamical systems, it is necessary and useful to introduce time delays associated with transitions in the Petri net models. This intuition has led to the emergence of stochastic Petri nets.

2.2 Stochastic Petri Nets

Stochastic Petri nets are Petri nets where exponentially distributed firing time is attached to each transition. In Generalized Stochastic Petri nets (GSPN)[22], transitions are allowed to be either *timed* (exponentially distributed firing time, drawn as rectangular boxes) or *immediate* (zero firing time, represented by thin black bars). Immediate transitions always have priority over timed transitions to fire. If several immediate transitions compete for firing, a specified probability mass function is used to break the tie.

A marking of a GSPN is called *vanishing* if at least one immediate transition is enabled in the marking and *tangible* otherwise. GSPN also introduces *inhibitor arc* connecting a place to a transition. Inhibitor arcs have small hollow circles instead of arrows at their terminating ends. A transition with an inhibitor arc can not fire if the input place of the inhibitor arc contains equal to or more tokens than the multiplicity of the arc.

It has been shown that a unique CTMC corresponds to a given GSPN under condition that only a finite number of transitions can fire in finite time with non-zero probability [22].

The GSPN analysis can be decomposed into four steps [14]:

- Generating the extended reachablility graph which contains the markings of the reachablity set as nodes and some stochastic information attached to the arcs, thereby all the markings are related to each other with stochastic information.
- Eliminating the vanishing markings with zero sojourn times and the corresponding transitions. This procedure generates the CTMC.
- Analyzing the steady-state, transient and cumulative behavior of the CTMC.
- Outputing the measures, such as the average number of tokens in each place and the throughput of each timed transition.

When SPN is applied to performance evaluation of computer networks, *places* can be used to denote the number of packets or cells in the buffer or the number of active users or flows in the system, while the arrival and departure of packets, cells, users or flows can be represented by *transitions*.

Fig. 2 shows a GSPN model of an $M/E_m/1/N$ queueing system. Transition T_1 represents the arrival of a customer with firing rate λ. An inhibitor arc with multiplicity N from place P_{on} to T_1 represents the capacity of the queueing system.The transition T_1 is disabled when the number of tokens in the system equals N. The immediate transition t_1 fires when there is at least one token in place P_{on} and P_{ser} is empty, and m tokens will be deposited in place P_{ser} after the firing of t_1.

Fig. 2 also shows the extended reachability graph of the GSPN model, where (i, j) represents that the number of tokens in places P_{on} and P_{ser} are i and j, respectively. After the extended reachablility graph is generated, the vanishing markings will be deleted, which is mapped to a CTMC. In this example, there are N vanishing markings which are represented by rectangles, i.e., (1,0), (2,0), ..., (N,0). Then the CTMC can be evaluated with well-known numerical solutions.

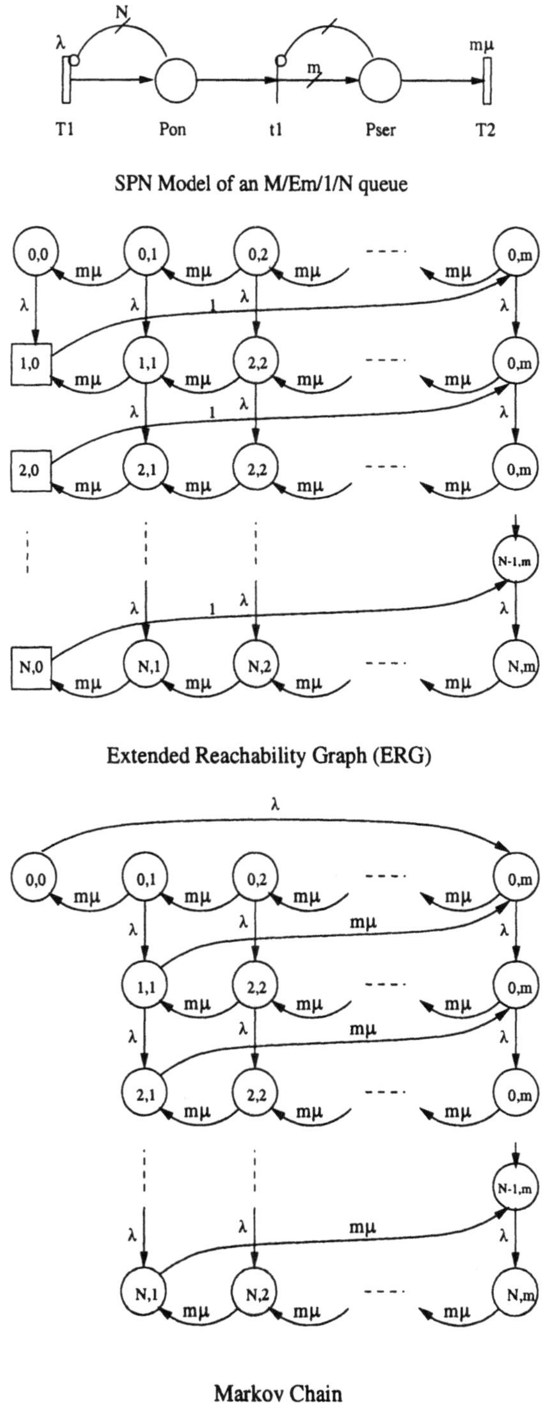

Fig. 2. An Example of GSPN

2.3 Stochastic reward nets

Stochastic reward nets (SRNs) are based on GSPN but extend it further [12]. In SRN, every tangible marking can be associated with a reward rate. It can be shown that an SRN can be mapped into a Markov reward model. Thus a variety of performance measures can be specified and calculated using a very convenient formalism. SRN also allows several other features that makes specification convenient:

- Each transition may have an enabling function (also called a guard) so that a transition is enabled only if its marking-dependent enabling function is true. This feature provides a powerful means to simplify the graphical representation and to make SRNs easier to be understood.
- Marking dependent arc multiplicities are allowed. This feature can be applied when the number of tokens to be transferred depends on the current marking.
- Marking dependent firing rates are allowed. This feature allows the firing rate of the transitions to be specified as a function of the number of tokens in any place of the Petri net.
- Transitions can be assigned different priorities, and a transition is enabled only if no other transition with a higher priority is enabled.
- Besides the traditional output measures obtained from a GSPN, such as throughput of a transition and the mean number of tokens in a place, more complex reward function can be defined.

We will show in the following sections how to use these features to build concise models.

3 Example 1: SRN Model for an Ethernet/ATM Bridge

3.1 Problem Description

Consider an Ethernet/ATM bridge, which has N Ethernet interfaces and one *ATM* interface. The packet arrivals from one Ehternet interface follow a 2-state Markov Modulated Poisson Process (MMPP). Assume that the traffic on the Ethernet interfaces are homogeneous, then the aggregation of the traffic may be described by an N-state MMPP. The aggregated packet arrival process is Poisson with rate r_i while it is at state i. The transitions between these states form a homogeneous, continuous time Markov chain. The rate from state i to $(i+1)$ is $i\delta$; to $(i-1)$ is $(N-i)\gamma$.

After the frames arrive into the buffer, they will be segmented into cells and each cell is serviced with a deterministic service time, i.e., 2.7 microseconds (we assume the port rate of the ATM interface is 155.520 Mb/s). From the previous measurements of real networks, it has been observed that the frame sizes of data traffic in Ethernet have three predominant values: 46 bytes (with probability c_1), 144 bytes (with probability c_2) and 1500 bytes (with probability c_3), which corresponds to 2 cells, 4 cells and 32 cells, respectively, after including the *LANE* (LAN Emulation) overhead (see [33,24]). It is found that $c_1 = 0.342$, $c_2 = 0.093$, $c_3 = 0.565$.

These three different sizes are the consequence of three different application classes. Short frames are transmitted during terminal-to-host communication, whereas applications based on network-file system protocol (NFS) generate short frames in one direction followed by medium-sized frames in the reverse direction. The maximum frame size in Ethernet traffic is 1512 bytes, used mostly during file transfer applications. In other words, the data frames in the LAN consist of large packets, medium packets and small packets. This observation allows us to analyze and design the buffer more effectively.

3.2 SRN Model

The traffic is modelled by an N-state MMPP, which is represented by the subnet in the dotted rectangle in Fig. 3. The firing rate of T_{on} depends on the number of tokens in place P_{on}, i.e., $(N - \#P_{on})\gamma$. The firing rate of T_{off} depends on the number of tokens in place P_{on} as well , i.e., $\#P_{on}\delta$. The firing rate of transition $T_{arrival}$ depends on the number of tokens in P_{on}, say $\lambda\#P_{on}$.

The frames generated according to the MMPP have three possible sizes, 2 cells (with probability c_1), 4 cells (with probability c_2), and 32 cells (with probability c_3). The split between the large, medium and small frame is represented by the immediate transitions t_1,t_2, and t_3. When a packet finds there is enough space in the buffer whose capacity is M cells, the packet enters the buffer, which is represented by immediate transitions t_7, t_8, and t_9. If the packet finds that there is not enough space in the buffer to accomodate the entire packet, the immediate transitions t_7, t_8, and t_9 are inhibited by inhibitor arcs from P_{buffer} to t_7, t_8, and t_9. Then the entire packet will be discarded, which is represented by immediate transitions t_4, t_5, and t_6. In this paper, discard is frame-based, not cell-based. Therefore, partial packet loss will not exist in our study, and loss amplification caused by partial packet loss is prevented [4]. The deterministic service time of the multiplexer, which can accomodate M cells, is approximated with an n-stage Erlang distribution.

The marking dependent firing rates for the timed transitions are listed in Table 1. The guards for the immediate transitions are listed in Table 2.

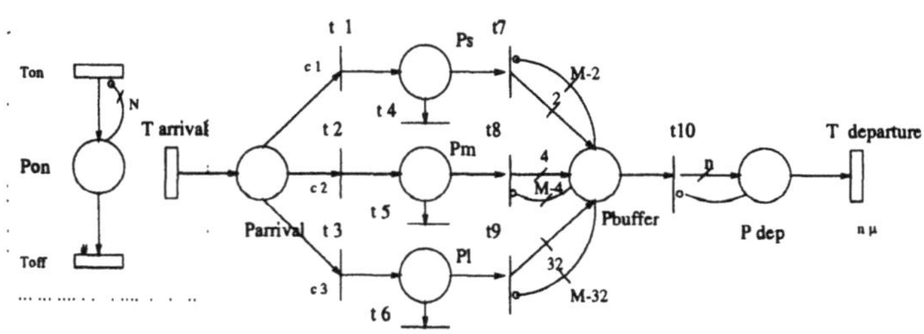

Fig. 3. Stochastic Reward Net Model for Ethernet/ATM Bridge

Table 1. Rates for the timed transitions in Fig. 3

Transition	Rate
T_{on}	$(N - \#P_{on})\gamma$
T_{off}	$\#P_{on}\delta$
$T_{arrival}$	$\#P_{on}\lambda$
$T_{departure}$	$n\mu$

Table 2. Rates and Guards for the immediate transitions in Fig. 3

Transition	Guard
t_4	$\#P_{buffer} + \#P_{dep} > M - 2$
t_5	$\#P_{buffer} + \#P_{dep} > M - 4$
t_6	$\#P_{buffer} + \#P_{dep} > M - 32$

4 Example 2: RF Recovery in Wireless Communication Systems[29]

4.1 Problem Description

In wireless networks, an RF (Radio Frequency) channel is assigned to a call either during the call set-up process when a new call is initiated or during the handoff process when an ongoing call subscriber roams into the cell.

A common assumption in the previous studies has been that the channel in use never fails. However, in a practical environment, wireless networks, like any other physical system, are subject to failures, especially RF failure. There are many factors that cause RF failure, such as base repeater power failure, base repeater RF amplifier failure, etc. With the increasing penetration of wireless communications, a disruption in service could cause severe consequences in both economic and social sense. Thus providing restoration subsequent to channel failures has become an important issue in ensuring network integrity.

4.2 Performance SRN Model (No RF Failure)

Before proposing our channel recovery scheme, we first present a pure performance model under the assumption that the channels in a wireless network never fail.

In cellular networks, a given geographical area is divided into a certain number of cells. When a new call (NC) is attempted in a cell covered by a base station (BS), the NC is connected if an idle channel is available in the cell. Otherwise, the call is blocked. When a mobile station (MS) travels across the cell boundaries, the channel in the old serving cell is released, and an idle channel is required in the target cell, which would be the new serving cell. This process is called *handoff*. If an idle channel exists in the target cell, the handoff call (HC) continues nearly transparently to the user. Otherwise, the HC is dropped.

The dropping of a handoff call (HC) is considered more severe than the blocking of a new call (NC). One method to reduce the dropping probability of HCs is to reserve a fixed number of channels exclusively for HCs. These exclusively reserved channels are referred to as *guard channels*. For example, if the total number of RF channels is C and the number of the channels in the reserved channel pool is g, then the number of RF channels available for new calls is $C - g$.

Fig. 4 shows an SRN of a performance model for channel allocation with guard channels.

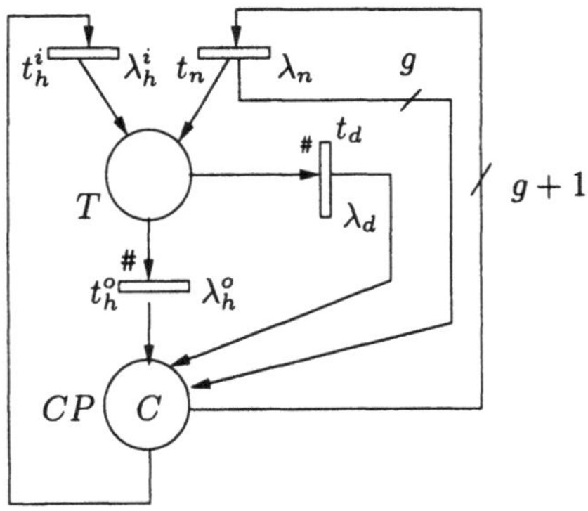

Fig. 4. SRN of a performance model for channel allocation.

In Fig. 4, place CP is the channel pool for the cell. Initially, there are C idle channels which are accessible for both the NCs and the HCs. Transitions t_n and t_h^i represent the arrivals of NCs and HCs respectively. Transition t_h^i is enabled with at least one idle channel in place CP. Otherwise, it is blocked. Transition t_n is disabled if there are less than $g + 1$ channels in place CP. This is represented by the multiple input arc from place CP to transition t_n and the multiple output arc from transition t_n to place CP. The resulting effect is that when transition t_n fires, only one token is moved from place CP to place T. The number of tokens in place T is the number of channels currently being utilized in the cell. Transitions t_d and t_h^o respectively represent the departure of a call, either due to the termination of the call or due to the MS leaving the cell. The clearing rate for a *single* call is λ_d. The rate at which an MS leaves the cell is λ_h^o. Notice that transitions t_d and t_h^o have marking dependent firing rates. The *actual* firing rates for transitions t_d and t_h^o are $k\lambda_d$ and $k\lambda_h^o$ respectively, where k is the number of tokens in place T. The marking dependency is indicated by the # signs next to the transitions in Fig. 4.

Let T_n denote the number of tokens in place T and consequently let $m = \{T_n, CP_n\}$ denote the marking of the SRN in Fig. 4. The continuous time Markov chain (CTMC) for the SRN of the performance model is shown in Fig. 5, where $\lambda_t = \lambda_n + \lambda_h^i$ and $\lambda_o = \lambda_d + \lambda_h^o$.

4.3 Performability SRN Model (with RF Failure and Recovery)

With RF failure and recovery, a failed channel (FCh) is automatically switched by an idle channel, if one is available. Otherwise, the call with a failed channel is queued until an idle channel is available. Since the spectrum is scarce, no spare channels are reserved exclusively for calls with failed channels. However, calls with failed channels are treated with the same priority as the HCs in the sense that both of them can access any available channel in the reserved channel pool. We assume

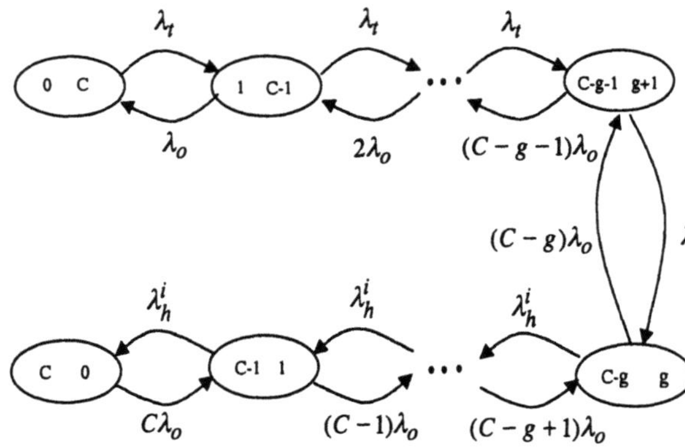

Fig. 5. CTMC for the SRN model in Fig. 4.

that an idle channel is always in perfect condition for service. In other words, a channel can only fail when it is in service. A call fails when the channel it holds fails.

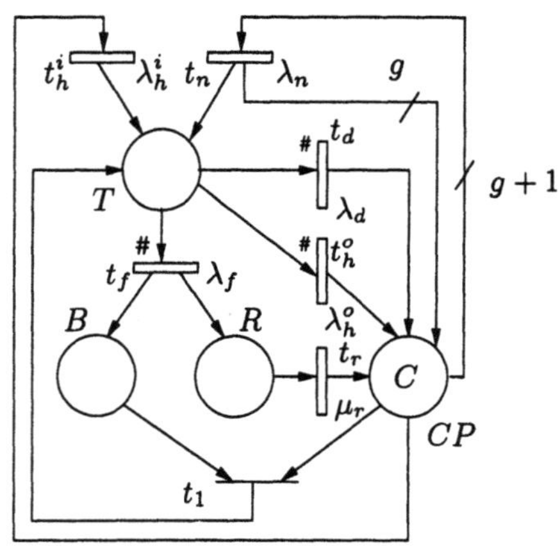

Fig. 6. Performability SRN Model

In Fig. 6, we show the SRN model. Compared with the pure performance model in Fig. 4, two places (B and R), two timed transitions (t_f and t_r) and one immediate transition (t_1) are added in Fig. 6. Transition t_f represents the failure of a channel while it is in use. The failure rate for a *single* channel is λ_f. When a channel fails, the FC is switched to an idle channel if one is available. In this case, the FC is restored to service immediately. When the channel pool is empty, the FC is queued in the buffer B. As soon as an idle channel is available, an FC is restored instantly. The queued FCs are served by the first-in/first-out (FIFO) policy. The above process is represented by the immediate transition t_1 in Fig. 6. In order to fire transition t_1, at least one token is required in both places CP (an idle channel) and B (an FC). In the mean time, the FChs are being recovered in place R under the FIFO policy with a single recovery facility. This is represented by transition t_r and the recovery rate is μ_r.

Let T_n denote the number of tokens in place T and consequently let $m = \{T_n, B_n, R_n, CP_n\}$ denote the marking of the SRN in Fig. 6. Then Fig. 7 shows the $\mathcal{ERG}$ obtained from the initial marking shown in Fig. 6, where $C = 3$ and $g = 1$. Vanishing markings are represented by rectangles and ovals are used to represent the tangible markings. The corresponding CTMC is shown in Fig. 8.

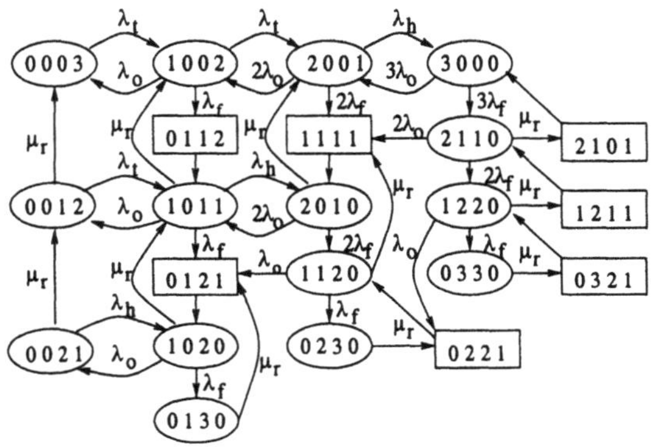

Fig. 7. ERG for the SRN model in Fig. 6.

5 Recent Research on SPN

5.1 Tools Development

Concurrent with the stochastic evolution of Petri nets, effort has been devoted to defining and implementing automatic solution tools for PN models. For example,

- SPNP (Stochastic Petri Net Package) and its GUI version iSPN can provide steady state, transient and cumulative measures, and advanced features such as marking-dependent rates, marking-dependent arc multiplicities and guards [11,13].
- GreatSPN provides graphical input, steady state and transient solution to GSPN[8].
- DSPNexpress can deal with Deterministic and Stochastic Petri Nets (DSPN) under the assumption that only one deterministic transitions can be enabled[5,6]. The solution to DSPN with two concurrently enabled deterministic transitions has been presented recently.
- ESP implements an approximate solution based on the use of phase type distributions[1].

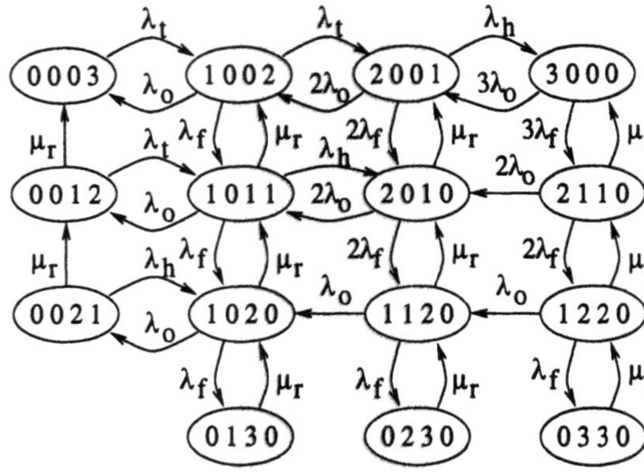

Fig. 8. CTMC for the SRN in Fig. 6.

5.2 Solutions Dealing with Largeness

Since SPNs are based on the solution of CTMC, the state space and reachable markings grows exponentially as the number of transitions, places and tokens increase in the SPNs, which poses a higher requirement to the memory and capacity of the computers, and limits the applicability of SPNs to deal with real life applications. For example, in the Ethernet/ATM bridge, the number of states is about nMN. If we choose 4-stage Erlang to approximate the deterministic transition, and assume that the bridge has 8 Ethernet interfaces and the ATM port can accomodate 20,000 cells (i.e., about 1Mbytes), then the number of states is about 640,000. A large effort has been devoted to overcome or alleviate the *largeness*. For example, we can decompose the SPN model into multiple smaller models which can be solved separately [18,9]. Actually, the decomposition is based on representing the infinitesimal generator matrix in a compact form as a combination of smaller component matrices with Kronecker operators [28,23]. However, sometimes the decomposition might not be very "clean", i.e., there are interactions among the submodels: the input parameters in submodel A depend on the behavior of submodel B, and the input parameters in submodel B depend on the behavior of the submodel A. In this case, fixed-point iteration can be used [31,20]. For a proof of existence of a fixed point, see [31]. The proof of uniqueness, however, needs to be worked out on a case by case basis. The software package, SPNP, facilitates such fixed point iteration schemes.

5.3 Solutions Dealing with Stiffness

The wide difference among the rates of the transitions in SPNs is called stiffness, which will cause numerical difficulties while solving the Markov chain. In *ATM* networks, there are various processes operating on widely different time scales.

For example, the cells may be transmitted within 3 microseconds with the *ATM* port rate is 155Mb/s, while the cells arrive in bursts whose length is at order of magnitude of millisecond, and the connection duration in *ATM* networks might be from several seconds to tens of minutes. Then if we want to study the performance of *ATM* networks at cell level, bursty level and connection level within one model, we have to deal with the stiffness in the model. One method to deal with the stiffeness is based on hierarchical decomposition [18,2], which decomposes the stiff model into multiple submodels so that there might be large difference in the magnitudes of the transition rates in different submodels, but the transition rates within each submodel do not differ significantly. The submodels can be solved separately and reward functions can be defined to get the final performance indices. Fixed-point iteration is still useful if the decompostion is not very "clean".

Besides the hierarchical decomposition, uniformization has been recently improved to handle stiff problems [17] and implemented in SPNP.

5.4 Extensions to SRN

Non-Markovian Stochastic Petri Nets The analytical tractability of SRNs and underlying CTMC are based on the exponential assumption of the distribution of the holding time in a given state. However, many activities in real life systems do not have an exponentially distributed duration, e.g., the transmission of cell in *ATM* networks has deterministic delay, self-similarity and heavy-tailed distributions were found in computer networks which demonstrates rather different properties from the Markovian processes [27,21,32]. Although non-exponential distributions can be approximated by phase type distributions, the price for introducing phase type distribution is an enlargement of the state space. In recent years, several classes of SPN models have been elaborated which incorporate some non-exponential characteritics in their definition [16,7,25,3,6]. SPNP supports simulations of general non-Markovian SPNs.

Fluid Stochastic Petri Nets [15] Fluid Stochastic Petri Nets (FSPN) extend the SPNs by introducing real (positive) tokens to special continuous places. The places are partitioned into a set of discrete places containing an integer number of tokens and a set of fluid (or continuous) places containing a real fluid level. The state space of an FSPN is partially discrete and partially continuous. The discrete part is an integer vector accounting for the number of tokens in the discrete places. The continuous part is a vector of real numbers accounting for the fluid levels in the continuous places. Conceptually, FSPN is based on stochastic fluid flow models which have been used extensively to evaluate the performance of high-speed networks. In high-speed networks, the stochastic processes can be viewed as continuous-state as the network speed increases (e.g., the cell transmission delay is 3 microseconds in 155Mb/s *ATM* networks). Therefore *FSPN* is very useful in high-speed networks. Besides, *FSPN* can be used to deal with the problem of largeness.

Numerical analysis of FSPNs with single fluid place and discrete event simulation with multiple fluid places have been integrated in SPNP [10].

6 Conclusion

As a high-level description language, SPN is very concise in its specification and its form is closer to a designer intuition about what a model should look like. Software packages developed by the researchers enable the automated generation and solution of Markovian stochastic systems represented by SPNs. It really relieves the performance analyst from the tedious task of building Markov chains and solving the associated linear equations. The aim of this paper is to encourage the researchers in communication area to use SPNs as a tool of performance analysis, therefore we omitted rigorous mathematical proofs for the procedures of automated generation and solution to CTMC. We focused on how to use SPNs as a tool of performance analysis and two examples in communication networks were presented and studied in this paper. We hope the "transition" from traditional methods to SPNs will be "enable"ed eventually. We do believe the "transition" is "deterministic" , if not "immediate".

References

1. A. Cumani, "ESP- A Package for the Evaluation of Stochastic Petri Nets with Phase-type Distributed Transition Times", *Proc. IEEE Int. Workshop on Timed Petri Nets*, pp.433-440, 1995.
2. A. Bobbio, K. S. Trivedi, "An Aggregation Technique for the Transient Analysis of Stiff Markov Chain," *IEEE Trans. on Computers*, vol. 35, pp. 803-814, 1986.
3. A. Bobbio, M. Telek, "A Benchmark for PH Estimation Algorithm: Results for Acyclic-PH", *Stochastic Models*, vol. 10, pp. 661-667, 1994.
4. A. Romanow, S. Floyd, "Dynamics of TCP Traffic over ATM Networks", *IEEE J -SAC*, pp. 633-641, May 1995.
5. C. Lindemann, "DSPNexpress: A Software Package for the Efficient Solution of Deterministic and Stochastic Petri Nets", *Performance Evaluation*, vol. 22, pp. 3-21, 1995.
6. C. Lindemann, *Performance Modelling with Deterministic and Stochastic Petri Nets*, John Wiley, 1998.
7. D. Logothetis, K. S. Trivedi, "Transient Analysis of the Leaky Bucket Rate Control Scheme Under Poisson and ON-OFF Sources", *Proceedings of the IEEE INFOCOM 94*, Toronto, Canada, June 1994.
8. G. Chiola, " GreatSPN 1.5 Software Architecture", *Computer Performance Evaluation*, pp. 121-136, Elsevier Science Publisher, 1992.
9. G. Ciardo, K. S. Trivedi, "A decomposition approach for stochastic Petri net models", *Proceedings of the Fourth International Workshop on Petri Nets and Performance Models (PNPM91)*, pp. 74–83, 1991.
10. G. Ciardo, D. Nicol, K. S. Trivedi, "Discrete-event Simulation of Fluid Stochastic Petri Nets",*IEEE Trans. Software Engineering* , vol. 25, no. 2, pp. 207–217, March-April, 1999.
11. G. Ciardo, J. Muppala and K. Trivedi, "SPNP: Stochastic Petri Net Package", *International Conference on Petri Nets and Performance Models*, Kyoto, Japan, December 1989.
12. G. Ciardo, J. Muppala, K. S. Trivedi, "Analyzing Concurrent and Fault-tolerant Software Using Stochastic Reward Nets", *Journal of Parallel and Distributed Computing*, 15, p255-269, 1992.

13. G. Ciardo, K. S. Trivedi, "Manual for SPNP: Stochastic Petri Net Package", version 6.0, CACC, ECE Department, Duke University, 1996.
14. G. Bolch, S. Greiner and H. de Meer, Kishor S. Trivedi, *Queueing Networks and Markov Chains,*John Wiley, 1998.
15. G. Horton, V. Kulkarni, D. Nicol, K. S. Trivedi, "Fluid stochastic Petri nets: Theory, application, and solution", *European Journal of Operations Research,* vol. 105, no. 1, pp. 184–201, Feb. 1998.
16. H. Choi, V. Kulkarni, K. S. Trivedi, "Markov Regenerative Stochastic Petri Nets", *Performance Evaluation*, vol. 20, no. 1-3, pp. 337-357, 1994.
17. J. Muppala, M. Malhotra, K. S. Trivedi,"Stiffness-Tolerant Methods for Transient Analysis of Stiff Markov Chains", *Microelectronics and Reliability,* vol. 34, no. 11, pp. 1825-1841, 1994.
18. J. K. Muppala, K. S. Trivedi, "Composite Performance and Availability Analysis using a Hierarchy of Stochastic Reward Nets", *Proc. of the Fifth International Conference on Modelling Techniques and Tools for Computer Performance Evaluation*, Torino, 1991.
19. K. S. Trivedi, *Probability and Statistics with Reliability, Queueing, and Computer Science Applications*, Prentice-Hall, Englewood Cliffs, NJ, 1982.
20. L. Tomek,K. S. Trivedi, "Fixed-Point Iteration in Availability Modeling", *Informatik-Fachberichte, Vol. 283: Fehlertolerierende Rechensysteme,* M. Dal Cin (ed.), pp. 229-240, Springer-Verlag, Berlin, 1991.
21. M. E. Crovella, A. Bestavros, "Self-similarity in World Wide Web Traffic- Evidence and Possible Causes", *Sigcomm'96*, pp. 160-169.
22. M. Ajmone Marsan, G. Balbo, and G. Conte, "A Class of Generalized Stochastic Petri Nets for the Performance Evaluation of Multiprocessor Systems", *ACM Transactions on Computer Systems*, vol. 2, pp. 93–122, May 1984.
23. P. Kemper, "Numerical Analysis of Superposed GSPNs", *IEEE Trans. on Software Engineering*, vol. 22, 1996.
24. Raif O. Onvural, *Asynchronous Transfer Mode Networks: Performance Issues,* Artech House, 1994.
25. R. German, C. Lindermann, "Analysis of Stochastic Petri Nets by the Method of Supplementary Variables", *Performance Evaluation*, vol. 20, pp. 317-335, 1994.
26. R. German, C. Kelling, A. Zimmerman, G. Hommel, " TimeNET - A Toolkit for Evaluating Non-Markovian Stochastic Petri Nets", *Performance Evaluation*, 1998
27. S. Deng, "Empirical Model of WWW Document Arrivals at Access Link", *ICC'96*, pp1797-1802.
28. S. Donatelli, " Superposed Stochastic Automata: A Class of Stochastic Petri Nets Amenable to Parallel Solution", *Performance Evaluation*, voo. 18, pp. 21-36, 1993.
29. Yue Ma, James J. Han, K. S. Trivedi, "Channel Allocation with Recovery Strategy in Wireless Networks", *European Transactions on Telecommunications (ETT)*, vol. 11, no. 4, pp. 395-406, July-August 2000.
30. T. Murata, "Petri Nets: Properties, Analysis and Applications", *Proc. of IEEE*, vol. 77, no. 4, pp. 541-560, 1989.
31. V. Mainkar, K. S. Trivedi, "Sufficient Conditions for the Existence of a Fixed Point in Stochastic Reward Net-Based Iterative Models", IEEE Trans. on Soft. Eng., vol. 22, Sept., pp. 640-653, 1996.

32. V. Paxson, S. Floyd, "Wide Area Traffic: The Failure of Poisson Modeling," *IEEE/ACM Trans. on Networking*, pp. 227-444, June 1995.
33. "LAN Emulation over ATM Specification Version 1.0", *ATM Forum, af-lane-0021.000*, Jan. 1995.

Part VIII

Performance Analysis

Dynamic Routing and Wavelength Assignment Using First Policy Iteration, Inhomogeneous Traffic Case

Esa Hyytiä and Jorma Virtamo

Helsinki University of Technology,
Laboratory of Telecommunications Technology
P.O.Box 3000, FIN-02150 HUT, Finland

Abstract. The routing and wavelength assignment problem (RWA) in WDM network can be viewed as a Markov Decision Process (MDP). The problem, however, defies calculation of the exact solution because of the huge size of the state space. Several heuristic algorithms have been presented in the literature. Generally, these algorithms, however, do not take into account the available extra information about the traffic, e.g. inhomogeneous arrival rates. In this paper we propose an approach where, starting from a given heuristic algorithm, one obtains a better algorithm by the first policy iteration. At each decision epoch a decision analysis is made where the costs of all the alternative actions are estimated by simulations on the fly. Being computationally intensive, this method can be used in real time only for systems with slow dynamics. Off-line it can be used to assess how close the heuristic algorithms come to the optimal policy. Numerical examples are given about the policy improvement.

1 Introduction

The wavelength division multiplexing (WDM) is a promising technology for the future all-optical networks. In WDM several optical signals using different wavelengths share the same fibre. The capacity of such fibre links can be huge, even terabits per second. The routing in network nodes is based on wavelengths of incoming signals [1–3].

Generally, the routing and wavelength assignment (RWA) problem in WDM networks consists of choosing a route and a wavelength for each connection so that no two connections using the same wavelength share the same fibre [4,5]. For example a simple form of RWA problem is a static traffic case with single fibre links. If the nodes are incapable to perform wavelength translations, an assumption made throughout this work, the problem can be mapped to a node colouring problem, once routing is fixed (see e.g. [4]).

When the traffic is not static, lightpath requests arrive randomly following some traffic process. Connection requests between a given source destination pair constitute a *traffic class*, which we index by k, $k \in \mathcal{K}$, where $\mathcal{K}$ is the set of all source destination pairs. Usually the RWA algorithm configures lightpaths in the network unless there are not enough resources available and the request is blocked (see e.g. [6–8]).

The possible schemes considered under dynamical traffic can be divided into two cases. If it is possible to reconfigure the whole network when blocking would occur, then the blocking probability can be considerably reduced. Such an operation, however, interrupts all (or at least many) active lightpaths and requires a lot of coordination between all the nodes. In large networks the reconfiguration seems infeasible. In any case the reconfiguration algorithm should try to minize the number of reconfigured lightpaths [9].

The other case is when active lightpaths may not be reconfigured. In this case it is important to decide which route and wavelength are assigned to an arriving connection request in order to balance the load and minimize the future congestion in the network.

Several heuristic algorithms have been proposed and studied (see e.g. [6–8]). These algorithms, however, do not take into account the traffic characteristics like unequal costs of different requests or inhomogeneous arrival rates. In this paper we study this problem in the setting of Markov Decision Processes (MDP) and propose an approach, where we try to improve any given heuristic algorithm by the first policy iteration [10,11]. The policy iteration, indeed, is known to lead to a new policy with better performance. In order to avoid dealing with the huge size of the state space in calculating the relative state costs needed in the policy improvement step, we suggest estimating these costs on the fly by simulations for the limited set of states that are relevant at any given decision epoch, i.e. when the route and wavelength assignment for an arriving call has to be made. This approach was introduced in [12]. In this paper we focus on the performance improvement in heterogeneous traffic environment, while the traffic characteristics in [12] were assumed to be uniform.

The rest of the paper is organized as follows. In Section 2 we briefly review Markov Decision Processes and policy iteration in general, and the first policy iteration, in particular. In Section 3, we consider the relative costs of states and how they are used in the policy iteration, and in the following Section 4 we study how theses state costs can be estimated by simulations. Some heuristic RWA algorithms are presented in Section 5. These are used as a starting point for policy iteration, and, in Section 6 some numerical results obtained by simulations are presented. Finally, Section 7 contains conclusions.

2 Policy Iteration

Routing and wavelength assignment constitutes a typical decision making problem. When certain events occur, one has to decide on some action. In the RWA problem, in particular, upon arrival of a request for new connection one has to decide whether or not to accept the request, and if accepted, which resources to allocate for it, i.e. which of the available routes and wavelengths are used for that connection.

In general, one is interested in the optimal policy which maximizes or minimizes the expectation (infinite time horizon) of a given objective function. Here we assume that the objective is defined in terms of minimizing some cost function. The cost may represent e.g. the loss of revenue due to blocked calls, where different revenue may be associated to each type of call.

When the arrival process of type k calls is a Poisson process with intensity λ_k, the holding times of those calls are distributed exponentially with mean $1/\mu_k$ and the expected revenue per carried call is w_k, then the system constitutes a Markov Process and the problem of determining the optimal policy belongs to the class of Markov Decision Processes (MDP) described e.g. in [10] and [11].

Three main approaches for solving the optimal policy in the MDP setting are the policy iteration, value iteration and linear programming approach. In this paper, we concentrate on the policy iteration, where, as the name says, one tries to find the optimal policy by starting from some policy and iteratively improving it. This policy iteration is known to converge rather quickly to the optimal policy. Even the first iteration often yields a new policy which is rather close to the optimal one. In practice, it is seldom possible to go beyond the first iteration. Also in this work, we will restrict ourselves to the first policy iteration.

At each decision epoch, i.e. arrival of a new request, there is a finite set of possible actions: either reject the call or accept it and assign a feasible combination of route and wavelength (RW) to it. A feasible RW combination is such that along the route from the source to destination the wavelength is not being used on any of the links. If no feasible RW combination exists, the call is unconditionally rejected.

A *policy* defines for each possible state of the system and for each class k of an arriving call which action is taken among all the possible actions in that state Many heuristic policies have been proposed in the literature such as the first-fit wavelength and most-used wavelength policies combined with shortest path routing or near shortest path routing [6–8]. Some of them work reasonably well. Common to all these heuristic policies is that they are simple. The choice of the action to be taken at each decision epoch can usually be described in simple terms and does not require much computation. We take one of the heuristic policies as a starting point and call it the *standard policy*. The policy resulting from the first policy iteration we refer to as the *iteration policy*.

By doing the first policy iteration we have two goals in mind: 1) Finding a better RWA algorithm which, being computationally intensive, may or may not be calculable in real time, depending on the time scale of the dynamics of the system. 2) Even in the case the algorithm is not calculable in real time, estimating how far the performance of the heuristic algorithm is from the optimal one.

Briefly, as explained in more detail below, our idea in the policy iteration is the following: at each decision epoch we make a decision analysis of all the alternative actions. For each of the possible actions, i.e. decision alternatives, we estimate the future costs by simulation. Thus, assuming that a given action is taken we let the system proceed from the state where it is after that action and use the standard policy to make all the subsequent decisions in the simulation. The iteration policy is the policy which is obtained when at each decision epoch the action is chosen for which the estimated cost is the minimum. It can be shown that the iteration policy is always better or at least as good a policy as the standard policy, and as said, it often comes rather close to the optimal policy.

3 Relative costs of states

In the MDP theory, the first policy iteration consists of the following steps: With the standard policy one solves the Howard equations (see e.g. [10,11]) to give the so called relative costs of the states, C_i, which for each possible state i of the system describe the difference in the expected cumulative cost from time 0 to infinity, given that the system starts from state i rather than from the equilibrium. Then, given that the current state of the system is j and a class-k call is offered, one calculates the cost $C_j + w_k$ for the action that the call is rejected, and the cost C_i, $i \in \mathcal{A}(j,k)$, for the case the call is accepted, where $\mathcal{A}(j,k)$ is the set of states reachable from state j by assigning call-k a feasible RW pair. By choosing always the action which minimizes the cost, one gets the iteration policy, i.e. the policy resulting from the first policy iteration.

Though the Howard equations are just a set of linear equations for relative costs C_i and the average cost rate c of the standard policy (see below), their solution cannot be obtained because of the prohibitive size of the state space of any realistic system. However, at any decision epoch the relative costs C_i are needed only for the current state j and a small set of states $\mathcal{A}(j,k)$ reachable from the current state. We propose estimating these values on the fly by means of simulations. To this end, it is useful to consider the physical interpretation of the relative costs C_i.

Given that the system starts from state i at time 0 and the standard policy is applied for all the decisions, the cumulative costs are accrued at the expected rate $c_t(i)$ at time t,

$$c_t(i) = \sum \lambda_k w_k \mathrm{P}\{I_t \in \mathcal{B}_k | I_0 = i\}, \tag{1}$$

i.e. the expected rate of lost revenue, where $\mathrm{P}\{I_t \in \mathcal{B}_k\}$ is the probability that at time t the state I_t of the system is a blocking state for class-k calls. When $I_t \in \mathcal{B}_k$ class-k calls arriving at time t are blocked by the standard policy because either no feasible RW pair exists or the policy otherwise deems the blocking to be advantageous in the long run. The expected cost rate $c_t(i)$ depends on the initial state i. However, no matter what the initial state is, as t tends to infinity, the expected cost rate tends to a constant c, which is specific to the standard policy, and corresponds to (1) with steady state blocking probabilities $\mathrm{P}\{I_t \in \mathcal{B}_k\}$.

The behaviour of the function $c_t(i)$ is depicted in Fig. 1 for two different initial values i_1 and i_2. The relative cost C_i is defined as the integral

$$C_i = \int_0^\infty (c_t(i) - c)\, dt,$$

i.e. the area between the curve $c_t(i)$ and the line at level c. So we are interested in the transient behaviour of $c_t(i)$; after the transient no contribution comes to the integral. The length of the transient is of the order $1/\mu$, where $1/\mu$ is the maximum over $\{1/\mu_k\}$, $k \in \mathcal{K}$. After that the system essentially forgets the information about the initial state, as most of the initial connections have died out. So we can restrict ourselves to an appropriately chosen finite interval $(0, T)$. The actual choice of T is a tradeoff between different considerations as will be discussed later.

One easily sees that in the policy improvement step only the differences of the values C_i between different states are important. Therefore, we can neglect

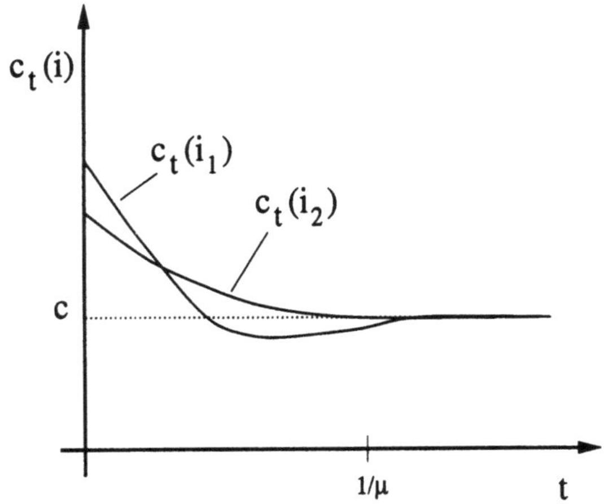

Fig. 1. Expected costs with different initial choices as a function of time.

the constant c in the integral, as it is common to all states, and end up for thus redefined C_i,

$$C_i \approx C_i(T) = \int_0^T c_t(i)\, dt, \tag{2}$$

which is simply the expected cumulative cost in interval $(0, T)$ starting from the initial state i.

4 Estimation of the state costs by simulation

In practice, it is not feasible to calculate the cost rate function $c_t(i)$ analytically even for the simplest policies. Therefore, we estimate the state costs C_i by simulations. In each simulation the system is initially set in state i and then the evolution of the system is followed for the period of length T, making all the RWA decisions according to the standard policy.

4.1 Statistics collection: blocking time vs. blocking events

In collecting the statistics one has two alternatives. Either one records the time intervals when the system is in a blocking state of class-k calls, for all $k \in \mathcal{K}$. If the cumulative time within interval $(0, T)$ when the system is in the blocking state of class-k calls is denoted by $\tau_k(i)$, then the integral is simply

$$\hat{C}_i = \sum \lambda_k w_k \tau_k(i). \tag{3}$$

Alternatively, one records the number $\nu_k(i)$ of blocked calls of type k in interval $(0, T)$. Then we have

$$\hat{C}_i = \sum w_k \nu_k(i). \tag{4}$$

In these equations we have written explicitly $\tau_k(i)$ and $\nu_k(i)$ in order to emphasize that the system starts from the state i. Both (3) and (4) give an unbiased estimate for C_i. In either case, the simulation has to be repeated a number of times in order to get an estimator with small enough a confidence interval.

Denote the estimates of future costs obtained in the jth simulation run by $\hat{C}_i^{(j)}$, using (3) or (4) as the case may be. Then our final estimator for C_i is

$$\hat{C}_i = \frac{1}{N}\sum_{j=1}^{N} \hat{C}_i^{(j)}, \tag{5}$$

where N is the number of simulation runs. In fact, for the policy improvement the interesting quantity is the difference

$$E_{i_1,i_2} = C_{i_2} - C_{i_1},$$

for which we have the obvious estimate

$$\hat{E}_{i_1,i_2} = \hat{C}_{i_2} - \hat{C}_{i_1}. \tag{6}$$

From the samples $\hat{C}_{i_1}^{(j)}$ and $\hat{C}_{i_2}^{(j)}$, $j = 1, \dots, N$, we can also derive an estimate for the variance $\hat{\sigma}_{i_1,i_2}^2$ of the estimator $\hat{E}_{i_1,i_2}$

$$\hat{\sigma}_{i_1,i_2}^2 = \frac{N\sum_j(\hat{C}_{i_2}^{(j)} - \hat{C}_{i_1}^{(j)})^2 - \left(\sum_j \hat{C}_{i_2}^{(j)} - \hat{C}_{i_1}^{(j)}\right)^2}{N^2(N-1)} = \frac{\hat{S}_{i_1,i_2}^2 - (\hat{E}_{i_1,i_2})^2}{N-1},$$

where $\hat{S}_{i_1,i_2}^2 = \frac{1}{N}\sum_j \left(\hat{C}_{i_2}^{(j)} - \hat{C}_{i_1}^{(j)}\right)^2$.

The choice between the alternative statistics collection methods is based on technical considerations. Though estimator (3) (blocking time) has a lower variance per one simulation run, it requires much more bookkeeping and the variance obtained with a given amount of computational effort may be lower for estimator (4) (blocking events).

4.2 Policy iteration with uncertain state costs

The important parameters of the simulation are the length of the simulation period T and the number of simulation runs N used for the estimation of each C_i. In practice, we are interested in the smallest possible values of T and N in order to minimize the simulation time. However, making T and N too small increases the simulation noise, i.e. error in the estimates $\hat{C}_i$, occasionally leading to decisions that differ from that of the true iteration policy, consequently deteriorating the performance of the resulting algorithm.

No matter how the parameters are chosen, some uncertainty in the estimators $\hat{C}_i$ is unavoidable. In order to deal with this uncertainty of the estimators $\hat{C}_i$, we do not blindly accept the action with the smallest estimated cost, but give a special status for the decision which would be chosen by the standard policy. Let us give this action the index 0. Based on the simulations we form estimates $\hat{E}_{0,i}$ for each

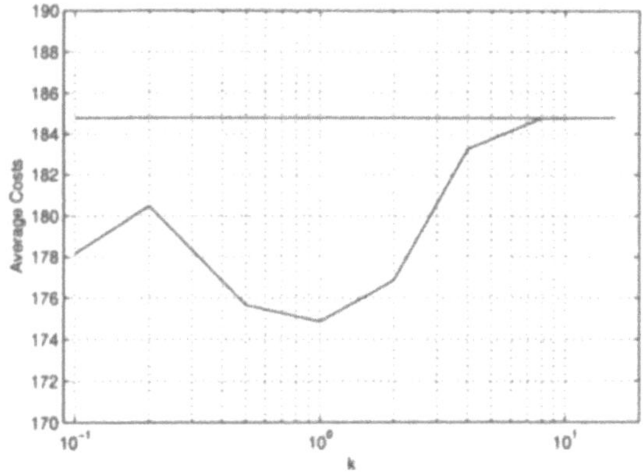

Fig. 2. Average cost rate c of the iteration policy as a function of the parameter k. The horizontal line represents the cost rate of the standard policy. The minimum lies in the range $k = 0.5 \ldots 2$ in this case. The set of routes was specifed with $\Delta l = 0$ and $rmax = 4$ as explained in Section 6.1.

possible action i. Then, as the decision we choose the action which minimizes the quantity

$$\hat{E}_{0,i} + k \cdot \hat{\sigma}_{0,i}, \tag{7}$$

where k is an adjustable parameter. Note that for $i = 0$ this quantity is equal to 0. Thus, in order for another action i to replace the action 0 of the standard policy, we must have $\hat{E}_{0,i} < -k \cdot \hat{\sigma}_{0,i}$, i.e. we require a minimum level of confidence for the hypothesis $C_i < C_0$. An appropriate value for k has to be determined experimentally.

If k is too small, wrong decisions are made more frequently. On the other hand too high a value of k prevents the choice of other alternative actions totally. In Fig. 2 the performance behaviour of a certain system is depicted as a function of k. The horizontal line represents the costs obtained with the standard policy. From the figure it can be seen that once k is higher than about 10 the iteration policy reduces to the standard policy.

4.3 Time complexity of iteration approach

Clearly the simulation of the future, even for a limited period T, at each decision epoch makes this algorithm very time consuming in comparison with the standard policy. Assume that a single decision of the standard policy takes a constant time u. Let N be the number of the simulations that are run for each alternative action, A average number of alternative actions per decision (possible RW pairs), λ the total arrival rate to the network (assuming uniform load for simplicity), and T the period covered by one simulation run. Then, the running time of each decision is on the average

$$u_i = A \cdot N \cdot (\lambda T)u = \lambda ANT \cdot u,$$

so the running time is λANT times longer than with the underlying algorithm. Neither λ nor A are parameters of the algorithm. Hence, the tradeoff between the goodness of solution and the running time is defined by choosing the value for the product $N \cdot T$.

For example, to get decent results with a simple 11 node network (Fig. 3) with moderate load ($\mu = 1$ and $\lambda_k = 0.4$ for all k), about 100 samples (simulation runs) are required, each $1/\mu$ time units long. So the increase in running time is of the order of $10^3 - 10^4$. It is clearly essential that the decisions of the underlying standard policy can be determined quickly.

5 Heuristic Algorithms

Several quick heuristic algorithms for the dynamic RWA problem have been proposed in the literature. Here we briefly present some of them and later study how the iteration approach works with them. The first set of algorithms assumes that a fixed set of possible routes for each connection is given in advance. Some papers refer to this as alternate routing strategy. In practice this set usually consists shortest or nearly shortest path of routes. These algorithms are greedy and accept the first feasible RW pair they find (first-fit).

- *basic* algorithm goes through all the routes in a fixed order and for each route tries all the wavelengths in a fixed order. The routes are sorted in the shortest route first order.
- *porder* algorithm is similar to *basic*-algorithm but it goes through all the wavelengths in a fixed order and for each wavelength tries all the routes in a fixed order.
- *pcolor* algorithm works like *porder* but wavelengths are searched in the order of the current usage instead of a fixed order, so that the most used wavelength is tried first.
- *lpcolor* algorithm is the smartest algorithm. It packs colours, but the primary target is to minimize the number of used links. So the algorithm first tries the most used wavelength with all the shortest routes, then the next often used wavelength and so on. If no wavelength works, the set of routes is expanded to include routes having one link more and wavelengths are tried again in the same order.

Another set of heuristic algorithms, adaptive unconstrained routing (AUR) algorithms, are described in [7]. These use dynamic routing instead of fixed set of routes, and are thus a little bit slower.

- *aurpack* is similar to *pcolor*, but without the limitations of a fixed set of routes.
- *aurexhaustive* finds a route with each wavelength (if possible) and chooses the shortest among them, i.e. it is identical to *lpcolor* except that the set of possible routes is not limited.

Thus AUR-algorithms will search for a free route dynamically based on the current state of the network. There is no need to store possible routes (which without any limitations can form a very large set) in advance. This approach is in slight conflict with the iteration approach, where the set of possible actions (RW pairs) is expected to be known in advance.

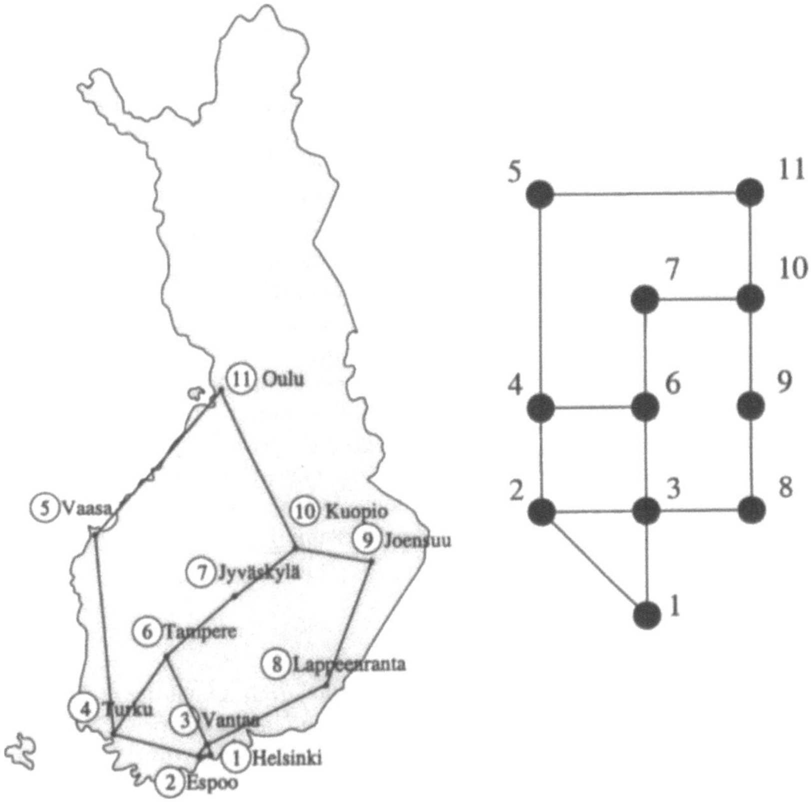

Fig. 3. Hypothetical WDM-network residing in Finland.

Also other heuristics are given in [7], e.g. *random* (tries wavelengths in random order) and *spread* (tries least used wavelength first), but they were reported to work worse than the ones described above, and are not further discussed here.

A deficiency of all the presented heuristic algorithms is that they do not take into account the possible additional information about the arrival rates, the distribution of holding times or the priorities of traffic classes (different costs/revenues). Also the duration of the call when it arrives could be known (for example one channel is reserved for certain event which lasts exactly two days). We could of course try to come up with better heuristics which somehow take into account additional information, but that means that we would need a new heuristic policy for each case. On the other hand the first policy iteration automatically adapts to the new situation, because the simulation automatically takes into account all the peculiarities of the system. So, even if the approach is very simple, it is very powerful due to its flexibility.

Table 1. The test scenarios: case I is a reference point where everything is uniform and in other two cases either arrival rates or costs are non-uniform.

Set	Description
I	*Uniform case.* The offered traffic between each node pair is uniform with the rate of $\lambda = 0.4$ and the cost of each missed request is equal to 1.0. The average duration of connection is $1/\mu = 1.0$.
II	*Non-uniform costs.* The offered load between each node pair is still uniform, but cost of missed calls differ. A missed connection request to/from node 2 costs 3.0, and connections 1-3, 1-6, 1-7, 6-7 and 6-10 have weight 1.0, and rest of the connections have weight 0.5. So the total cost arrival rate is the same as in the uniform case. The average duration of connections is the same 1.0.
III	*Non-uniform arrival rates.* Similarly here we set the arrival rates of the connections where the other end is node 2 to 1.2, and arrival rates between 1-3, 1-6, 1-7, 6-7 and 6-10 are set to 0.4, while rest of the connection have arrival rate of 0.2. Thus, the total arrival rate of the connection requests to the network stays again the same. The average duration of connections is the same 1.0.

6 Simulation results

In this section we present some numerical results from simulations. Three test scenarios were created each having different kind of characteristics. Every traffic scenario was based on the small network shown in Fig. 3. The network was assumed to have 8 wavelengths available on each link and all the links contained one fibre. The test scenarios used in the simulations are listed in Table 1. In the first case we have uniform traffic and it is used as a reference point. In the other two cases we have given a special status to the node 2. The special status could arise e.g. in the case where the node represents a gateway to international connections. To facilitate comparison with the uniform traffic case also rates λ_k and expected revenues per call w_k were adjusted so that the offered income rate ,$W = \sum_k \lambda_k w_k$, was kept constant.

It should be recognized that the results for the iteration policy were obtained by two levels of nested simulations. In order to assess the performance of the policy, an outer simulation is run, where connections arrive and leave the network and blocking times or events are recorded. Upon each arrival in this outer simulation, a number of inner simulations are launched from the current state in order to make a comparison between different decision alternatives. Based on this comparison one alternative is chosen and used in the outer simulation, which then continues until upon the next arrival the decision analysis by the inner simulations is again started.

6.1 Selection of routes

The possible routes per node pair (or traffic class) were calculated beforehand. Generally the set of routes is enormous, so some way of pruning it is needed. In this study the set of routes was specified with parameters Δl and $rmax$. Parameter Δl sets the maximum number of extra additional links a route can contain when compared to the shortest route. The second parameter *rmax* defines the maximum number of routes per traffic class, i.e. only the *rmax* first routes are included in set. For example, with Δl=0 and *rmax*=10 only the shortest routes are included, and if there are more than 10 equally short routes for some node pair, then only the first 10 routes found are included.

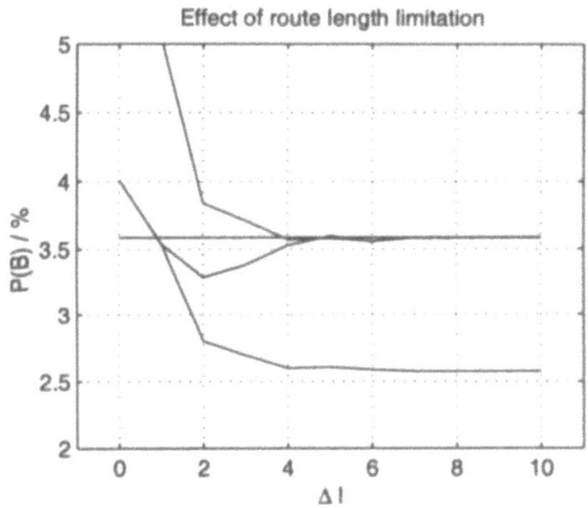

Fig. 4. Blocking probability as a function of the parameter Δl. The other routing parameter *rmax* set no limit on the number of routes. At $\Delta l = 3$ the algorithms from worst to best are *basic*, *aurpack*, *pcolor* and *lpcolor*.

Third possible parameter to reduce the running time of the first iteration approach is *maxtest*, which defines the number of alternative actions evaluated against the standard policy. This effectively limits the running time when the load is low in the network and there are plenty of RW pairs available.

In Fig. 4 the performances of a few heuristic algorithms are presented as a function of the routing parameter Δl. The other routing parameter *rmax* was chosen to be high enough in order not to cause any restriction on the set of routes. The y-axis represents the blocking probability under a uniform load. Clearly too small a set of routes limits the performance but, as can be seen from figure, also too large a set of routes can be a problem for some algorithms. Here the problematic algorithm is *pcolor* which needlessly favors the most used colours at the expense of longer routes, leading to degraded performance.

6.2 Estimation of optimal simulation period

Usually the longer the simulation period T is the better results are obtained. Here we are, however, interested in how current decision affects the results, when the standard policy is used for all later decisions. As was explained before, after a transient period the cost rate $c_t(i)$ is very near to the long time average c of the standard policy. Simulating over a period longer than the duration of the transient

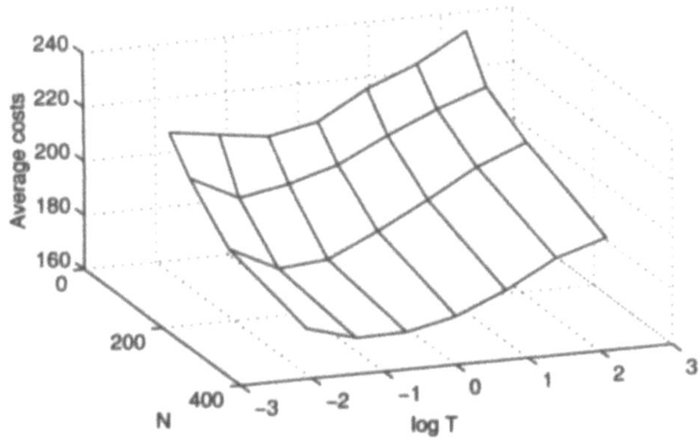

Fig. 5. The performance of the algorithm with different simulations periods T and number of simulation runs N. The traffic is uniform and the routing parameters are Δl=0 and *rmax*=4. *basic* is used as the standard policy. Load is $a = 0.4$ for each traffic class. For the reference, the standard policy gives an average cost of 260 for the same arrivals.

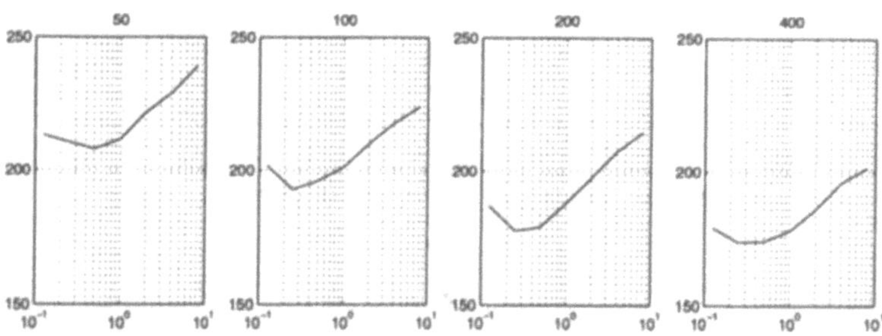

Fig. 6. The performance of the algorithm with different simulations periods T with constant number of simulation runs N. The setup is the same as in Fig. 5 so the graphs represent cuts from the 3D-surface of Fig. 5

thus gives no new information but actually only increases the noise resulting from the stochastic nature of the simulation.

The average costs are depicted in Figs. 5 and 6 using the first policy iteration with different simulation periods and number of simulations. The offered load to the network is $a = 0.4$ for each traffic class. In the mesh Fig. 5 the x-axis is $\log_{10}$ of the simulation period T and y-axis is the number of simulation runs N. The z-axis represents the average costs.

In Fig. 6 each subfigure has a fixed number of simulation runs (50, 100, 200 or 400). The x-axis represents the length of simulation period T and y-axis the average costs. As can be seen from Figs. 5 and 6 the results get worse as the simulation period T grows longer than $0.5 \ldots 1.0$ average holding times. This suggests that the optimal simulation period is about $0.5 \cdot 1/\mu$ in this case.

6.3 Performance of the iteration algorithm under non-uniform traffic

In this section we investigate the performance of the policy iteration starting from different heuristic policies. The parameters used in the simulation are given in Table 2. The choice of these parameters represents a tradeoff between the performance and the running time, and was done on the basis of the considerations of the previous sections.

Table 2. The running parameters used in test cases.

param.	value	description
W	22	total offered cost rate to network, uniform in cases I and II
λ_{tot}	22	total offered load to network, uniform in cases I and III
μ	1.0	duration of connection, $\sim$Exp(μ)
k	2.0	constant at decision making, defines "certainty"
N	200	number of simulations run in iteration approach
T_1	$0.25 \cdot 1/\mu$	length of simulation run in iteration approach, first run
T_2	$0.50 \cdot 1/\mu$	length of simulation run in iteration approach, second run

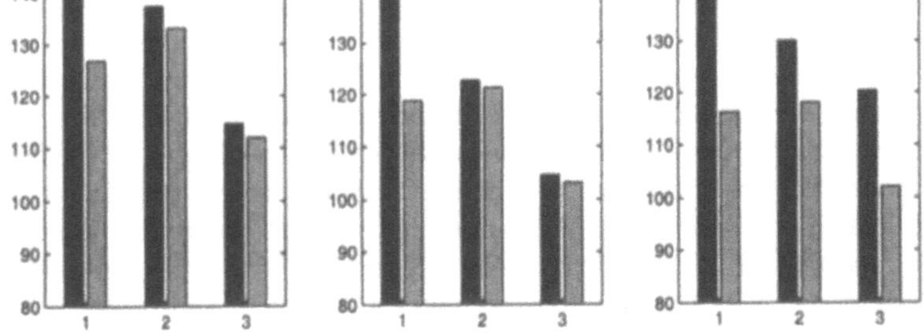

Fig. 7. The performance of the algorithm in different test scenarios with *basic*, *pcolor* and *lpcolor* as standard policy. The simulation period is $0.25 \cdot 1/\mu$. Each pair of bars relates to one of these policies, the left bar is obtained with the standard policy and the right bar with the corresponding iteration policy.

As suggested by Fig. 6 the number of simulation runs N was chosen to be 200 and the simulation period T to be 1/4 or 1/2 times the average holding time. The improvement when the number of samples was increased from 200 to 400 was not significant. Note that the previous simulations were ran with a much smaller set

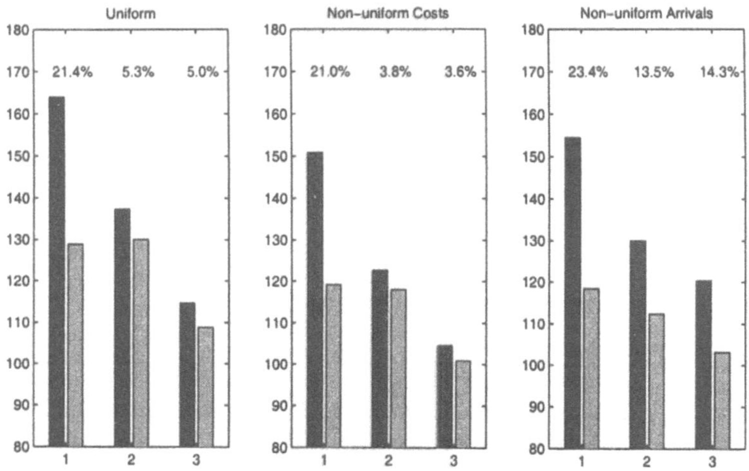

Fig. 8. The performance of the algorithm in different test cases with *basic*, *pcolor* and *lpcolor* as standard policy. The simulation period is $0.50 \cdot 1/\mu$.

of possible routes per node pair. New routing parameters $\Delta l = 3$ and $rmax = 30$ allow routes that are longer than shortest path in the search space. While the number of routes in these simulations is much larger than in the ones from which the parameters were obtained, it is expected that the same parameters still work quite well. As the number of possible actions increased, the value of k was chosen to be 2.0 to be on the safer side (see Fig. 2).

The first policy approach was studied with these parameters in three different traffic scenarios presented in the beginning of this section. The results with different parameters and algorithms are presented in Figs. 7 and 8. The y-axis represents the average costs obtained from the outer simulation. In Fig. 7 the simulation period T was 1/4 times the average holding time while in Fig. 8 the period was extended to 1/2 times the average holding time. The results of standard policies are naturally the same in both figures. Each pane represents one traffic scenario. In the first pane the traffic is uniform, in the second pane the costs are non-uniform and in the last pane the traffic is non-uniform. The bars in figures represent average costs in each case. In each pair of bars the bar on the left represents the result obtained with the standard policy and the bar on the right is the one obtained with the policy iteration. The standard policies from left to right are *basic*, *pcolor* and *lpcolor*. The length of the outer simulation from which the average costs were collected is 200 holding times.

Figures 7 and 8 show that in each case the iteration leads to a better policy, and that the improvement is really notable in the case where *basic* was used as the standard policy. Since *basic* is the worst algorithm, this improvement is not a surprise. Note that in the uniform case the average costs are generally higher than in the other cases. This is probably because a large part of the "important" connections were quite short making the non-uniform case easier to control. It is also worth noting that *basic* policy improved with the policy iteration gave results roughly equal to *pcolor* with the iteration.

In the non-uniform cost case the improvements were generally much less when *pcolor* or *lpcolor* were used as standard policy. Still the difference between *pcolor* and *lpcolor* performance is clear and this suggests that the first policy iteration was not able to come to very close to the optimal policy.

The non-uniform arrival case on the other hand was very favorable for the first iteration approach. The improvement over any heuristic policy was around 10% or higher. This can be explained by the fact that as we are actually sampling the possible future realizations, the important (i.e. more probable) ones are automatically more frequently chosen. So we get a better grasp of the future with fewer samples. In the case where costs differ, similar favoring of more probable realizations does not occur naturally. It would probably be useful to study the applicability of importance sampling method to make more important paths more frequent in the simulations.

7 Conclusions

In this paper we have studied the RWA problem in the optical networks where the offered traffic is non-uniform or different traffic classes have different revenues. The problem was handled in the framework of Markov decisions processes. In particular, we have studied the applicability of the first policy iteration where the relative costs of states are estimated as they are needed by simulations on the fly.

The problem with heuristic algorithms presented in the literature is that they do not take into account the non-uniform traffic or other peculiarities of the system. The first policy iteration is expected to some extent to come over these deficiences, and this was actually the case in both of the non-uniform test scenarios. The performance improvement obtained by the policy iteration depends on the standard policy one starts with, and in these tests the average cost rate was reduced by about 10% to 20% in most cases. The improvement was not so high in the case where *pcolor* heuristics was used as the standard policy.

The running time of first iteration approach is probably too long for systems where decisions must be made in few seconds. But for slower systems it indeed can be used as an improvement to heuristic algorithms in real time. Even when a long running time makes the approach infeasible in real time, this method can still be used to assess how close the performance of an arbitary heuristic algorithm comes to the optimal policy.

References

1. B. Mukherjee (1997) Optical Communication Networks, McGraw-Hill series on computer communications, McGraw-Hill
2. R. Ramaswami and K. Sivarajan (1998) Optical Networks, A Practical Perspective, Morgan Kaufmann Series in Networking, Morgan Kaufmann Publishers
3. A. Willner (1997) Mining the optical bandwidth for a terabit per second. IEEE Spectrum 34(3):32–41
4. D. Banaree and B. Mukherjee (1996) A practical approach for routing and wavelength assignment in large wavelength-routed optical networks. IEEE JSAC 14:903–908
5. S. Baroni (1998) Routing and wavelength allocation in WDM optical networks, PhD thesis, University College London
6. E. Karasan and E. Ayanoglu (1998) Effects of wavelength routing and selection algorithms on wavelength conversion gain in wdm optical networks. IEEE/ACM Trans Networking 6:186–196
7. A. Mokhtar and M. Azizoğlu (1998) Adaptive wavelength routing in all-optical networks, IEEE/ACM Trans Networking 6:197–206
8. R. Ramaswami and K. Sivarajan (1995) Routing and wavelength assignment in all-optical networks, IEEE/ACM Trans Networking 3:489–500
9. G. Mohan and S. Murthy (1999) A time optimal wavelength rerouting algorithm for dynamic trafffic in wdm networks, IEEE J Lightwave Technol 17:406–417
10. H. C. Tijms (1994) Stochastic Models, An Algorithmic Approach, John Wiley & Sons Ltd
11. Z. Dziong (1997) ATM Network resource management, McGraw-Hill
12. E. Hyytid and J. Virtamo (2000) Dynamic routing and wavelength assignment using first policy iteration, ISCC'2000, the 5th IEEE Symposium on Computers and Communications, Antibes, Juan les Pins, France, 4–6 July, 2000

Performability Analysis of TDMA Cellular Systems Based on Composite and Hierarchical Markov Chain Models

Yonghuan Cao, Hairong Sun, and Kishor S. Trivedi

Center for Advanced Computing and Communications
Department of Electrical and Computer Engineering
Duke University
Durham, NC 27708-0291
E-mail: {ycao, hairong, kst}@ee.duke.edu

Abstract. Composite Markov chain models are built to study the performability of wireless TDMA system with and without automatic protection switching for the control channel. The models are then decomposed hierarchically. Measures of interest are explicitly given in closed form for the approximate hierarchical models, that are proven to be more robust and less time-consuming without losing accuracy. The presented models are of great interest in wireless network design and operation.

1 Introduction

With the increasing popularity of cellular communications systems nowadays, customers are expecting the same level of service, availability and performance from the wireless communications systems as provided by the traditional wire-line networks. The high degree of mobility enjoyed in wireless networks is in turn the cause of inherent unreliability. Compared with wired networks, wireless networks need to deal with disconnects due to the handoffs, noise and interference, blocked and weak signals and run-down batteries, etc. In addition, the performance and availability of a wireless system is affected by the outage-and-recovery of its supporting function units. Unplanned as well as planned outages of the equipments also contribute to the degradation of the system's availability and performance. From the designer and operator's point of view of the wireless network, it would be of great importance to take these factors into account integratively. Thus, a comprehensive model accounting for both performance and availability would be very useful in network design and operation. This motivates us to study performability of wireless systems.

In this paper, we consider a typical cellular system. Such a system consists of several operational areas, called *cells*. Cells are assumed to be statistically identical in this study. A cell has multiple base repeaters. Each base repeater provides a number of time-division-multiplexed channels for mobile stations to communicate with the system. Normally, one of the channels is dedicated to transmitting control messages. Such a channel is called the *control channel.* Failure of the control channel will cause the whole system to fail. To avoid this undesirable situation, an automatic protection switching (APS) scheme is suggested in [1] so that the system automatically selects a channel from the rest of the available channels to substitute the failed control channel. If all channels are in use (talking), then one of them is forcefully terminated and is used as the control channel.

Two kinds of calls may arrive to a cell: *new calls* and *handoff calls* (from neighboring cells). A call is accepted only when the cell can find a channel not in use; otherwise, the call is dropped. Since dropping handoff calls is considered less desirable than blocking a new call, a guard channel scheme (GCS) ([2–4]) that reserves a number of channels for handoff calls is often used. Changing the number of guard channels results in different new call blocking probabilities and handoff call dropping probabilities.

A cell as a whole is subject to failures which will make all channels inaccessible, causing a full outage. In practice, this type of failures may occur when the communication links between base station controller and base repeaters do not function properly, or critical function units (such as base station controller) fail. In this paper, we will refer to this type of failure as the *platform failure.* Each base repeater is also subject to failure which disables the channels that it provides. In a system without APS, if a failed base repeater happens to be the one hosting the control channel, it results in a full outage, same as the situation caused by a platform failure. In this paper, we consider both systems: with and without APS. If APS is implemented, a failure of the base repeater hosting the control channel will only cause a partial outage. Readers interested in the impact of the control channel recovery scheme are encouraged to consult our previous work [1] based on stochastic Petri net models.

Performance analysis of the system has been carried out by several authors ([4,1,5]). In particular, [5] derived the "wireless Erlang-B formulae" and revealed their important properties. However, pure performance analysis tends to be optimistic since it ignores the failure-repair or transient failure-recovery behavior in the wireless communication networks. In this paper, our objective is to present comprehensive yet analytically tractable performability models in which not only performance but also availability are considered. We will first present accurate composite Markov chain models. We then apply decomposition technique [6] and use a two-level hierarchical model to approximate the composite models.

The remainder of the paper is organized as follows. In Section 2, the system specification is made for the performability analysis of the TDMA wireless system. In Section 3, exact composite models are developed for cellular systems with and without APS. In Section 4, the corresponding hierarchical models are presented to approximate the composite models. The numerical results from both the exact and approximate models are presented in Section 5. Section 6 concludes the paper.

2 Model Description

We consider a cell with N_b base repeaters. Each base repeater has M channels. Therefore, a total number of N_bM channels are available when the whole system is working properly. Since one of the channels has to be used as the control channel, the total number of available talking channels is N_bM-1. We also assume that the control channel is selected randomly out of N_bM channels.

A channel can be either idle or occupied (talking). As mentioned earlier, when a handoff call arrives and an idle channel is available in the channel pool, the call is accepted and a channel is assigned to it. Otherwise, the handoff call is dropped. Let g be the number of guard channels. When a new call arrives, it is accepted if $g+1$ or more idle channels are available in the channel pool; otherwise, the new call is blocked. We assume that the arrival stream of new calls and handoff calls are independent Poisson processes with rates λ_1 and λ_2, respectively. Ongoing call (new or handoff) completion times are exponentially distributed with parameter μ_1 and the time at which the mobile station engaged in the call departs the cell are exponentially distributed with parameter μ_2 which is independent of call completion times.

All failure events are assumed to be mutually independent. Times to platform failures and repair are assumed to be exponentially distributed with mean $1/\lambda_s$ and $1/\mu_s$, respectively. Also assumed is that times to base repeater failures and repair are exponentially distributed with mean $1/\lambda_b$ and $1/\mu_b$ respectively, and that a single repair facility is shared by all the base repeaters.

The underlying stochastic process is thus a homogeneous continuous time Markov chain (CTMC), which we describe in the next section.

3 Exact Composite Model

In order to develop the Markov chain of the system, we need to know the distribution of the number of ongoing talking channels on a base repeater when it fails. Consider b $(0 < b \le N_b)$ base repeaters available in the system. Let i denote the number of talking channels on the failing base repeater. Also let k denote the number of talking channels in the system, and for expository simplicity, we also use j $(j+k = bM-1)$, the number of idle channels in the system. Assuming the channels are allocated randomly to arriving calls, we give the probability of i talking channels residing in the failing base repeater in the following two cases:

1. If the failing base repeater is not hosting the control channel, the probability, a_i, that i $(0 \le i \le M)$ talking channels are on the failing base repeater, is given as follows.

   - if $k < M$,

$$a_i = \begin{cases} 0, & \text{if } i > k \\ \dfrac{\binom{(b-1)M-1}{k-i}\binom{M}{i}}{\binom{bM-1}{k}}, & \text{if } i \le k \end{cases} \tag{1}$$

- if $M \leq k < (b-1)M - 1$,

$$a_i = \frac{\binom{(b-1)M-1}{k-i}\binom{M}{i}}{\binom{bM-1}{k}} \tag{2}$$

- if $k \geq (b-1)M - 1$,

$$a_i = \begin{cases} 0, & \text{if } i < M - j \\ \dfrac{\binom{(b-1)M-1}{j-M+i}\binom{M}{M-i}}{\binom{bM-1}{j}}, & \text{if } i \geq M - j \end{cases} \tag{3}$$

2. If the failing base repeater is hosting the control channel, the probability, a'_i, that i $(0 \leq i \leq M-1)$ talking channels are on the failing base repeater, is given in the similar way.
   - if $k < M - 1$,

$$a'_i = \begin{cases} 0, & \text{if } i > k \\ \dfrac{\binom{(b-1)M}{k-i}\binom{M-1}{i}}{\binom{bM-1}{k}}, & \text{if } i \leq k \end{cases} \tag{4}$$

   - if $M - 1 \leq k < (b-1)M - 1$,

$$a'_i = \frac{\binom{(b-1)M}{k-i}\binom{M-1}{i}}{\binom{bM-1}{k}} \tag{5}$$

   - if $k \geq (b-1)M - 1$,

$$a'_i = \begin{cases} 0, & \text{if } i < M - 1 - j \\ \dfrac{\binom{(b-1)M}{j-M+i}\binom{M-1}{M-i}}{\binom{bM-1}{j}}, & \text{if } i \geq M - 1 - j \end{cases} \tag{6}$$

3.1 System without APS

In a system without APS, if a base repeater fails and this base repeater happens to be hosting the control channel, then this failure will cause the whole system to go down. Let c denote this state in which the control channel is down. Also let s denote the state in which the system is down due to a platform failure.

We use a 2-tuple (b, k) to represent a state in which there is no platform failure or control channel failure. Clearly, $0 < b \leq N_b$[1] and $0 \leq k \leq bM - 1$. In addition to states s and c, they consist of the state space of the underlying Markov chain. It follows that the size of state space is $2 + M\dfrac{N_b(N_b+1)}{2}$.

The state transitions of the Markov chain can be completely explored by iteratively enumerating *only* all outgoing transitions from all states. Hence, for a state, we only list its outgoing transitions. The outgoing transitions from state (b, k) are shown in Figure 1. From state (b, k), the following transitions can occur.

[1] For $b = 0$, *i.e.*, there is no base repeater functioning in the system, a control channel failure must have occurred.

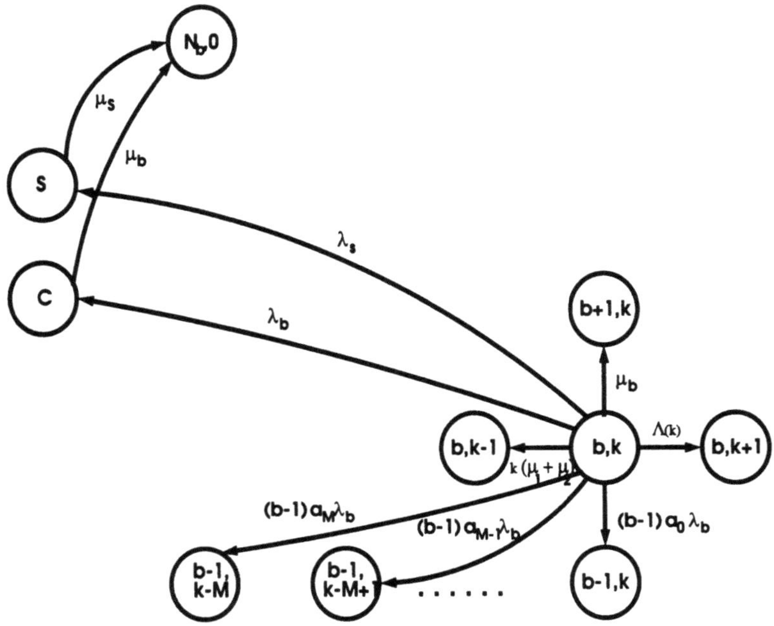

Fig. 1. State diagram of a system without APS

1. If $k < bM - 1$, the system is not fully-loaded and can accept handoff calls (and new calls if the number of idle channels in the channel pool is more than reserved channels for handoff calls). This corresponds to the transition from (b, k) to $(b, k + 1)$ with rate $\Lambda(k)$, where $\Lambda(k) = \lambda_1 + \lambda_2$ if $k < bM - 1 - g$ or $\Lambda(k) = \lambda_2$ if $bM - 1 - g \leq k < bM - 1$.
2. If $k > 0$, ongoing new calls and handoff calls will depart from the cell at rate $k(\mu_1 + \mu_2)$, *i.e.*, transition from (b, k) to $(b, k - 1)$. We note that the call departure rate depends on the number of talking channels, k, in the system.
3. If $b > 0$, a base repeater may fail at rate $b\lambda_b$. Let i $(0 \leq i \leq M)$ be the number of ongoing calls on the failing base repeater. If the base repeater is not hosting the control channel, transition from (b, k) to $(b - 1, k - i)$ will occur at rate $b\lambda_b a_i \left(1 - \frac{1}{b}\right) = (b - 1)a_i\lambda_b$. Here $\frac{1}{b}$ is the probability that the failing base repeater hosts the control channel and the probability a_i can be calculated by (1), (2) and (3). If the base repeater happens to be hosting the control channel, transition from (b, k) to c occurs with rate $b\lambda_b\frac{1}{b} = \lambda_b$.
4. If $b < N_b$, a failed base repeater not hosting the control channel may be recovered at rate μ_b. This leads to the transition from (b, k) to $(b + 1, k)$ with rate μ_b in view of the assumption that a single repair facility is shared.
5. For state (b, k), a platform failure may also occur in the system causing a full outage at rate λ_s. This corresponds to the transition from (b, k) to s with λ_s.

There is only one transition from state c, that is the failed base repeater hosting the control channel being recovered at rate μ_b, *i.e.*, the transition from c to $(N_b, 0)$. Similarly, only one transition from state s can occur, that is the platform failure is recovered at rate μ_s. This corresponds to the transition from s to $(N_b, 0)$. This completes the description of the underlying Markov chain of the non-APS system.

3.2 System with APS

An APS-capable system is able to switch the control channel from a failing base repeater to a functioning one. In case that the system is fully-loaded with currently available channels, one talking channel will be forced to terminate and is used as the control channel. Using the same notation introduced in Section 3.1, the underlying Markov chain, depicted in Figure 2, is very similar to that of a system without APS. The only differences are listed below.

1. If $b > 1$, a base repeater failure may occur at rate $b\lambda_b$. If the failing base repeater is not hosting the control channel, the same transition from (b, k) to $(b-1, k-i)$ will occur at rate $b\lambda_b a_i \left(1 - \frac{1}{b}\right) = (b-1)a_i\lambda_b$ as described in Section 3.1. However, due to the APS mechanism, if the failing repeater is the one hosting the control channel, the transitions are different: because in this case the base repeater provides only $M-1$ channels, instead of entering the control failure state c, the chain will move into state $(b-1, k-i)$ for $0 \le i \le M-1$ at rate $b\lambda_b a'_i \frac{1}{b} = a'_i\lambda_b$. Here a'_i is determined by (4), (5) and (6). Superposing the two cases, transition from (b, k) to $(b-1, k-i)$ has rate $(b-1)a_i\lambda_b + a'_i\lambda_b = [a_i(b-1) + a'_i]\lambda_b$ for $0 \le i \le M-1$ or $a_i(b-1)\lambda_b$ for $i = M$.
2. If $b = 1$, a base repeater failure will disable the control channel on the last base repeater and leads to state c. This corresponds to the transition from (b, k) to c with rate λ_b. We note that $(1, k)$ are the only states that can enter state c in a system with APS.

From the above discussion, it is clear that the gain by APS is achieved by the control channel switching ability upon failures to the hosting base repeater, lessening the chance of entering the control failure state, c.

3.3 Performability Indices

Three steady-state performability indices are considered in evaluating a cellular system: (I) the system unavailability ($\overline{A}$), (II) the overall new call blocking probability (P_b^o) and (III) the overall handoff call dropping probability (P_d^o). Let p_c, p_s, and $p(b, k)$ be the steady-state state probabilities of the Markov chains described

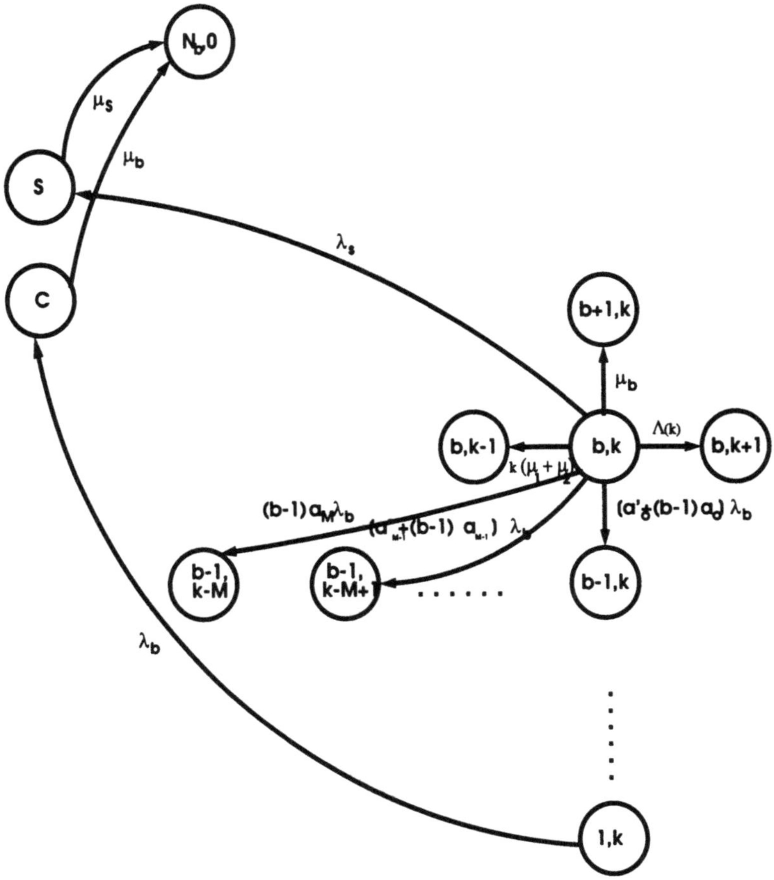

Fig. 2. State diagram of a system with APS

in Section 3.1 and 3.2. The three measures can be given as follows.

$$\overline{A}(N_b) = p_c + p_s \tag{7}$$

$$\begin{aligned} P_b^o(N_b, M, g) &= p_c + p_s + \sum_{b=1}^{N_b} \sum_{k=bM-1-g}^{bM-1} p(b,k) \\ &= \overline{A}(N_b) + \sum_{b=1}^{N_b} \sum_{k=bM-1-g}^{bM-1} p(b,k) \end{aligned} \tag{8}$$

$$\begin{aligned} P_d^o(N_b, M, g) &= p_c + p_s + \sum_{b=1}^{N_b} p(b, bM-1) \\ &= \overline{A}(N_b) + \sum_{b=1}^{N_b} p(b, bM-1) \end{aligned} \tag{9}$$

It is clear from the equations that, for both systems with and without APS, (1) the system unavailability $\overline{A}$ consists of one part of the overall new call blocking probability P_b^o and handoff call dropping probability P_d^o; (2) for system without GCS, *i.e.*, $g = 0$, $P_b^o(N_b, M, g) = P_d(N_b, M, g)$; and (3) $P_b^o(N_b, M, g) > P_d(N_b, M, g)$ holds for systems with one or more guard channels.

4 Approximate Hierarchical Model

For systems with large number of base repeaters, constructing the underlying Markov chain and seeking the solution is not trivial. Furthermore, the large difference between the time scales of reliability and performance parameters may cause the generator matrix to be highly ill-conditioned and may impose great *stiffness* problem in the steady-state solution. Hence it would be desirable to have well-behaved, less time-consuming and yet accurate approximate models.

We therefore use the decomposition method [6] to build a two-level performability model [7]: we first present an availability model which accounts for the failure-repair behavior of platform and base repeaters; second, we use a performance model to compute performance indices (new call blocking probability and handoff call dropping probability) given the number of non-failed base repeaters; finally, we combine them together and give performability measures of interest in closed forms.

4.1 Availability Model

Let $s \in S = \{0, 1\}$ denote a binary value indicating whether or not the system is down due to a platform failure (0: system down due to a platform failure; 1: no platform failure has occurred). Also let $k \in B = \{0, 1, \cdots, N_b\}$ denote the number of non-failed base repeaters. The 2-tuple $(s, k), s \in S, k \in B$ defines a state in which the system is undergoing a (no) platform failure if $s = 0$ (if $s = 1$) and k base repeaters are up. The underlying stochastic process is a homogeneous continuous time Markov chain (CTMC) with state space $S \times B$. Let $P(s, k; N_b)$ be the corresponding steady state probability. The state diagram of this irreducible CTMC is depicted in Figure 3.

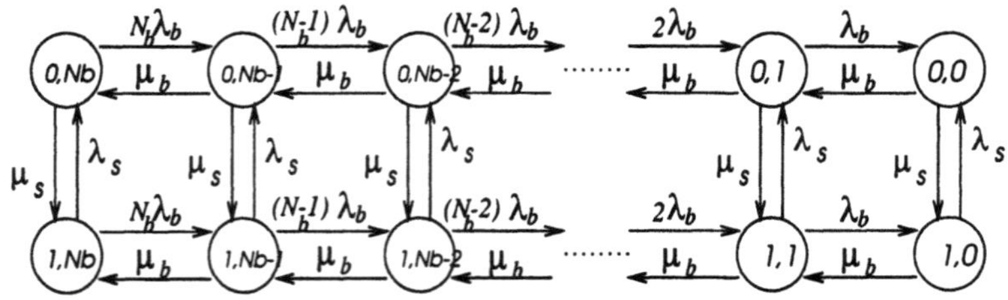

Fig. 3. Markov chain of Availability Model

Solving the Markov chain, we have

$$P(s,k;N_b) = \begin{cases} \dfrac{\lambda_s}{\lambda_s+\mu_s}\dfrac{1}{k!}\left(\dfrac{\mu_b}{\lambda_b}\right)^k\left[1+\sum_{j=1}^{N_b}\dfrac{1}{j!}\left(\dfrac{\mu_b}{\lambda_b}\right)^j\right]^{-1}, & \text{if } s=0, \\ \dfrac{\mu_s}{\lambda_s+\mu_s}\dfrac{1}{k!}\left(\dfrac{\mu_b}{\lambda_b}\right)^k\left[1+\sum_{j=1}^{N_b}\dfrac{1}{j!}\left(\dfrac{\mu_b}{\lambda_b}\right)^j\right]^{-1}, & \text{if } s=1. \end{cases} \tag{10}$$

The system unavailability corresponds to all the states in which either the system has a platform failure that brings the whole system down, or in a system without APS, a base repeater hosting the control channel fails, or the system even without platform failure has no working base repeater left. For a system without APS, the probability that one of the (N_b-k) failed base repeaters happens to host the control channel is $(N_b-k)/N_b$. Let $\overline{A}(N_b)$ denote the steady state system unavailability. For both systems with and without APS, we thus write unavailability as

$$\overline{A}(N_b) = \begin{cases} \sum_{k=0}^{N_b} P(0,k;N_b) + \sum_{k=0}^{N_b} P(1,k;N_b)\dfrac{N_b-k}{N_b}, & \text{w/o APS} \\ \sum_{k=0}^{N_b} P(0,k;N_b) + P(1,0;N_b), & \text{w/ APS.} \end{cases} \tag{11}$$

4.2 Performance Model

For each of the states of the availability model of Figure 3, we now seek to obtain key performance indices. Performance indices of interests are the steady state new call blocking probability and handoff call dropping probability. Given the number of available channels, the previous work in [5,1] provided formulae for these indices. We recall the results here. For a system having k non-failed channels and g guard channels, based on a birth-and-death process, the new call blocking probability is given as

$$P_b(k,g) = \frac{\sum_{n=k-g}^{k} \dfrac{A^{k-g}}{n\,!}A_1^{n-(k-g)}}{\sum_{n=0}^{k-g-1}\dfrac{A^n}{n\,!} + \sum_{n=k-g}^{k}\dfrac{A^{k-g}}{n\,!}A_1^{n-(k-g)}} \tag{12}$$

and the handoff call dropping probability is given as

$$P_d(k,g) = \frac{\dfrac{A^{k-g}}{k\,!}A_1^{g}}{\sum_{n=0}^{k-g-1}\dfrac{A^n}{n\,!} + \sum_{n=k-g}^{k}\dfrac{A^{k-g}}{n\,!}A_1^{n-(k-g)}}, \tag{13}$$

where $A = \frac{\lambda_1 + \lambda_2}{\mu_1 + \mu_2}$, $A_1 = \frac{\lambda_2}{\mu_1 + \mu_2}$. It should be noted that $P_b(k,g) > P_d(k,g)$ holds for $g \geq 1$ and when $g = 0$, $P_b(k,0) = P_d(k,0)$ becomes the Erlang B formula. In practice, the incoming handoff rate λ_2 is an unknown parameter which needs to be determined as a function of k, g, λ_1, μ_1 and μ_2. For a generic cell, assuming all cells are statistically identical, λ_2 can be determined by the following fixed point: the steady-state throughput of incoming handoff calls should be equal to the throughput of outgoing handoff calls. The existence and uniqueness of the fixed point are proved in [5].

4.3 Performability

From the last two sections, we notice that calls can be blocked (or dropped) due to system *being down* or *being full*. The former type of loss is captured by the pure availability model while the latter type of loss is captured by the pure performance model. We now wish to combine the two types of losses. The primary vehicle for doing this is to determine pure performance losses for each of the availability model states and attach these loss probabilities as reward rates (or weights) to these states. Such a Markov reward model has been called a performability model ([7,6]). We list

State (s,k)	Reward rate	
	New call blocking	Handoff call dropping
$(0,k)$, for $k = 0, \cdots, N_b$	1	1
$(1,0)$	1	1
$(1,k)$, for $k = 1, \cdots, N_b$	1, if $kM - 1 \leq g$ $\frac{N_b - k}{N_b} + P_b(kM-1,g)\frac{k}{N_b}$, o.w.	$\frac{N_b - k}{N_b} + P_d(kM-1,g)\frac{k}{N_b}$

Table 1. Reward rates for systems without APS

reward rates for the states of the availability model in Table 1 for systems without APS and Table 2 for system with APS. Let us first consider states of system being down.

Clearly, for both systems without and with APS, a cell is not able to accept any new calls or handoff calls if it has platform failure which corresponds to the states $(0,k)$ for $k = 0, \cdots, N_b$, or all base repeaters are down which corresponds to the state $(1,0)$, Therefore, reward rates of both overall new call blocking and handoff call dropping are 1's.

In addition, for a system without APS, control channel down may occur in states $(1,k)$ for $k = 1, \cdots, N_b$ with probability $(N_b - k)/N_b$ and cause new call blocking and handoff call dropping. This corresponds to the rates with $(N_b - k)/N_b$ in the last row of Table 1.

All cases mentioned above contribute to system unavailability, $\overline{A}(N_b)$, discussed in Section 4.1. Hence, system unavailability, $\overline{A}(N_b)$, also consists of one of the parts of the overall new call blocking probability and handoff call dropping probability.

State (s,k)	Reward rate	
	New call blocking	Handoff call dropping
$(0,k)$, for $k = 0,\cdots,N_b$	1	1
$(1,0)$	1	1
$(1,k)$, for $k = 1,\cdots,N_b$	1, if $kM-1 \le g$ $P_b(kM-1,g)$, o.w.	$P_d(kM-1,g)$

Table 2. Reward rates for systems with APS

We now consider states in which the system is not undergoing a full outage caused by failures of platform, control channel (if system w/o APS) or all base repeaters being down.

The corresponding states are $(1,k)$ for $k = 1,\cdots,N_b$. The total number of available channels for state $(1,k)$ is $kM-1$. From Section 4.3, new call blocking probability and handoff call dropping probability in these states are $P_b(kM-1,g)$ and $P_d(kM-1,g)$, respectively. Thus, these probabilities are used as reward rates to these states for overall new call blocking and handoff call dropping.

For a system without APS, we note that the probability of not having the control channel down in state $(1,k)$ for $k>0$ is k/N_b. Therefore, the reward rates, $P_b(kM-1,g)$ and $P_d(kM-1,g)$, are also weighted by k/N_b (shown in the last row of Table 1).

Also, in case that the number of idle channels is less than the number of guard channels, *i.e.*, $kM-1<g$ for states $(1,k)$, $k=1,\cdots,N_b$, a cell is not able set up any new calls because all available channels are reserved for handoff calls. Hence, the reward rates for new call blocking assigned to the corresponding states are 1's.

Now let $G = \lfloor (g+1)/M \rfloor$. Summarizing Table 1 and Table 2, the overall call blocking probability can be written as the expected steady state reward rate,

$$P_b^o(N_b,M,g) = \overline{A}(N_b) + \begin{cases} \mathbf{1}(G>0)\sum_{k=1}^{G} P(1,k;N_b)\left(\frac{k}{N_b}\right) \\ \quad + \sum_{k=G+1}^{N_b} P(1,k;N_b)P_b(kM-1,\ g)\left(\frac{k}{N_b}\right), & \text{w/o APS} \\ \mathbf{1}(G>0)\sum_{k=1}^{G} P(1,k;N_b) \\ \quad + \sum_{k=G+1}^{N_b} P(1,k;N_b)P_b(kM-1,\ g), & \text{w/ APS} \end{cases} \tag{14}$$

where $\mathbf{1}(e)$ is the indicator function: $\mathbf{1}(e) = 1$ if expression e is true; $\mathbf{1}(e) = 0$, otherwise. Similarly the overall handoff call dropping probability can be given as

$$P_d^o(N_b, M, g) = \overline{A}(N_b) + \begin{cases} \sum_{k=1}^{N_b} P(1,k;N_b) P_d(kM-1,\ g) \frac{k}{N_b}, & \text{w/o APS} \\ \sum_{k=1}^{N_b} P(1,k;N_b) P_d(kM-1,\ g), & \text{w/ APS.} \end{cases} \tag{15}$$

5 Numerical Results and Discussion

We present numerical results in this section. Table 3 summarizes parameters used.

Parameter	Meaning	Value
N_b	Number of base repeaters	10
M	Number of channels/base repeater	8
λ_1	New call arrival rate	20 calls/minute
$1/\mu_1$	Mean call holding time	2.5 minutes
$1/\mu_2$	Mean time to handout	1.25 minutes
λ_s	Platform failure rate	1/year
μ_s	Mean repair time of platform	8 hours
λ_b	Base repeater failure rate	2/year
μ_b	Mean repair time of base repeater	2 hours

Table 3. Parameters used in numerical study

5.1 Accuracy of the hierarchical models

We tabulate some results from composite models and hierarchical models to show the accuraracy of hierarchical models. Tables 4, 5 and 6 compare system unavailability $\overline{A}$, overall new call blocking probability P_b^o and overall handoff call dropping probability P_d^o, respectively, from both models. The presented results from the two models show negligible difference, with the maximum relative error for all three measures less than 0.2%.

The proven high accuracy of hierarchical models with respect to accurate composite models allows us to carry out a variety of experiments with much higher computational efficiency. The following study is therefore conducted by solving hierarchical models.

5.2 Improvement by APS

In Figure 4 we plot the overall new call blocking probability, P_b^o, and handoff call dropping probability, P_d^o, against new call arrival rate, λ_1, for both systems with

Parameters	APS	$\overline{A}_C$ Composite	$\overline{A}_H$ Hierarchical	$100\frac{\overline{A}_H - \overline{A}_C}{\overline{A}_C}$ (%)
$N_b = 8$	no	0.00136799	0.00136987	+0.1374
$N_b = 8$	yes	0.00091280	0.00091241	-0.0427
$N_b = 10$	no	0.00136799	0.00137028	+0.1674
$N_b = 10$	yes	0.00091241	0.00091241	0
$N_b = 12$	no	0.00136799	0.00137070	+0.1981
$N_b = 12$	yes	0.00091241	0.00091241	0

Table 4. Comparison of $\overline{A}$ from composite and hierarchical models

Parameters	APS	$P^o_{b\,C}$ Composite	$P^o_{b\,H}$ Hierarchical	$100\frac{P^o_{b\,H} - P^o_{b\,C}}{P^o_{b\,C}}$ (%)
$\lambda_1 = 12$	no	0.00136799	0.00137028	+0.1674
$\lambda_1 = 12$	yes	0.00091241	0.00091241	0
$\lambda_1 = 20$	no	0.00174871	0.00175125	+0.1452
$\lambda_1 = 20$	yes	0.00129612	0.00129638	+0.0201
$\lambda_1 = 32$	no	0.13844626	0.13845285	+0.0048
$\lambda_1 = 32$	yes	0.13807912	0.13808954	+0.0075

Table 5. Comparison of P^o_b from composite and hierarchical models

Parameters	APS	$P^o_{d\,C}$ Composite	$P^o_{d\,H}$ Hierarchical	$100\frac{P^o_{d\,H} - P^o_{d\,C}}{P^o_{d\,C}}$ (%)
$\lambda_1 = 12$	no	0.00136799	0.00137028	+0.1674
$\lambda_1 = 12$	yes	0.00091241	0.00091241	0
$\lambda_1 = 20$	no	0.00137566	0.00137796	+0.1672
$\lambda_1 = 20$	yes	0.00092016	0.00092017	+0.0011
$\lambda_1 = 32$	no	0.00846951	0.00847205	+0.0300
$\lambda_1 = 32$	yes	0.00801872	0.00801933	+0.0076

Table 6. Comparison of P^o_d from composite and hierarchical models

APS and without APS. The plots show that both probabilities increase but stay nearly flat when new call traffic is low (< 20 calls/minute). The probabilities then increase sharply after λ_1 exceeds 20 calls/minute. The improvement by APS can be seen as reductions of P^o_b and P^o_d. Improvement remains steady (a > 30% relative reduction of both P^o_b and P^o_d) given low traffic but diminishes rapidly as traffic load becomes heavier.

5.3 Percentage of $\overline{A}$ in P^o_b and P^o_d

In the same figure, we also plot the percentage of unavailability $\overline{A}$ in P^o_b and P^o_d to see how much failures of platform, base repeaters and control channel (*i.e.*, system being down) contribute to overall performability measures. It can be seen from the plots that system unavailability dominates under light traffic and becomes a less important factor under intense traffic when heavy traffic with limited system capacity becomes the major factor causing blocking and dropping.

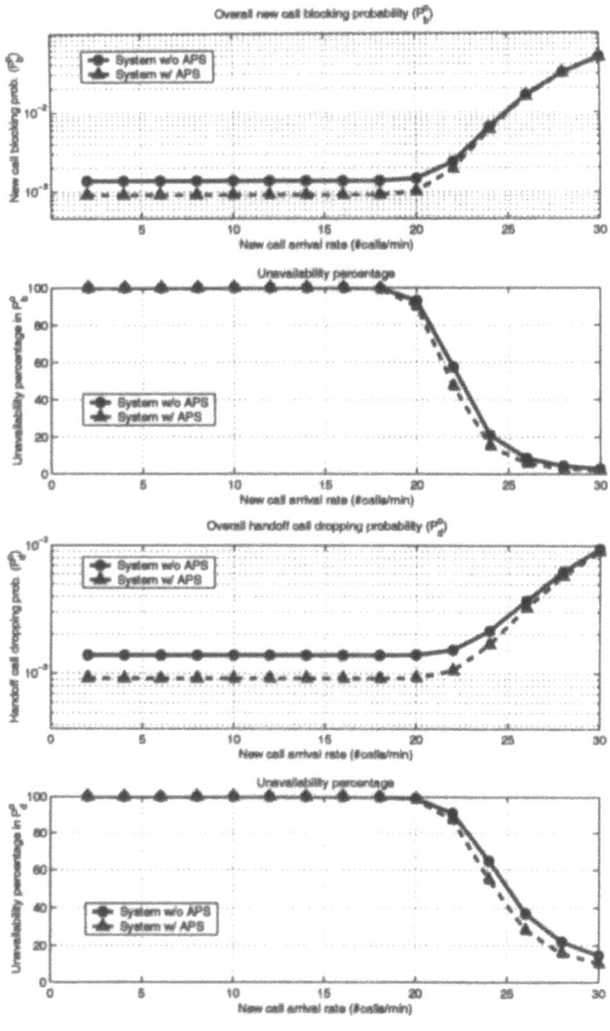

Fig. 4. $P_b^o(N_b, M, g)$ (1st from top) and $P_d^o(N_b, M, g)$ (3rd) versus g for systems w/o APS and w/ APS; Percentage of unavailability $\overline{A}(N_b)$ in $P_b^o(N_b, M, g)$ (2nd) and $P_d^o(N_b, M, g)(4th)$

5.4 Optimization of N_b and g

We also plot curves of both P_b^o and P_d^o against the number of guard channels, g, for $N_b = 8, 9$ and 10 for an APS-capable system[2] in Figure 5. The figure shows that, for each N_b, (I) each curve of P_b^o and P_d^o starts with the same values, *i.e.*, $P_b^o = P_d^o$ when $g = 0$ and (II) increasing g results in a decrease in P_d^o and an increase in P_b^o. From Figure 5, it is also clear that when increasing the number of base repeaters, N_b, both curves of P_b^o and P_d^o move down, indicating the performability improvement.

[2] Similar curves can also be plotted for systems w/o APS (see [8]).

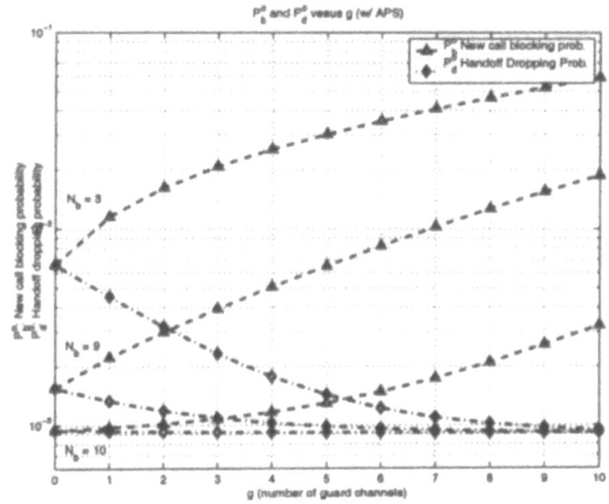

Fig. 5. P_b^o and P_d^o versus N_b and g

In practice, we may face problems to minimize the overall new call blocking probability P_b^o and handoff call dropping probability P_d^o. The number of channels M that a base repeater provides is normally fixed after the repeater is manufactured. Given that the reliability parameters $(\lambda_s, \mu_s, \lambda_b, \mu_b)$ and traffic parameters (A, A_1) are fixed, the decision variables are the number of guard channels, g, and the number of base repeaters, N_b. For example, the following optimization problem may occur to a network designer,

O: Given reliability parameters $(\lambda_s, \mu_s, \lambda_b, \mu_b)$, traffic parameters (A, A_1) and the number of channels M on a base repeaters, determine the optimal integer values of N_b and g so as to

$$\text{minimize } N_b$$

$$\text{such that } \begin{cases} P_b^o(N_b, M, g) \leq P_{b0} \\ P_d^o(N_b, M, g) \leq P_{d0}. \end{cases}$$

We now show how the plot in Figure 5 also provides a graphical way to solve the optimization problems concerning N_b and g. We may draw two lines, $P_b^o = P_{b0}$ and $P_d^o = P_{d0}$. Pairs of triangle marks ($\triangle$) for P_b^o under line $P_b^o = P_{b0}$ and diamond marks ($\diamond$) for P_d^o under line $P_d^o = P_{d0}$ consist of the set of possible solutions. We then choose the minimum N_b. For $P_{b0} = 0.003$ and $P_{d0} = 0.002$, $N_b^* = 9$ and $g^* = 0$ or 1.

6 Conclusion

We have presented the exact composite Markov chain models for performbability study of TDMA wireless systems with and without automatic protection switch. We then have followed a reward-assigning approach to develop the two-level hierarchical performability models. Measures of interest, such as system unavailability, overall new call blocking probability and handoff call dropping probability are explicitly given in closed form. This enables the approximate models to possess excellent scalability. Compared with composite models, the more robust and less time-consuming hierarchical models are proven to be able to provide high accuracy. Numerical results are given under realistic parameter settings. It is expected that the models presented in this paper will be useful in wireless network design and operation.

Acknowledgement

The authors would like to thank anonymous reviewers for their constructive comments. The authors also wish to thank Dr. S. Dharmaraja and Dong Chen for their valuable reviews. This work was in part supported by a fellowship from Motorola Inc. to Duke University and by the Air Force Office of Scientific Research under the MURI Grant No. F49620-00-1-0327.

References

1. H.-R. Sun, Y. Cao, and K. Trivedi, "Availability and performance evaluation for automatic protection switching in TDMA wireless system," *Pacific Rim International Symposium on Dependable Computing (PRDC'99)*, 1999.
2. R. Guerin, "Queueing-blocking system with two arrival streams and guard channels," *IEEE Transactions on Communication*, vol. 32, no. 2, pp. 153–163, February 1988.
3. D. Hong and S. Rappaport, "Traffic model and performance analysis for cellular mobile radio telephone systems with prioritized and non-prioritized handoff procedures," *IEEE Transactions on Vehicular Technology*, vol. VT-35, no. 3, pp. 77–92, August 1986.
4. Y.-B. Lin, S. Mohan, and A. Noerpel, "Queueing priority channel assignment strategies for PCS hand-off and initial access," *IEEE Trans. on Vehi. Tech.*, vol. 43, no. 3, pp. 704–712, Aug. 1994.
5. G. Haring, R. Marie, R. Puigjaner, and K. Trivedi, "Loss formulae and their optimization for cellular networks," *IEEE Transactions on Vehicular Technology (to appear)*, 1999.
6. K. Trivedi, J. Muppala, S. Woolet, and B. Haverkort, "Composite performance and dependability analysis," *Performance Evaluation*, vol. 14, no. 3-4, pp. 197–216, February 1992.
7. J. F. Meyer, "On evaluating the performability of degradable computing systems," *IEEE Transactions on Computers*, vol. C-29, no. 8, pp. 720–731, August 1980.
8. Y. Cao, H.-S. Sun, and K. S. Trivedi, "Performability analysis and optimization for cellular networks (technical report)," Tech. Rep., CACC, ECE Department, Duke University, Feburary 2000.

The MM CPP/GE/c/L G-Queue at equilibrium

P.G. Harrison[1] and R. Chakka[2]

[1] Imperial College, London SW7 2BZ, UK. pgh@doc.ic.ac.uk
[2] Middlesex University, London N11 2NQ, UK. ram2@mdx.ac.uk

Abstract. The MM CPP/GE/c/L G-Queue is a Markov modulated queue with compound Poisson arrivals of both positive and negative customers and generalised exponential service times at c parallel servers. The system considered has either finite or infinite ($L = \infty$) capacity and customers in service cannot be killed by a negative arrival (immune servicing). The equilibrium queue length probabilities are derived as well as the Laplace transform of the response time probability density function of successful customers. This model can form the basis of a building block for networks with bursty, correlated traffic and with load balancing and unreliable servers.

1 Introduction

Various models have been proposed for describing the traffic that arises in today's telecommunications systems such as ATM and the Internet. In particular, such traffic often exhibits burstiness, i.e. batches of transmission units (e.g. packets) arrive together, and correlation between interarrival times. The compound Poisson process (CPP) assumes that batch arrivals are Poisson with geometric batch size – equivalently, that interarrival times have generalised exponential (GE) probability distribution [12]. The Markov modulated Poisson process (MMPP) and self-similar traffic models such as Fractional Brownian Motion (FBM) [14] can represent correlated traffic and, in the latter case, 'heavy tails' which are a consequence of long range correlation. A CPP traffic model often gives a good representation of burstiness, e.g. [5,?], but cannot describe auto-correlation. Conversely, the MMPP models can capture auto-correlation but not burstiness. Although the self-similar models such as FBM can account for both auto-correlation and burstiness, they are analytically intractable in a queueing context.

We introduce a new queueing/traffic model, called the Markov Modulated CPP/GE/c/L G-queue[1], specified fully in section 2. This is a multi-server queue with GE service times and with both positive and negative arrival streams, each of which is a CPP. In addition, all three GE distributions (for positive and negative interarrival times and for service time) are modulated jointly by a continuous

[1] The name G-queue has been adopted for queues with negative customers in acknowledgement of Gelenbe who first introduced them [6].

time Markov phase process. Negative customers remove (positive) customers in the queue and have been used to model random neural networks, task termination in speculative parallelism and faulty components in manufacturing systems [6,4,3]. This queueing model can account for burstiness and correlation, but in addition, the negative customers can represent additional behaviours such as server breakdowns, killing signals and load balancing.

With the negative customers introduced, several customer removal strategies are possible. The one that is considered in this paper is the 'remove customer from the end of the queue' (RCE) killing discipline with immune servicing, where customers actually being served are immune and cannot be removed by a negative arrival. This killing strategy is appropriate for modelling load balancing where a customer in service would never be moved to another server. We first derive, in section 3 the steady state probability distribution for the length of this queue using the method of spectral expansion. Obviously, the MMCPP/GE/c/L, MMCPP/MMGE/c/L and MMPP/M/c/L queues [2], with only positive customers, are rather special cases of this more general queueing model. The system is modelled as a two-dimensional Markov process and solved for the equilibrium queue length probability distribution, from which several steady state performance measures can be worked out.

The Laplace transform of the sojourn time probability density function for successful (not killed) customers is then derived in section 4. Sojourn time distributions in G-queues were first considered by Harrison and Pitel who considered single Markovian queues with negative arrivals in [8] and tandem networks thereof in [9], with numerically tractable results in the former and in certain special cases of the latter. The results presented here are initially obtained conditional on the state of the system seen by a new positive arrival and then deconditioned using the steady state queue length probability distribution.

The paper concludes in section 6 with a discussion of the implications of our results on the modelling of ATM neworks; the analysis at least provides the basis of a building block for networks of switches described in terms of their internal arrival processes.

2 System description

2.1 Modulation

The entire system is modulated by a continuous time, irreducible Markov phase process with N states. Let Q be the generator matrix of this process, given by

$$Q = \begin{bmatrix} -q_1 & q_{1,2} & \cdots & q_{1,N} \\ q_{2,1} & -q_2 & \cdots & q_{2,N} \\ \vdots & \vdots & \ddots & \vdots \\ q_{N,1} & q_{N,2} & \cdots & -q_N \end{bmatrix},$$

where $q_{i,k}(i \neq k)$ is the instantaneous transition rate from phase i to phase k, and

$$q_i = \sum_{j=1}^{N} q_{i,j} \quad , \quad q_{i,i} = 0 \qquad (i = 1, \dots, N)$$

Let $\mathbf{r} = (r_1, r_2, \dots, r_N)$ be the vector of equilibrium probabilities of the modulating phases. Then, $\mathbf{r}$ is uniquely determined by the equations:

$$\mathbf{r}Q = 0 \quad ; \quad \mathbf{r}\mathbf{e}_N = 1 \, .$$

where $\mathbf{e}_N$ stands for the column vector with N elements, each of which is unity.

2.2 The arrival process

The arrival process is the superposition of two CPP arrival streams in each of the modulating phases. One of these CPP processes is of positive customers and the other of negative customers. The parameters of the GE inter-arrival time distribution for the positive customers in phase i are (σ_i, θ_i), and (ρ_i, δ_i) are those of the negative customers. That is, the inter-arrival time probability distribution function is $1 - (1 - \theta_i)e^{-\sigma_i t}$, in phase i, for the positive customers and $1 - (1 - \delta_i)e^{-\rho_i t}$ for the negative customers. Thus, the arrival *point*-processes are Poisson, with batches arriving at each point having geometric size distribution. Specifically, the probability that a batch is of size s is $(1 - \theta_i)\theta_i^{s-1}$, in phase i, for the positive customers, and $(1 - \delta_i)\delta_i^{s-1}$ for the negative customers.

The overall average arrival rates of the positive customers ($\overline{\sigma}$) and negative customers ($\overline{\rho}$) are given by,

$$\overline{\sigma} = \sum_{i=1}^{N} \frac{r_i \sigma_i}{1 - \theta_i} \qquad ; \qquad \overline{\rho} = \sum_{i=1}^{N} \frac{r_i \rho_i}{1 - \delta_i} \, .$$

2.3 The GE multi-server

The service facility has c homogeneous servers, each with GE-distributed service times with parameters (μ_i, ϕ_i) in phase i. The service discipline is FCFS and each server serves at most one positive customer at any given time. Negative customers neither wait in the queue, nor are served. The operation of the GE server is similar to that described for the CPP arrival processes above. However, the batch size associated with a service completion is bounded by one more than the number of customers waiting to commence service at the departure instant. For queues of length $c \leq j < L+1$ (including any customers in service), the maximum batch size at a departure instant is $j-c+1$, only one server being able to complete a service period at any one instant under the assumption of exponentially distributed batch-service times. Thus, the probability that a departing batch has size s is $(1-\phi_i)\phi_i^{s-1}$ for $1 \leq s \leq j-c$ and ϕ_i^{j-c} for $s = j-c+1$. In particular, when $j = c$, the departing batch has size 1 with probability one, and this is also the case for all $1 \leq j \leq c$ since each customer is already engaged by a server and there are then no customers waiting to commence service.

2.4 Negative customer semantics

A negative customer removes a positive customer in the queue, according to a specified *killing discipline*. We consider here a variant of the RCE killing discipline (removal of the customer from the end of the queue), where the most recent positive arrival is removed [8], but which does *not* allow a customer actually in service to be removed: a negative customer that arrives when there are no positive customers waiting to start service has no effect. We may say that customers in service are *immune* to negative customers or that the service itself is *immune servicing*. Such a killing discipline is suitable for the modelling of load balancing where work is transferred from overloaded queues but never work that is actually in progress.

When a batch of negative customers of size $l\,(1 \leq l < j-c)$ arrives, l positive customers are removed from the end of the queue leaving the remaining $j-l$ positive customers in the system. If $l \geq j-c \geq 1$, then $j-c$ positive customers are removed, leaving none waiting to commence service (queue length equal to c). If $j \leq c$, the negative arrivals have no effect.

2.5 The queueing capacity

L is the queueing capacity, in all phases, including the customers in service, if any. L can be finite or infinite. We assume, when the number of customers is j and the arriving batch size of positive customers is greater than $L-j$ (assuming finite L), that only $L-j$ customers are taken in and the rest are rejected.

2.6 Condition for stability

When L is finite, the system is ergodic since the representing Markov process is irreducible. Otherwise, when $L = \infty$, the overall average departure rate is maximum when the queue length is large, and is given by,

$$\overline{\mu} = c\sum_{i=1}^{N} \frac{r_i \mu_i}{1-\phi_i}. \tag{1}$$

Hence, we conjecture the necessary and sufficient condition for the existence of steady state probabilities is

$$\overline{\sigma} < \overline{p} + \overline{\mu}\,. \tag{2}$$

3 Steady state queue length distribution

3.1 Balance equations

The state of the system at any time t can be specified completely by two integer-valued random variables, $I(t)$ and $J(t)$. $I(t)$ varies from 1 to N, representing the phase of the arrival process, and $0 \leq J(t) < L+1$ represents the number of (positive) customers in the system at time t, including any in service. The system is now modelled by a continuous time discrete state Markov process, $\overline{Y}$ (or Y if L is infinite), on a rectangular lattice strip. For convenience, let $I(t)$, the phase, vary in the horizontal direction and $J(t)$, the level, in the vertical direction. We denote the steady state probabilities, when they exist, by $\{p_{i,j} \mid 1 \leq i \leq N, 0 \leq j < L+1\}$, where $p_{i,j} = \lim_{t\to\infty} \text{Prob}(I(t) = i, J(t) = j)$, and we write $\mathbf{v}_j = (p_{1,j}, \dots, p_{N,j})$.

The process $\overline{Y}$ evolves with the following instantaneous transitions whose rates are:

(a) $Q(i,k)$ – purely lateral transition rate, from state (i,j) to state (k,j) for all $j \geq 0$ $(1 \leq i,k \leq N\ ;\ i \neq k)$, caused by a phase transition in the modulating Markov phase process;

(b) $B_{i,j,j+s}$ – s-step upward transition rate, from state (i,j) to state $(i,j+s)$, for all phases i, caused by a new batch arrival of size s of positive customers $(1 \leq s \leq L-j)$. Thus, for each j, s can be seen as bounded when L is finite and unbounded when L is infinite;

(c) $C_{i,j,j-s}$ – s-step downward transition rate, from state (i,j) to state $(i,j-s)$, $(j-s \geq c+1)$, for all phases i, caused by either a batch service completion of size s or a batch arrival of negative customers of size s;

(d) $C_{i,c+s,c}$ – s-step downward transition rate, from state $(i,c+s)$ to state (i,c), for all phases i, caused by a batch arrival of negative customers of size $\geq s$ or a batch service completion of size $s\,(1 \leq s \leq L-c)$;

(e) $C_{i,c-1+s,c-1}$ – s-step downward transition rate, from state $(i,c-1+s)$ to state $(i,c-1)$, for all phases i, caused by a batch departure of size $s\,(1 \leq s \leq L-c+1)$;

(f) $C_{i,j+1,j}$ – 1-step downward transition rate, from state $(i,j+1)$ to state (i,j), $(c \geq 2\ ;\ 0 \leq j \leq c-2)$, for all phases i, caused by a single departure;

where

$$B_{i,j-s,j} = (1-\theta_i)\theta_i^{s-1}\sigma_i \qquad (\forall i\ ;\ 0 \leq j-s \leq L-2\ ;\ j-s < j < L)\ ;$$

$$B_{i,j,L} = \sum_{s=L-j}^{\infty} (1-\theta_i)\theta_i^{s-1}\sigma_i = \theta_i^{L-j-1}\sigma_i \qquad (\forall i\ ;\ j \leq L-1)\ ;$$

$$\begin{aligned} C_{i,j+s,j} &= (1-\phi_i)\phi_i^{s-1}c\mu_i + (1-\delta_i)\delta_i^{s-1}\rho_i \\ &\qquad\qquad (\forall i\ ;\ c+1 \leq j \leq L-1\ ;\ s = 1,2,\dots,L-j)\ ; \\ &= (1-\phi_i)\phi_i^{s-1}c\mu_i + \delta_i^{s-1}\rho_i \quad (\forall i\ ;\ j = c\ ;\ s = 1,2,\dots,L-c)\ ; \\ &= \phi_i^{s-1}c\mu_i \qquad (\forall i\ ;\ j = c-1\ ;\ s = 1,2,\dots,L-c+1)\ ; \\ &= 0 \qquad (\forall i\ ;\ c \geq 2\ ;\ 0 \leq j \leq c-2\ ;\ s \geq 2)\ ; \\ &= (j+1)\mu_i \qquad (\forall i\ ;\ c \geq 2\ ;\ 0 \leq j \leq c-2\ ;\ s = 1)\ . \end{aligned}$$

Now define the matrices,

$$\begin{aligned}
B_{j-s,j} &= \text{Diag}\,[B_{1,j-s,j}, B_{2,j-s,j}, \dots, B_{N,j-s,j}] \\
&\qquad\qquad\qquad\qquad (j-s<L\,;\ j-s<j\le L)\,; \\
B_s &= B_{j-s,j}\ \ (j<L)\ \ ;\ \ B = B_1 = (E-\Theta)\Sigma \\
\Sigma &= \text{Diag}\,[\sigma_1, \sigma_2, \dots, \sigma_N]\ \ ;\ \ \Theta = \text{Diag}\,[\theta_1, \theta_2, \dots, \theta_N]\ \ ; \\
R &= \text{Diag}\,[\rho_1, \rho_2, \dots, \rho_N]\ \ ;\ \ \Delta = \text{Diag}\,[\delta_1, \delta_2, \dots, \delta_N]\ \ ; \\
M &= \text{Diag}\,[\mu_1, \mu_2, \dots, \mu_N]\ \ ;\ \ \Phi = \text{Diag}\,[\phi_1, \phi_2, \dots, \phi_N]\ \ ; \\
C_j &= jM \qquad (1 \le j \le c)\,; \\
&= cM = C \qquad (j \ge c)\,; \\
C_{j+s,j} &= \text{Diag}\,[C_{1,j+s,j}, C_{2,j+s,j}, \dots, C_{N,j+s,j}]\ .
\end{aligned}$$

where $E = \text{Diag}(\mathbf{e}_N)$ is the unit matrix of size $N \times N$. Then, we get,

$$\begin{aligned}
B_s &= \Theta^{s-1}B = \Theta^{s-1}(E-\Theta)\Sigma\ ;\ \sum_{s=1}^{\infty} B_s = \Sigma\ ;\ B_{L-s,L} = \Theta^{s-1}\Sigma\,; \\
C_{j+s,j} &= C(E-\Phi)\Phi^{s-1} + R(E-\Delta)\Delta^{s-1} \\
&\qquad\qquad (c+1 \le j \le L-1\,;\ s = 1, 2, \dots, L-j)\,; \\
&= C(E-\Phi)\Phi^{s-1} + R\Delta^{s-1} \qquad (j = c\,;\ s = 1, 2, \dots, L-c)\,; \\
&= C\Phi^{s-1} \qquad (j = c-1\,;\ s = 1, 2, \dots, L-c+1)\,; \\
&= 0 \qquad (c \ge 2\,;\ 0 \le j \le c-2\,;\ s \ge 2)\,; \\
&= C_{j+1} \qquad (c \ge 2\,;\ 0 \le j \le c-2\,;\ s = 1)\,;
\end{aligned}$$

Proposition 1 *The balance equations of the* Markov modulated CPP/GE/c/L *queue with positive and negative customers and with immune servicing can be expressed in the form (for $L > c+4$)*[2]*:*

[2] If $L \le c+4$, then the resulting Markov process with $(L+1)N$ states can be solved directly, e.g. [17].

(i) For $c+2 \le j \le L-3$,

$$\mathbf{v}_{j-1}Q_0 + \mathbf{v}_j Q_1 + \mathbf{v}_{j+1}Q_2 + \mathbf{v}_{j+2}Q_3 = 0\,; \tag{3}$$

(ii) For $j = L, L-1, L-2$,

$$\sum_{s=1}^{L} \mathbf{v}_{L-s}\Theta^{s-1}\Sigma + \mathbf{v}_L(Q-C-R) = 0\,; \tag{4}$$

$$\begin{aligned}&\mathbf{v}_{L-2}\left[\Sigma - (Q-C-R)\Theta\right]\\ &\quad + \mathbf{v}_{L-1}\left[Q - \Sigma - C(E+\Theta-\Theta\Phi) - R(E+\Theta-\Theta\Delta)\right]\\ &\quad + \mathbf{v}_L\left[C(E-\Phi)(E-\Theta\Phi) + R(E-\Delta)(E-\Theta\Delta)\right] = 0\,;\end{aligned} \tag{5}$$

$$\begin{aligned}&\mathbf{v}_{L-3}\left[\Sigma - (Q-C-R)\Theta\right]\\ &\quad + \mathbf{v}_{L-2}\left[Q - \Sigma - C(E+\Theta-\Theta\Phi) - R(E+\Theta-\Theta\Delta)\right]\\ &\quad + \mathbf{v}_{L-1}\left[C(E-\Phi)(E-\Theta\Phi) + R(E-\Delta)(E-\Theta\Delta)\right]\\ &\quad + \mathbf{v}_L\left[C(E-\Phi)\Phi(E-\Theta\Phi) + R(E-\Delta)\Delta(E-\Theta\Delta)\right] = 0\,;\end{aligned} \tag{6}$$

(iii) For $j = c+1, c, c-1$,

$$\begin{aligned}&\mathbf{v}_c\left[\Sigma - (Q-C)\Theta\right] + \mathbf{v}_{c+1}\left[Q - \Sigma - C(E+\Theta-\Theta\Phi) - R(E+\Theta)\right]\\ &+ \sum_{s=1}^{L-c-1} \mathbf{v}_{c+1+s}\left[C(E-\Phi)\Phi^{s-1}(E-\Theta\Phi) + R\Delta^{s-1}(E-\Delta-\Theta\Delta)\right]\\ &\qquad = 0\,; \qquad (\textit{for } c \ge 1)\end{aligned} \tag{7}$$

$$\begin{aligned}&\mathbf{v}_{c-1}\left[\Sigma - (Q-C_{c-1})\Theta\right] + \mathbf{v}_c\left[Q - \Sigma - C(E+\Theta)\right]\\ &\quad + \sum_{s=1}^{L-c} \mathbf{v}_{c+s}\left[C\Phi^{s-1}(E-\Phi-\Theta\Phi) + R\Delta^{s-1}\right] = 0\,; \quad (\textit{for } c \ge 2)\end{aligned} \tag{8}$$

$$\begin{aligned}&\mathbf{v}_{c-2}\left[\Sigma - (Q-C_{c-2})\Theta\right] + \mathbf{v}_{c-1}\left[Q - \Sigma - C_{c-1}(E+\Theta)\right]\\ &\qquad + \sum_{s=1}^{L-c+1} \mathbf{v}_{c-1+s}\left[C\Phi^{s-1}\right] = 0\,; \qquad (\textit{for } c \ge 3)\end{aligned} \tag{9}$$

(iv) For $c-2 \ge j \ge 2$,

$$\mathbf{v}_{j-1}\left[\Sigma - (Q-C_{j-1})\Theta\right] + \mathbf{v}_j\left[Q - \Sigma - C_j(E+\Theta)\right] + \mathbf{v}_{j+1}C_{j+1} = 0\,; \tag{10}$$

(v) For $j = 1$

$$\mathbf{v}_0\left[\Sigma - Q\Theta\right] + \mathbf{v}_1\left[Q - \Sigma - C_1(E+\Theta)\right] + \mathbf{v}_2 C_2 = 0\,; \qquad (\textit{for } c \ge 3) \tag{11}$$

$$\mathbf{v}_0\left[Q - \Sigma\Theta\right] + \mathbf{v}_1\left[Q - \Sigma - C_1(E+\Theta)\right] + \sum_{s=1}^{L-1} \mathbf{v}_{1+s} C\Phi^{s-1} = 0\,; \qquad (\textit{for } c = 2) \tag{12}$$

$$\begin{aligned}&\mathbf{v}_0\left[E - Q\Theta\right] + \mathbf{v}_1\left[Q - \Sigma - C_1(E+\Theta)\right]\\ &\qquad + \sum_{s=1}^{L-1}\left[C\Phi^{s-1}(E-\Phi-\Theta\Phi) + R\Delta^{s-1}\right] = 0\,; \quad (\textit{for } c = 1)\end{aligned} \tag{13}$$

(vi) For $j = 0$

$$\mathbf{v}_0 [Q - \Sigma] + \mathbf{v}_1 C_1 = 0 \; ; \qquad (\textit{for } c \geq 2) \tag{14}$$

$$\mathbf{v}_0 [Q - \Sigma] + \sum_{s=1}^{L} \mathbf{v}_s C \Phi^{s-1} = 0 \; ; \qquad (\textit{for } c = 1) \tag{15}$$

(vi) Normalisation

$$\sum_{j=0}^{L} \mathbf{v}_j \mathbf{e}_N = 1 \; ; \tag{16}$$

where

$$\begin{aligned}
Q_0 &= \Sigma - (Q - C - R)\Theta \; ; \\
Q_1 &= Q(E + \Theta\Phi + \Theta\Delta) - \Sigma(E + \Delta + \Phi) - C(E + \Theta + \Theta\Delta) \\
&\quad - R(E + \Theta + \Theta\Phi) \; ; \\
Q_2 &= -Q(\Phi + \Delta + \Theta\Phi\Delta) + \Sigma(\Phi + \Delta + \Delta\Phi) + C(E + \Delta + \Theta\Delta) \\
&\quad + R(E + \Phi + \Theta\Phi) \; ; \\
Q_3 &= Q\Phi\Delta - \Sigma\Phi\Delta - C\Delta - R\Phi \; ; \\
I_{k>l} &= 1 \qquad (k > l) \; ; \\
&= 0 \qquad (k \leq l) \; ;
\end{aligned}$$

Proof The balance equations are:

$$\sum_{s=1}^{L} \mathbf{v}_{L-s} B_{L-s,L} + \mathbf{v}_L (Q - C - R) = 0 \tag{17}$$

for the Lth row, and, for $0 \leq j \leq L-1$,

$$\sum_{s=1}^{j} \mathbf{v}_{j-s} B_s + \mathbf{v}_j \left[Q - \Sigma - C_j - R I_{j>c}\right] + \sum_{s=1}^{L-j} \mathbf{v}_{j+s} C_{j+s,j} = 0 \tag{18}$$

Substituting $B_{L-s,L} = \Theta^{s-1}\Sigma$, the balance equation for the Lth row becomes equation (4). Substituting $B_s = \Theta^{s-1}(E-\Theta)\Sigma$, the balance equations for the levels j and $j-1$, $(0 \leq j-1, j \leq L-1)$, respectively are:

$$\sum_{s=1}^{j} \mathbf{v}_{j-s} \Theta^{s-1}(E-\Theta)\Sigma + \mathbf{v}_j \left[Q - \Sigma - C_j - R I_{j>c}\right] + \sum_{s=1}^{L-j} \mathbf{v}_{j+s} C_{j+s,j} = 0 \qquad (0 \leq j \leq L-1) \tag{19}$$

$$\sum_{s=1}^{j-1} \mathbf{v}_{j-1-s} \Theta^{s-1}(E-\Theta)\Sigma + \mathbf{v}_{j-1} \left[Q - \Sigma - C_{j-1} - R I_{j-1>c}\right] \tag{20}$$

$$+ \sum_{s=1}^{L-j+1} \mathbf{v}_{j-1+s} C_{j-1+s,j-1} = 0 \qquad (0 \leq j-1 \leq L-1 \text{ or } 1 \leq j \leq L)$$

We modify the balance equation for level j as follows. Post-multiply (20) by Θ and subtract the resulting equation from (19) to get the modified equation for level j $(1 \leq j \leq L-1)$ as:

$$\mathbf{v}_{j-1}\left[\Sigma - (Q - C_{j-1} - RI_{j-1>c})\Theta\right] + \mathbf{v}_j\left[Q - \Sigma - C_j - RI_{j>c} - C_{j,j-1}\Theta\right] + \sum_{s=1}^{L-j} \mathbf{v}_{j+s}\left[C_{j+s,j} - C_{j+s,j-1}\Theta\right] = 0 \qquad (1 \leq j \leq L-1) \quad (21)$$

We now partition the levels and consider five cases: the first (above the threshold-level $j = c+1$) and the rest, which are similar but not considered here. In addition, since the sum of the steady state probabilities over all the states is 1, we have equation (16).

Case 1: $c + 2 \leq j \leq L - 3$ Consider the balance equations for the levels, $c + 2 \leq j \leq L - 1$. For this range, we have, $C_{j-1} = C$; $C_j = C$. Also, for $(c+2 \leq j \leq L-1)$, we have

$$\begin{aligned} C_{j,j-1} &= C(E - \Phi) + R(E - \Delta) \ ; \\ C_{j+s,j} &= C(E - \Phi)\Phi^{s-1} + R(E - \Delta)\Delta^{s-1} \ ; \\ C_{j+s,j-1} &= C(E - \Phi)\Phi^{s} + R(E - \Delta)\Delta^{s} \ . \end{aligned}$$

Substituting the above in (21), the balance equation for level j becomes,

$$\begin{aligned} &\mathbf{v}_{j-1}\left[\Sigma-(Q-C-R)\Theta\right] \\ &+\mathbf{v}_j\left[Q-\Sigma-C(E+\Theta-\Theta\Phi)-R(E+\Theta-\Theta\Delta)\right] \\ &+\sum_{s=1}^{L-j}\mathbf{v}_{j+s}\left[C(E-\Phi)\Phi^{s-1}(E-\Theta\Phi)+R(E-\Delta)\Delta^{s-1}(E-\Theta\Delta)\right]=0 \\ &\qquad (c+2\leq j\leq L-1) \end{aligned} \tag{22}$$

Using the above, the balance equation for level $j+1$ $(c+2\leq j+1\leq L-1)$ is obtained by replacing j by $j+1$ in (22), giving

$$\begin{aligned} &\mathbf{v}_{j}\left[\Sigma-(Q-C-R)\Theta\right] \\ &+\mathbf{v}_{j+1}\left[Q-\Sigma-C(E+\Theta-\Theta\Phi)-R(E+\Theta-\Theta\Delta)\right] \\ &+\sum_{s=1}^{L-j-1}\mathbf{v}_{j+1+s}\left[C(E-\Phi)\Phi^{s-1}(E-\Theta\Phi)+R(E-\Delta)\Delta^{s-1}(E-\Theta\Delta)\right] \\ &=0 \qquad (c+2\leq j+1\leq L-1 \text{ or } c+1\leq j\leq L-2) \end{aligned} \tag{23}$$

We further modify the balance equation (22) for level j by subtracting from it equation (23) post-multiplied by Φ, to get,

$$\begin{aligned}
&\mathbf{v}_{j-1}\left[\Sigma - (Q - C - R)\Theta\right] \\
&+ \mathbf{v}_{j}\left[Q(E + \Theta\Phi) - \Sigma(E + \Phi) - C(E + \Theta) - R(E + \Theta + \Theta\Phi - \Theta\Delta)\right] \\
&+ \mathbf{v}_{j+1}\left[-Q\Phi + \Sigma\Phi + C + R(E + (\Phi - \Delta)(E + \Theta - \Theta\Delta))\right] \\
&+ \sum_{s=2}^{L-j} \mathbf{v}_{j+s}\left[R(E - \Delta)\Delta^{s-2}(E - \Theta\Delta)(\Delta - \Phi)\right] = 0 \\
&\qquad\qquad\qquad\qquad (c + 2 \le j \le L - 2)
\end{aligned} \tag{24}$$

Using the above equation, the modified equation pertaining to level $j+1$ ($c+2 \le j+1 \le L-2$ or $c+1 \le j \le L-3$) is obtained by replacing j by $j+1$ in (24), giving

$$\begin{aligned}
&\mathbf{v}_{j}\left[\Sigma - (Q - C - R)\Theta\right] \\
&+ \mathbf{v}_{j+1}\left[Q(E + \Theta\Phi) - \Sigma(E + \Phi) - C(E + \Theta) - R(E + \Theta + \Theta\Phi - \Theta\Delta)\right] \\
&+ \mathbf{v}_{j+2}\left[-Q\Phi + \Sigma\Phi + C + R(E + (\Phi - \Delta)(E + \Theta - \Theta\Delta))\right] \\
&+ \sum_{s=2}^{L-j-1} \mathbf{v}_{j+1+s}\left[R(E - \Delta)\Delta^{s-2}(E - \Theta\Delta)(\Delta - \Phi)\right] = 0
\end{aligned} \tag{25}$$

Further modifying (24) by subtracting from it equation (25) post-multiplied by Δ, we get equations (3) which are essentially the modified balance equations for levels $c+2, c+3, \ldots, L-3$. Notice that the coefficient matrices Q_0, Q_1, Q_2, Q_3 are *independent* of j.

3.2 Spectral solution of the balance equations

The set of equations (3) for the levels $c+2$ to $L-3$, have the coefficient matrices Q_0, Q_1, Q_2, Q_3 that are *j-independent* and hence have an efficient solution by the spectral expansion method [16,1]. Define the matrix polynomials $Z(\lambda)$ and $\overline{Z}(\xi)$ as,

$$Z(\lambda) = Q_0 + Q_1\lambda + Q_2\lambda^2 + Q_3\lambda^3 \; ; \qquad \overline{Z}(\xi) = Q_3 + Q_2\xi + Q_1\xi^2 + Q_0\xi^3 \tag{26}$$

Then, the spectral solution for $\mathbf{v}_j$ $(c+1 \le j \le L-1)$ is given by,

$$\mathbf{v}_j = \sum_{k=1}^{N} a_k \boldsymbol{\psi}_k \lambda_k^{j-c-1} + \sum_{k=1}^{2N} b_k \boldsymbol{\gamma}_k \xi_k^{L-1-j} \qquad (c+1 \le j \le L-1) \tag{27}$$

where λ_k $(k = 1, 2, \ldots, N)$ are the N eigenvalues of least absolute value of the matrix polynomial $Z(\lambda)$ and ξ_k $(k = 1, 2, \ldots, 2N)$ are the $2N$ eigenvalues of least absolute value of the matrix polynomial $\overline{Z}(\xi)$. $\boldsymbol{\psi}_k$ and $\boldsymbol{\gamma}_k$ are the left-eigenvectors of $Z(\lambda)$ and $\overline{Z}(\xi)$ respectively, corresponding to the eigenvalues λ_k and ξ_k respectively. a_k, b_k are certain arbitrary constants determined by the other balance equations (see below).

Observe that the matrix $Q_0 + Q_1 + Q_2 + Q_3$ is singular, hence $\lambda = 1$ is an eigenvalue that is on the unit-circle for both $Z(\lambda)$ and $\overline{Z}(\xi)$. If (2) is satisfied, the number of eigenvalues of $Z(\lambda)$ that are strictly within the unit circle is N. If (2) is not satisfied, that number is $N-1$. The properties of these eigenvalues and eigenvectors, together with the relevant spectral analysis, are dealt with in [1,16]. Some of them are summarised below. Let the rank of Q_0 be $N - n_0$ and that of Q_3 be $N - n_3$. Then,

(a) $Z(\lambda)$ would have $d = 3N - n_3$ eigenvalues of which n_0 are null eigenvalues (also referred as zero-eigenvalues), whereas $\overline{Z}(\xi)$ would have n_3 zero-eigenvalues and $3N - n_0 - n_3$ non-zero eigenvalues.
(b) If $(\lambda, \boldsymbol{\psi})$ is a non-zero eigenvalue-eigenvector pair of $Z(\lambda)$, then there exists a corresponding non-zero eigenvalue-eigenvector pair, $(\xi = \frac{1}{\lambda}, \boldsymbol{\gamma} = \boldsymbol{\phi})$ for $\overline{Z}(\xi)$. Thus, the non-zero eigenvalues of these two matrix polynomials are mutually reciprocal.
(c) The N eigenvalues of least absolute value of $Z(\lambda)$ and the $2N$ eigenvalues of least absolute value of $\overline{Z}(\xi)$ lie either strictly inside, or on, their respective unit-circles, but not outside.
(d) We assume, if multiple eigenvalues exist, they also have corresponding independent eigenvectors. This has been invariably observed in our numerical studies. Hence, for $Z(\lambda)$ each pair $(\lambda, \boldsymbol{\psi})$ is *distinct*. A similar result is valid for $\overline{Z}(\xi)$.

If the unknowns a_k, b_k can be determined in such a way that all the balance equations are satisfied, then that would yield the unique steady state solution. This is done as follows.

Using the spectral expansion solution for (27), the vectors $\mathbf{v}_j$ $(j = L-3, L-2, L-1)$ are already expressed as linear sums of known vectors with the unknown scalar coefficients a_k, b_k. Using (6), $\mathbf{v}_L$ can also be expressed in the same way as:

$$\mathbf{v}_L = \mathbf{y}_L \left[C(E - \Phi)\Phi(E - \Theta\Phi) + R(E - \Delta)\Delta(E - \Theta\Delta) \right]^{-1} \tag{28}$$

where,

$$\begin{aligned}\mathbf{y}_L = &- \mathbf{v}_{L-3}\,[\Sigma - (Q - C - R)\Theta] \\ &- \mathbf{v}_{L-2}\,[Q - \Sigma - C(E + \Theta - \Theta\Phi) - R(E + \Theta - \Theta\Delta)] \\ &- \mathbf{v}_{L-1}\,[C(E - \Phi)(E - \Theta\Phi) + R(E - \Delta)(E - \Theta\Delta)]\end{aligned}$$

Now that each $\mathbf{v}_j$ $(c+1 \leq j \leq L)$ is expressed as a linear sum of known vectors, with a_k, b_k as scalar unknowns, using the equations (7,27), we can express $\mathbf{v}_c$ as:

$$\mathbf{v}_c = \mathbf{y}_c\,[\Sigma - (Q - C)\Theta]^{-1} \tag{29}$$

where

$$\begin{aligned}\mathbf{y}_c = &- \mathbf{v}_{c+1}\,[Q - \Sigma - C(E + \Theta - \Theta\Phi) - R(E + \Theta)] \\ &- \mathbf{v}_L\left[C(E - \Phi)\Phi^{L-c-2}(E - \Theta\Phi) + R\Delta^{L-c-2}(E - \Delta - \Theta\Delta)\right] \\ &- \sum_{s=1}^{L-c-2}\left[\sum_{k=1}^{N} a_k \boldsymbol{\psi}_k \lambda_k^s + \sum_{k=1}^{2N} b_k \boldsymbol{\gamma}_k \xi_k^{L-c-2-s}\right] \times \\ &\qquad [C(E - \Phi)\Phi^{s-1}(E - \Theta\Phi) + R\Delta^{s-1}(E - \Delta - \Theta\Delta)]\end{aligned}$$

The last term on the right hand side can be simplified as,

$$\begin{aligned}&-\sum_{k=1}^{N} a_k \boldsymbol{\psi}_k\,[C(E - \Phi)(E - \Theta\Phi)\alpha_{k,c} + R(E - \Delta - \Theta\Delta)\beta_{k,c}] \\ &\qquad -\sum_{k=1}^{2N} b_k \boldsymbol{\gamma}_k\,[C(E - \Phi)(E - \Theta\Phi)\alpha'_{k,c} + R(E - \Delta - \Theta\Delta)\beta'_{k,c}]\end{aligned}$$

where,

$$\alpha_{k,c} = \mathrm{Diag}\left[\frac{\lambda_k - \lambda_k^{L-c-1}\phi_1^{L-c-2}}{1-\lambda_k\phi_1}, \dots, \frac{\lambda_k - \lambda_k^{L-c-1}\phi_N^{L-c-2}}{1-\lambda_k\phi_N}\right]$$

$$\beta_{k,c} = \mathrm{Diag}\left[\frac{\lambda_k - \lambda_k^{L-c-1}\delta_1^{L-c-2}}{1-\lambda_k\delta_1}, \dots, \frac{\lambda_k - \lambda_k^{L-c-1}\delta_N^{L-c-2}}{1-\lambda_k\delta_N}\right]$$

$$\alpha'_{k,c} = \mathrm{Diag}\left[\frac{\xi_k^{L-c-3} - \xi_k^{-1}\phi_1^{L-c-2}}{1-\frac{\phi_1}{\xi_k}}, \dots, \frac{\xi_k^{L-c-3} - \xi_k^{-1}\phi_N^{L-c-2}}{1-\frac{\phi_N}{\xi_k}}\right]$$

$$\beta'_{k,c} = \mathrm{Diag}\left[\frac{\xi_k^{L-c-3} - \xi_k^{-1}\delta_1^{L-c-2}}{1-\frac{\delta_1}{\xi_k}}, \dots, \frac{\xi_k^{L-c-3} - \xi_k^{-1}\delta_N^{L-c-2}}{1-\frac{\delta_N}{\xi_k}}\right]$$

Following a similar procedure as above, $\mathbf{v}_{c-1}$ and $\mathbf{v}_{c-2}$ can be expressed as linear sums of known vectors with scalar unknowns a_k, b_k, using equations (8, 9) respectively, together with (27).

The next step is to express $\mathbf{v}_{c-3}, \dots, \mathbf{v}_1$ as similar linear sums, using (10), as:

$$\mathbf{v}_{j-1} = -\left[\mathbf{v}_{j+1}C_{j+1} + \mathbf{v}_j[Q - \Sigma - C_j(E+\Theta)]\right][\Sigma - (Q - C_{j-1})\Theta]^{-1} \qquad (j = c-2, \dots, 2)$$

Using the equations (11, 27) and a similar procedure as used for $\mathbf{v}_c$ in (29), $\mathbf{v}_0$ too can be expressed as a linear sum of known vectors with the same unknowns, a_k, b_k.

We still have 3 vector equations, (5, 4, 14) and a scalar equation (16). By substituting the expressions we already have for all the $\mathbf{v}_j$'s in these equations, we obtain $3N+1$ linear simultaneous scalar equations in $3N$ unknowns, the a_k, b_k's. Out of these, only $3N$ (including the summation equation) are independent. Hence, they can be solved for the a_k, b_k's.

Since the eigenvalues and eigenvectors are either real or complex conjugate pairs, so are the a_k, b_k's. An efficient procedure for computing the eigenvalue-eigenvectors, a_k, b_k's and the required steady state probabilities is given in [1].

3.3 System with infinite queueing capacity

So far the analysis has been for the case of finite L. A similar – in fact, simpler – analysis can be done for the case of infinite queueing capacity. When that is done, the equations (3) become valid for $j = c+1, c+2, \ldots$. Then, the spectral expansion solution is

$$\mathbf{v}_j = \sum_{k=1}^{N} a_k \boldsymbol{\psi}_k \lambda_k^{j-c} \quad (j = c+1, c+2, \ldots) . \tag{30}$$

Here, we need only the N relevant eigenvalue-eigenvectors of $Z(\lambda)$, and only the a_k's are to be determined. Notice that the equation (30) is the same as (27) when the limit $L \to \infty$ is taken, and that the computation time for this case is much less than that of the finite L case. Moreover, the terms $\alpha_{k,c}$, $\beta_{k,c}$ become simpler and the terms $\alpha'_{k,c}$, $\beta'_{k,c}$ become zero.

4 Response time distribution

To investigate sojourn – or response – time distribution, we consider the passage of a special "tagged" customer through the queue. Ultimately, this customer will either be served or killed by a negative customer; we require the probability density function for the sojourn time of customers that are not killed, i.e. the time elapsed between the arrival instant and the completion of service. Consider then the Markov process $X(t) = \{I(t), B(t), A(t)\}$ where the random variable $A(t)$ (respectively $B(t)$) denotes the number of customers ahead of (respectively behind) the tagged customer at time t. Thus, $J(t) = A(t) + B(t) + 1$ in the above notation. Let the random variable T denote the time remaining, at time t, up to the service completion of the tagged customer. For $i, j \geq 0$, we define the probability distributions $\mathbf{F}_{ij}(t) = (F_{1ij}(t), \ldots, F_{Nij}(t))$ where, for $1 \leq k \leq N$,

$$F_{kij}(t) = P(T \leq t \mid I(t) = k, B(t) = i, A(t) = j)$$

Now, when the state is (k, i, j), we consider the initial small interval of length h and derive an expression for $\mathbf{F}_{ij}(t + h)$ in terms of $\{\mathbf{F}_{ab}(t) \mid a, b \geq 0\}$. By the memoryless property of the exponential distribution, we can write, for $j \geq c$:

$$\begin{aligned}
\mathbf{F}_{ij}(t+h) = {} & (E + Qh - \Sigma h - \Delta h - cMh)\mathbf{F}_{ij}(t) \\
& + h\sum_{s=1}^{\infty} \Sigma(E-\Theta)\Theta^{s-1}\mathbf{F}_{i+s,j}(t) \\
& + h\sum_{s=1}^{i} \Delta(E-R)R^{s-1}\mathbf{F}_{i-s,j}(t) \\
& + h\Delta R^{i}.0 \\
& + h\sum_{s=1}^{j-c+1} cM(E-\Phi)\Phi^{s-1}\mathbf{F}_{i,j-s}(t) \\
& + hcM\Phi^{j-c+1}.\mathbf{e}_N + o(h)
\end{aligned} \tag{31}$$

Notice that if a batch of negative customers arrives with size $i+1$ or more, the tagged customer is killed and so does *not* complete service in time less than t – i.e. does so with probability 0. Similarly, if there is a batch service completion of $j-c+2$ or more customers (the one in service, $j-c$ queueing in front of the tagged customer and the tagged customer), the tagged customer completes service in time less than t with probability 1.

For $j \leq c-1$, the tagged customer is in service and cannot be killed by a negative arrival. In this case, the remaining sojourn time of any customer in service is independent of both arrival processes and of the service completions at other servers. Thus, for $0 \leq j \leq c-1$, $i \geq 0$, $\mathbf{F}_{ij}(t) = \mathbf{F}_{00}(t)$ and we have:

$$\mathbf{F}_{00}(t+h) = (E + Qh - Mh)\mathbf{F}_{00}(t) + hM.\mathbf{e}_N + o(h)$$

This yields the vector differential equation

$$\frac{d\mathbf{F}_{00}(t)}{dt} = (Q - M)\mathbf{F}_{00}(t) + M\mathbf{e}_N \tag{32}$$

with solution

$$\mathbf{F}_{00}(t) = \left(1 - e^{-(Q-M)t}\right)(Q-M)^{-1}M\mathbf{e}_N$$

since $\mathbf{F}_{00}(0) = 0$.[3]

We define the Laplace transform vector of the distribution functions $\mathbf{F}_{ij}(t)$ by:

$$\mathbf{L}_{ij}(s) = \left(\int_0^\infty e^{-st}F_{1ij}(t)\mathrm{d}t, \ldots, \int_0^\infty e^{-st}F_{Nij}(t)\mathrm{d}t\right)$$

Then, the Laplace transform of the derivative with respect to t, i.e. of the vector of probability density functions $\mathbf{F}'_{ij}(t) = (F'_{1ij}(t), \ldots, F'_{Nij}(t))$, is $s\mathbf{L}_{ij}(s)$ by a simple integration by parts, since $\mathbf{F}_{ij}(0) = 0$ for all $i, j \geq 0$.

We can now derive recurrence formulas for the $\mathbf{L}_{ij}(s)$ which we solve using the generating function method. The Laplace transform of the (unconditional) equilibrium sojourn time distribution then follows directly in terms of these generating functions. To this end, we define the generating function vector $\mathbf{G}$ (one component per phase) by:

$$\mathbf{G}(y, z, s) = \sum_{i=0}^{\infty}\sum_{j=c}^{\infty} \mathbf{L}_{ij}(s)y^i z^j$$

The following proposition determines this generating function.

Proposition 2 *The generating function $\mathbf{G}(y,z,s)$ is given by the equation*

$$V(y,z,s)\mathbf{G}(y,z,s) = cz^c M(E - z\Phi)^{-1}\left((E-\Phi)K(y,s) + \frac{\Phi\mathbf{e}_N}{s(1-y)}\right)$$
$$- \ \Sigma(E-\Theta)(yE-\Theta)^{-1}\mathbf{G}(\Theta, z, s)$$

[3] In state $X(t) = (0, 0, k)$ for any k, it is understood that the tagged customer is actually undergoing a non-zero service period, i.e. that the state is not instantaneous.

for all $y \neq \theta_i$ $(1 \leq i \leq N)$, *where*

$$\begin{aligned}
V(y,z,s) &= S - Q + \Sigma + \Delta + cM - \Sigma(E-\Theta)(yE-\Theta)^{-1} \\
&\quad - \; y\Delta(E-R)(E-yR)^{-1} - zcM(E-\Phi)(E-z\Phi)^{-1} \\
K(y,s) &= \frac{(S-Q+M)^{-1}M\mathbf{e}_N}{s(1-y)} \\
\mathbf{G}(\Theta,z,s) &= (G_1(\theta_1,z,s), G_2(\theta_2,z,s), \ldots, G_N(\theta_N,z,s)) \\
S &= sE
\end{aligned}$$

Proof Taking the Laplace transform of equation 31, we get

$$\begin{aligned}
(S - Q + \Sigma + \Delta + cM)\mathbf{L}_{ij} &= \Sigma(E-\Theta)\sum_{s=1}^{\infty}\Theta^{s-1}\mathbf{L}_{i+s,j} \\
&\quad + \; \Delta(E-R)\sum_{s=1}^{i}R^{s-1}\mathbf{L}_{i-s,j} \\
&\quad + \; cM(E-\Phi)\sum_{s=1}^{j-c+1}\Phi^{s-1}\mathbf{L}_{i,j-s} \\
&\quad + \; (cM/s)\Phi^{j-c+1}\mathbf{e}_N
\end{aligned} \tag{33}$$

Multiplying throughout by $y^i z^j$ and summing over the domain $i \geq 0$ and $j \geq c$, we obtain:

$$
\begin{aligned}
(S - Q + \Sigma + \Delta + cM)\mathbf{G}(y,z,s) = \\
& \Sigma(E-\Theta)\sum_{j=c}^{\infty}\sum_{i=0}^{\infty}\sum_{s=1}^{\infty}\Theta^{s-1}\mathbf{L}_{i+s,j}y^i z^j \\
& + \Delta(E-R)\sum_{j=c}^{\infty}\sum_{i=0}^{\infty}\sum_{s=1}^{i}R^{s-1}\mathbf{L}_{i-s,j}y^i z^j \\
& + cM(E-\Phi)\sum_{j=c}^{\infty}\sum_{i=0}^{\infty}\sum_{s=1}^{j-c+1}\Phi^{s-1}\mathbf{L}_{i,j-s}y^i z^j \\
& + (cM/s)\sum_{j=c}^{\infty}\sum_{i=0}^{\infty}\Phi^{j-c+1}y^i z^j.\mathbf{e}_N
\end{aligned}
\tag{34}
$$

For the first term on the right hand side, we change the summation variable i to $k = i + s$ and sum over the domain $1 \leq s \leq k$ and $k \geq 1$, leaving the domain of j unchanged, to get the term

$$
\Sigma(E-\Theta)y^{-1}\sum_{j=c}^{\infty}\sum_{k=1}^{\infty}\sum_{s=1}^{k}(\Theta/y)^{s-1}\mathbf{L}_{k,j}y^k z^j = \Sigma(E-\Theta)\sum_{j=c}^{\infty}\sum_{k=1}^{\infty}\frac{y^k E - \Theta^k}{yE - \Theta}\mathbf{L}_{k,j}z^j
$$

where the matrix $yE - \Theta$ in the denominator denotes multiplication by its inverse.[4] This simplifies to $\Sigma(E-\Theta)(yE-\Theta)^{-1}(\mathbf{G}(y,z,s) - \mathbf{G}(\Theta,z,s))$, as required for the Θ-terms.

The second term is handled similarly, changing the summation domain to $\sum_{j=c}^{\infty}\sum_{s=1}^{\infty}\sum_{i=s}^{\infty}$ and then changing the last summation variable from i to $i+s$ so that the sum over s can be separated out.

---

[4] Notice that by the hypothesis that $y \neq \theta_i$, the inverse exists. If $y = \theta_1$, say, the first component of the expression becomes

$$
\Sigma(E-\Theta)\sum_{j=c}^{\infty}\sum_{k=1}^{\infty}kL_{k,j,1}y^{k-1}z^j = \Sigma(E-\Theta)\frac{dG_1}{dy}
$$

For the third term, the summation domain is written $\sum_{i=0}^{\infty}\sum_{s=1}^{\infty}\sum_{j=s+c-1}^{\infty}$ and the last summation variable is changed from j to $j+s$ giving

$$cM(E-\Phi)\sum_{i=0}^{\infty}\sum_{s=1}^{\infty}\sum_{j=c-1}^{\infty} z(z\Phi)^{s-1}\mathbf{L}_{i,j}y^i z^j =$$

$$zcM(E-\Phi)(E-z\Phi)^{-1}\left(\mathbf{G}(y,z,s) - z^{c-1}\sum_{i=0}^{\infty}\mathbf{L}_{i,c-1}y^i\right)$$

Now, taking the Laplace transform of equation 32 yields $s\mathbf{L}_{00} = (Q-M)\mathbf{L}_{00} + M\mathbf{e}_N/s$ so that $\mathbf{L}_{00} = (S-Q+M)^{-1}M\mathbf{e}_N/s$, and since $\mathbf{L}_{ij} = \mathbf{L}_{00}$ for all $j \le c$ and $i \ge 0$, we have

$$\sum_{i=0}^{\infty}\mathbf{L}_{i,c-1}y^i = \frac{(S-Q+M)^{-1}M\mathbf{e}_N}{s(1-y)} = K(y,s)$$

The last term is straightforward. ♠

5 Equilibrium response time distribution

The result we seek for the unconditional response time of a randomly selected arrival, call it $L(s)$, requires the joint probability mass function vector $\mathbf{a}_{ij}$ for the number of customers behind, i, and ahead, j, seen by the tagged arrival in each phase corresponding to the components of the vector. Thus we require

$$L = \sum_{i=0}^{\infty}\sum_{j=0}^{\infty}\mathbf{a}_{ij}.\mathbf{L}_{ij}$$

which can be written

$$L = \sum_{i=0}^{\infty}\sum_{j=0}^{c-1}\mathbf{a}_{ij}.\mathbf{L}_{ij} + \sum_{i=0}^{\infty}\sum_{j=c}^{\infty}\mathbf{a}_{ij}.\mathbf{L}_{ij}$$

Now, an arriving batch is Poisson and so, by the Random Observer Property, sees the steady state distribution p_{kq} for the phase (k) and queue length (q), as per section 3.1, where we wrote $\mathbf{v}_q = (p_{1q},\ldots,p_{Nq})$. The only way the tagged customer can see arrivals behind is if they are in the same batch and similarly the total number seen ahead is the sum of the queue length already present and the number in front *within* the arriving batch. Moreover, because batch sizes are geometric, the numbers of customers in the same batch ahead of and behind the randomly selected tagged arrival are independent and distributed as the whole batch size. We assume in this section that service times are exponential, i.e. cannot be instantaneous, and therefore have

$$\mathbf{a}_{ij} = (E-\Theta)^2\Theta^{i+j}\sum_{q=0}^{j}\Theta^{-q}\mathbf{v}_q$$

For $j \geq c$ this may be written

$$\mathbf{a}_{ij} = (E-\Theta)^2\Theta^{i+j}\left[\sum_{q=0}^{c-1}\Theta^{-q}\mathbf{v}_q + \sum_{q=c}^{j}\sum_{k=1}^{N} a_k \boldsymbol{\psi}_k \Theta^{-q}\lambda_k^q\right]$$

We therefore obtain

$$\begin{aligned} L = {} & (E-\Theta)\mathbf{L}_{00}.\sum_{j=0}^{c-1}\Theta^j\sum_{q=0}^{j}\Theta^{-q}\mathbf{v}_q \\ & + (E-\Theta)^2\sum_{i=0}^{\infty}\sum_{j=c}^{\infty}\Theta^{i+j}\mathbf{L}_{ij}.\sum_{q=0}^{c-1}\Theta^{-q}\mathbf{v}_q \\ & + (E-\Theta)^2\sum_{i=0}^{\infty}\sum_{j=c}^{\infty}\Theta^{i+j}\mathbf{L}_{ij}.\sum_{q=c}^{j}\sum_{k=1}^{N} a_k \boldsymbol{\psi}_k \left(\lambda_k\Theta^{-1}\right)^q \end{aligned}$$

which simplifies, after some algebraic manipulation, to

$$\begin{aligned} L = {} & (E-\Theta)\mathbf{L}_{00}.\sum_{j=0}^{c-1}\Theta^j\sum_{q=0}^{j}\Theta^{-q}\mathbf{v}_q \; + \; (E-\Theta)^2\mathbf{G}(\Theta,\Theta).\sum_{q=0}^{c-1}\Theta^{-q}\mathbf{v}_q \; + \\ & (E-\Theta)^2\sum_{k=1}^{N} a_k\lambda_k\Theta^{-1}\left(E-\lambda_k\Theta^{-1}\right)^{-1}.\left[\left(\lambda_k\Theta^{-1}\right)^{c-1}G(\Theta,\Theta) - G(\Theta,\lambda_k)\right] \end{aligned}$$

It now remains to find the vector function $\mathbf{G}(y, z, s)$ at $y = \Theta$.[5] This can be done from proposition 2 using the analyticity of $\mathbf{G}$ inside the unit y–disk. We write the right hand side $\mathbf{W}(y, z, s, \mathbf{G}(\Theta, z, s))$ where $\mathbf{W}$ is defined by

$$\mathbf{W}(y,z,s,x) =$$
$$cz^c M(E - z\Phi)^{-1}\left((E-\Phi)K(y,s) + \frac{\Phi \mathbf{e}_N}{s(1-y)}\right) - \Sigma(E-\Theta)(yE-\Theta)^{-1}x$$

The proposition can now be written $V\mathbf{G} = \mathbf{W}$ where the fourth argument of $\mathbf{W}$ is understood to be $\mathbf{G}(\Theta, z, s)$, i.e. $\mathbf{G} = V^{-1}\mathbf{W}$. This is a set of N equations for any particular choice of y, z, s. Thus, if V is singular for any choice of y, z with $|y| < 1, |z| < 1$, there must be a linear dependence amongst the equations on both sides, i.e. there must exist a vector $\boldsymbol{\eta}$ such that $\boldsymbol{\eta}.V = \mathbf{0}$ and $\boldsymbol{\eta}.\mathbf{W} = 0$, for the equations to be consistent. The vector $\boldsymbol{\eta}$ depends solely on V – essentially it is its left eigenvector for eigenvalue 0. This approach is a generalisation of that used in the scalar situation where we would have $G = W/V$ so that we must have $W = 0$ whenever $V = 0$ for any values of y, z inside their unit disks. In this way $G(\theta, z, s)$ would be determined for scalar θ. In our case, we have N unknowns, $\{G_i(\theta_i, z, s) \mid 1 \le i \le N\}$ and one equation for each (y, z) pair in the unit disks that renders V singular.

We therefore solve the equation $|V| = 0$ for y as a function of z. Suppose we obtain solutions inside the unit disks $y_i(z)$ for $i = 1, \ldots, N$. This yields corresponding vectors $\boldsymbol{\eta}_i(z)$ and, substituting y_i for y in $\mathbf{W}$, we obtain N linear equations in the required vector $\mathbf{G}(\Theta, z, s)$, *viz.* $\boldsymbol{\eta}_i(z).\mathbf{W} = 0$, which can be solved. The equation $|V| = 0$ in y is of degree $2N$ and we postulate that there are N solutions inside the unit disk in order that there exists a solution for G. This is the same argument as is used in the Spectral Analysis method [15] and has been observed to hold empirically for a 2-phase model. Notice that when there are no negative customers, so that $R = \Delta = 0$, the equation $|V| = 0$ is of degree N and so rather simpler, permitting a closed-form solution for a two-phase modulating Markov chain.

6 Conclusions

We have derived exact results for the equilibrium queue length and sojourn time probability distributions of a Markov modulated, multi-server queue with generalised exponential service times and with compound Poisson arrivals of both positive and negative customers – the MM CPP/GE/c/L G-queue, or MM CPP/M/1/∞

[5] Note that we often omit the last argument s for the sake of brevity.

G-queue in the case of sojourn times in this paper. This is a highly representative queue that can account for many types of traffic and processing patterns, e.g. correlated, bursty traffic, environmentally sensitive service times, unreliable servers and load balancing. It generalises significantly both queues with MMPP arrivals [2] and with generalised exponential service times [11,13]. Moreover, it also generalises results on G-queues which had been restricted to exponential service times and bulk-Poisson arrivals [4] with constant rates, apart from the numerically intractable extension to the M/G/1 G-queue [10].

One of our most immediate extensions to this work is to determine the departure process of the queue, in particular the probability distribution of the inter-departure times. By considering the departure process of the queue, we would have the basis of a building block for analysing *networks* of such queues in terms of the internal arrival processes at each constituent queue. An interesting approximate approach is to consider all queues in isolation with positive arrival streams determined by the busy/idle status of the source nodes and the negative arrival streams determined by the equilibrium dynamic behaviour of certain queue lengths so as to facilitate load balancing. For example, if a certain queue's length passes a given threshold, customers will be transferred out of it until the length goes below a lower threshold. Similarly, the positive arrival rate at underutilised queues will increase correspondingly. Transfers of work could be represented by a combination of negative arrivals at one queue and extra positive arrivals at another when these queues are over- and under-utilised respectively. The dynamics of the utilisation levels can be represented approximately in the modulating CTMC.

That the equilibrium behaviour of both queue lengths and response times can be determined in a tractable way renders the MM CPP/GE/c G-queue a viable building block for the approximate analysis of queueing networks with bursty, correlated traffic, incorporating load balancing and node-failures.

References

1. R. Chakka, *Performance and Reliability Modelling of Computing Systems Using Spectral Expansion*, Ph.D. Thesis, University of Newcastle upon Tyne, UK, 1995.
2. R. Chakka and P.G. Harrison, Analysis of MMPP/M/c/L queues, *Proceedings of the Twelfth UK Computer and Telecommunications Performance Engineering Workshop*, Edinburgh (1996) 117-128.
3. J.M. Fourneau and M. Hernandez, Modelling defective parts in a flow system using G-networks, *Second International Workshop on Performability Modelling of Computer Communication Systems*, Le Mont Saint-Michel, June 1993.
4. J.M. Fourneau, E. Gelenbe and R. Suros, G-networks with multiple classes of positive and negative customers, *Theoretical Computer Science*, Vol. 155, pp. 141–156, 1996.
5. R. Fretwell and D. Kouvatsos, ATM traffic burst lengths are geometrically bounded, *Proceedings of the 7 th IFIP Workshop on Performance Modelling and Evaluation of ATM & IP Networks*, Antwerp, Belgium (1999).
6. E. Gelenbe. Random neural networks with positive and negative signals and product form solution, *Neural Computation*, Vol. 1, No. 4, pp 502-510, 1989.

7. P.G. Harrison and A. de C. Pinto, An approximate analysis of a synchronous, packet-switched banyan networks with blocking, *Performance Evaluation* **19** (1994) 223-258.
8. P.G. Harrison and E. Pitel. Sojourn times in single server queues with negative customers. *Journal of Applied Probability* **30**, 943-963, 1993.
9. P.G. Harrison, E. Pitel. Response time distributions in tandem G-networks. *Journal of Applied Probability* **32**, 224-246, 1995.
10. P.G. Harrison and E. Pitel. The $M/G/1$ queue with negative customers. *Advances in Applied Probability* **28**, 540-566, 1996.
11. D.D. Kouvatsos, A maximum entropy analysis of the G/G/1 Queue at Equilibrium, *Journal of Operations Research Society*, **39**, 2 (1988) 183-200.
12. D. Kouvatsos, Entropy maximisation and queueing network models, *Annals of Operations Research*, **48** (1994), 63-126.
13. D. Kouvatsos, J. Wilkinson, P.G. Harrison and M.K. Bhabuta, Performance Analysis of Buffered Banyan ATM Switch Architectures, in *Performance Modelling and Evaluation of ATM Networks, Vol. II*, D. Kouvatsos, ed., Chapman-Hall (1996).
14. B.B. Mandelbrot and J.W.V. Ness, Fractional Brownian Motions, fractional noises and applications, *SIAM Review*, **10,4** (1968) 422-437.
15. I. Mitrani, *Probabilistic Modelling*, Cambridge University Press (1998).
16. I. Mitrani and R. Chakka, Spectral expansion solution for a class of Markov models: Application and comparison with the matrix-geometric method, *Performance Evaluation* **23** (1995) 241-260.
17. W.J. Stewart, *Introduction to Numerical Solution of Markov Chains*, Princeton University Press, 1994.

Performance Analysis of the IEEE 1394 Serial Bus

Takashi Norimatsu[1] and Hideaki Takagi[2]

[1] Doctoral Program in Policy and Planning Sciences, University of Tsukuba, Tsukuba-shi, Ibaraki, 305-8573, Japan, tnorimat@shako.sk.tsukuba.ac.jp
[2] Institute of Policy and Planning Sciences, University of Tsukuba, Tsukuba-shi, Ibaraki, 305-8573, Japan, takagi@shako.sk.tsukuba.ac.jp

Abstract. The IEEE 1394 is a standard for the high performance serial bus interface. This standard has the isochronous transfer mode that is suitable for real-time applications and the asynchronous transfer mode for delay-insensitive applications. It can be used to construct a small-size local area network. We propose a simple queueing model for a network with this standard under some assumptions, and calculate the average waiting time of an asynchronous packet in the buffer in the steady state. We give some numerical results, along with validation by simulation, in order to evaluate the performance.

1 Introduction

The IEEE 1394 is a high performance serial bus interface as a specification of the bus by which computers and I/O equipments are interconnected together [1]– [4]. It was standardized by IEEE in 1995 on the basis of the specification of a bus called *Fire Wire* that had been developed by Apple, Inc for a promising alternative to the Small Computer System Interface (SCSI).

The IEEE 1394 can be used not only as a bridge bus but also as an interconnection among personal computers, peripheral devices, video decks, digital video cameras, and so on. When one uses the IEEE 1394 to interconnect several devices, the topology of the resulting network must be in the form of a daisy-chain or a tree. However, it can be seen as a bus on a transport layer.

The IEEE 1394 has some useful features. Devices attached to the IEEE 1394 network configure the network status automatically without any task of a user each time the network topology changes. It means that users unfamiliar with communication networks can treat the IEEE 1394 network easily. In addition, the IP over IEEE 1394, a technology that enables IP transmission over the IEEE 1394 network, has already been released. This technology allows us to use Internet applications over the IEEE 1394 network. However, there are limits on the size of the network in terms of the number of attached devices and the distance. The IEEE 1394 cannot cover the network over 73 meters and/or with more than 63 attached devices. These abilities and limits lead us to construct a small-size local area network environment such as Small Office Home Office (SOHO). For example, the IEEE 1394 is considered to construct a home network in [5].

When it was standardized in 1995, the maximum transmission speed of the bus was 400 Mbps with copper wire or optical fiber. Then IEEE planned to specify a version of higher speed called *P1394b*. It supports the transmission speed of 800 Mbps, 1.6 Gbps, and 3.2 Gbps with optical fiber. It enhances the current IEEE 1394 to construct a large-size local area network in the office and other sites.

The IEEE 1394 has several characteristics that are different from any other LANs such as the Ethernet, the token-ring, etc. In particular, we should pay attention to the following two characteristics related to the performance of this bus.

One is that the IEEE 1394 uses an *arbitration method* for the media access control. The arbitration of IEEE 1394 is of centralized type such that there exists a special node that controls the access to the bus by all nodes in the network.

The other is that the IEEE 1394 has two kinds of data transfer modes called *isochronous transfer mode* (ITM) and *asynchronous transfer mode* (ATM). The IEEE 1394 guarantees almost exactly periodic data transmissions in the ITM. Therefore, the ITM is suitable for real-time applications. In the ATM, a node attached to the bus can send a packet when the bus is free. Besides, the node that wants to transmit a packet in the ATM must defer to other nodes transmitting packets in the ITM. It means that the transmission of a packet in the ITM has a priority over that in the ATM. In the IEEE 1394, time is divided into fixed-size frames called *cycles*. The duration of a cycle is 125 μseconds. At most 80% of a cycle is available to transmit packets in the ITM. The rest is available for the packets in the ATM. During the latter, the node that wants to transmit a packet in the ITM must defer to the nodes transmitting packets in the ATM. The capacity of transmitting packets in the ITM per cycle is independent of the traffic load in the ATM, while that of transmitting packets in the ATM per cycle depends on the traffic load in the ITM. Because of this asymmetry, the performance of the bus in the ATM is affected by the traffic condition in the ITM.

The transmission system we analyze in this paper is similar to the one with an integrated circuit and packet switching facility. Its transmission capacity is shared by two types of traffic. To implement this facility, the synchronous time-division multiplexed frame structure is used. Time is divided into frames of fixed duration. Each frame is subdivided into slots. The frame is partitioned into two regions by a boundary. One is for circuit switching type data like voice calls, and the other is for packet switching type data. The boundary may be fixed or movable. The use of a movable boundary leads to the dynamic resource allocation for the two different types of traffic in terms of slots. Such an integrated transmission system with a movable boundary is studied in [6]– [15].

Theoretically speaking, the transmission system we analyze can be thought of as the one with random inputs and scheduled periodical inputs. Such a system with scheduled secondary inputs has been analyzed in [16]. However, the scheme proposed in [16] cannot deal with dynamic ITM traffic. An approximate analysis has been done in [11] [13] by using a fluid model.

The purpose of this paper is to study the performance of the bus in terms of the average waiting time of an arbitrary asynchronous packet in the buffer in the steady state. Under some assumptions for analytical tractability, we derive the probability generating function for the number of asynchronous packets in the buffer by a Markov chain for each overhead of isochronous traffic. We then calculate the average waiting time of an asynchronous packet in the buffer as a performance index of the bus. Finally, we present numerical results based on our analysis as well as by simulation.

In Section 2, we explain how the isochronous and asynchronous transmission is done. We propose an analytical model in Section 3. Section 4 shows some numerical results based on the analytical model in Section 3 and simulation results. In Section 5, we summarize what is done in this paper and discuss what to do in the future.

2 Media Access Control

In this section, we describe the media access control in the two data transfer modes of IEEE 1394. See Fig. 1.

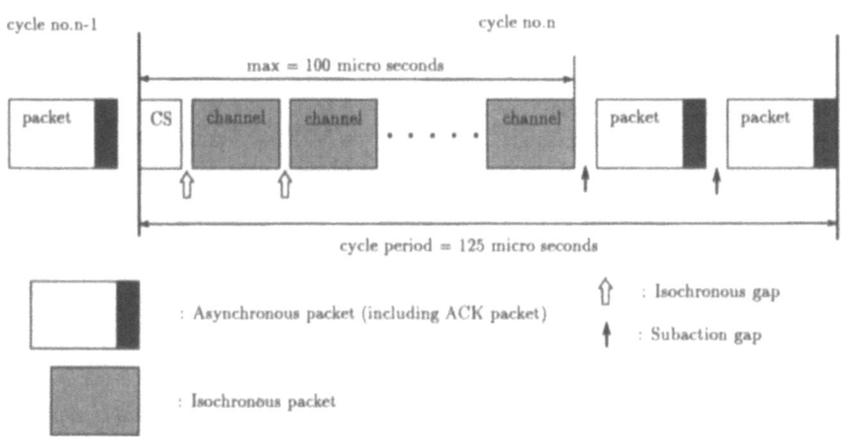

Fig. 1. The organization of a cycle in the IEEE 1394 [1].

Since the IEEE 1394 is a bus specification, no more than one node can transmit data simultaneously. Therefore, the IEEE 1394 needs some media access control as the Ethernet does. Unlike the Ethernet, it uses an access method called *arbitration*. There are two types of arbitration, namely, *isochronous arbitration* and *asynchronous arbitration* as described in the subsequent subsections.

2.1 Isochronous Transfer Mode

The transmission of a packet in the ITM is divided into three phases:

1. Isochronous arbitration (arb phase)
2. Data transfer (data phase)

3. Isochronous gap (gap phase)

According to [4, p.117], the isochronous arbitration done as follows.

a. All nodes attached to the bus set their clock by receiving a *cycle start* (CS) *packet* sent by the *root*. The root is a node that controls and manages access to the bus. The root sends a CS packet every 125 μseconds. The interval between consecutive CS packets is a cycle. An overall network cycle synchronization must be maintained.
b. A node that wishes to transmit a packet sends the root a request for transmission after it detects an *isochronous gap* (IG). The IG is a state in which no signal propagates on the bus during a short time.
c. The root assigns a certain channel to the node that has first sent a request, and this node transmits a packet. After that, this node is prohibited from transmitting until the next cycle. Since at most 100 μseconds are available to transmit packets in the ITM in a cycle, the transmission request is refused if the time taken for transmitting a packet exceeds 100 μseconds.
d. Steps a, b and c are repeated. In the case of the ITM, no ACK is returned. A node releases a channel assigned by the root when there are no ITM data left in its buffer. Otherwise, it transmits one data packet per cycle as ITM traffic.

2.2 Asynchronous Transfer Mode

The transmission of a packet in the ATM is divided into four phases:

1. Asynchronous arbitration (arb phase)
2. Data transfer (data phase)
3. Acknowledgment (ack phase)
4. Subaction gap or arbitration reset gap (gap phase)

According to [4, p.116], the asychronous arbitration is done as follows.

a. A node that wishes to transmit a packet sends a request for transmission to the root after it detects a *subaction gap* (SG) or an *arbitration reset gap* (ARG). The SG is a state in which no signal propagates on the bus for a short time. The ARG is a state in which no signal propagates on the bus for a period that is much longer than the SG.
b. The root allows the node that has first sent a request to transmit a packet. Then, this node transmits only one packet. After that, it is prohibited from transmitting until it detects an ARG.
c. The node that receives a packet returns an ACK. Then the bus gets in the state in which no signal propagates (SG).
d. Steps a, b and c are repeated until all nodes complete transmitting their packets at most once. After that, an ARG occurs, and every node can again send the root a request for another transmission in the same cycle.

The mechanism by which the transmission of a packet in the ITM has priority over that in the ATM works as follows. Assume that there are two nodes. One of them wishes to transmit a packet in the ITM, and the other wishes to transmit a packet in the ATM. At the beginning of a cycle, the root sends a CS packet. After that, the bus enters a state in which no signal propagates on the bus. After some time, both nodes identify this state as an IG since it is shorter than an SG and ARG. Thus the node that has a packet in the ITM transmits first.

3 Traffic Modeling and Analysis

3.1 Analytical Model

A discrete-time queueing system is proposed as a model of the IEEE 1394 network. First, some assumptions are made in order to make the analysis tractable. In particular, we assume that it takes a certain fixed duration to transmit ITM traffic every cycle. We then construct a Markov chain for the number of asynchronous packets in the buffer.

3.1.1 Modeling Assumptions

Our modeling assumptions are as follows.

- **Size of a packet:** The size of every packet is fixed.
- **Discrete time:** Time is divided into N fixed-length intervals, called *slots*, such that a slot is the time to transmit a packet. It implies that the arbitration, data transfer and the SG or ARG occur in one slot in the ATM.
- **Time to transmit a CS packet at the head of every cycle:** Ignored.
- **Time to transmit the ITM traffic in a cycle:** The number of slots used to transmit the ITM traffic in a cycle is fixed at K. According to the specification of the IEEE 1394, it must be that $K \leq \lfloor 0.8 \times N \rfloor$.
- **Arbitration in the ITM and ATM:** Not taken into consideration in our analytical model.
- **Queueing model:** We consider a single infinite-capacity FIFO queue for the asynchronous packets generated in all the nodes in the network.
- **Arrival stream of packets:** The arrival stream of asynchronous packets from each node is assumed to be Poisson. The total number of packets generated in a slot at each node has Poisson distribution with mean λ_{ATM}. Hereafter, asynchronous packets are called ATM packets.

Our model does not incorporate the effect of arbitration on the fairness of access. To do so, a very large multiple-dimensional Markov chain must be used, which makes analysis intractable. Therefore, our performance index shows no information on the fair access.

3.1.2 Markov Chain

A discrete-time Markov chain is used as a means of modeling the asynchronous traffic.

We introduce some random variables. Let Y_n be the number of ATM packets in the buffer at the beginning of the nth cycle, and Y_n^h; $1 \leq h \leq N$, be the number of ATM packets in the buffer at the end of the hth slot in the nth cycle. Also, let $Po(\lambda)$ be a random variable whose distribution is Poisson with mean λ. The deployment of Y_n and Y_n^h; $1 \leq h \leq N-1$, on the time axis is illustrated in Fig. 2.

The relationships for these random variables are given by

$$Y_n^h = \begin{cases} Po(\lambda_{\mathrm{ATM}}) + Y_n^{h-1}; & 1 \leq h \leq K \\ Po(\lambda_{\mathrm{ATM}}) + \left[Y_n^{h-1} - 1\right]^+; & K+1 \leq h \leq N, \end{cases} \tag{1}$$

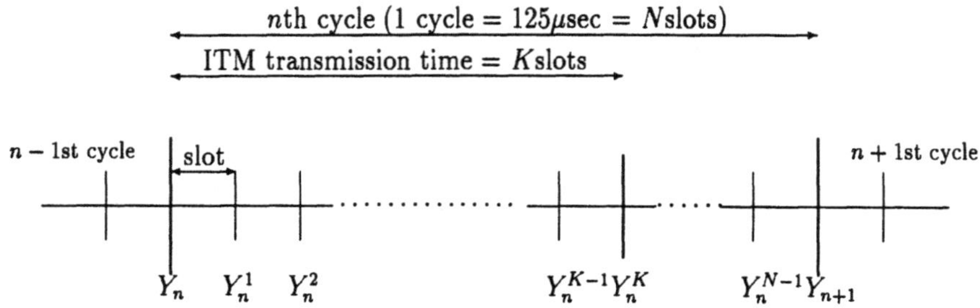

Fig. 2. Deployment of Y_n and $Y_n^h; 1 \leq h \leq N-1$.

where we assume

$$Y_{n+1} \equiv Y_n^N \equiv Y_{n+1}^0. \tag{2}$$

Note that Y_n^h consists of two parts. One is the number of ATM packets arriving in the hth slot in the nth cycle. The other is that of ATM packets that were in the buffer at the beginning of the hth slot and are still there at the end of the same slot.

From (1) and (2), Y_{n+1} is expressed as a function of Y_n:

$$\begin{aligned} Y_{n+1} &= Po(\lambda_{\text{ATM}}) + \left[Po(\lambda_{\text{ATM}}) + \left[\cdots [Po(K\lambda_{\text{ATM}}) + Y_n - 1]^+ \right.\right. \\ &\quad \left.\left. \cdots -1]^+ - 1\right]^+ . \end{aligned} \tag{3}$$

The number of nests in (3) is $N - K$, which is the number of slots available for transmission of ATM packets in each cycle.

3.2 Analysis

In this section, we obtain the probability generating function (PGF) for the number of ATM packets in the buffer in the steady state, and then get the average waiting time of an ATM packet as a performance measure. The stability condition is also derived.

3.2.1 Number of ATM Packets

Let us first derive the PGF for the number of ATM packets in the buffer in the steady state.

First, $Y_n^h(z)$; $1 \leq h \leq N$, is given by

$$Y_n^h(z) = \begin{cases} e^{\lambda_{ATM}(z-1)} Y_n^{h-1}(z); & 1 \leq h \leq K \\ \dfrac{e^{\lambda_{ATM}(z-1)}}{z} \Big[Y_n^{h-1}(z) & \\ \quad + P(Y_n^{h-1} = 0)(z-1) \Big]; & K+1 \leq h \leq N. \end{cases} \tag{4}$$

Using (4) iteratively, we can express $Y_n^h(z)$; $1 \le h \le N$, as in Theorem 1.

Theorem 1. $Y_n^h(z)$; $1 \le h \le N$, is given by

$$Y_n^h(z) = \begin{cases} e^{h\lambda_{ATM}(z-1)} Y_n(z); & 1 \le h \le K \\ \dfrac{1}{z^{h-K}} \Big[e^{h\lambda_{ATM}(z-1)} Y_n(z) + \\ (z-1) \displaystyle\sum_{i=K}^{h-1} T_n^i(0) e^{(h-i)\lambda_{ATM}(z-1)} z^{i-K} \Big]; & K+1 \le h \le N, \end{cases} \tag{5}$$

where $T_n^i(l)$; $K \le i \le N-1$, $0 \le l \le N-1-h$, is defined as

$$\begin{aligned} T_n^K(l) &= \sum_{j=0}^{l} \frac{(K\lambda_{\text{ATM}})^j}{j!} e^{-K\lambda_{\text{ATM}}} P(Y_n = l-j) \\ T_n^i(l) &= \sum_{j=0}^{l-1} \frac{\lambda_{\text{ATM}}^j}{j!} e^{-\lambda_{\text{ATM}}} T_n^{i-1}(l-j+1) + \\ &\quad \frac{\lambda_{\text{ATM}}^l}{l!} e^{-\lambda_{\text{ATM}}} \left[T_n^{i-1}(0) + T_n^{i-1}(1) \right]; \qquad K+1 \le i \le N-1. \end{aligned}$$

Note that $T_n^i(l); K \le i \le N-1$, is the probability that there are l ATM packets in the buffer at the end of the ith slot in the nth cycle. Such situation occurs when

- there are $l-j+1$ ATM packets in the buffer at the beginning of the ith slot, and $j(< l)$ ATM packets arrive in the same slot;

and

- there is zero or one ATM packet in the buffer at the beginning of the ith slot, and l ATM packets arrive in the same slot.

Proof. In the case of $1 \le h \le K$, the result is obvious from (1). So we prove the results in the case of $K+1 \le h \le N$ by mathematical induction. First, $Y_n^{K+1}(z)$ is shown from (1) as

$$\begin{aligned} Y_n^{K+1}(z) &= \frac{e^{\lambda_{ATM}(z-1)}}{z} \left[Y_n^K(z) + P(Y_n^K = 0)(z-1) \right] \\ &= \frac{e^{\lambda_{ATM}(z-1)}}{z} \left[e^{K\lambda_{ATM}(z-1)} Y_n(z) + P(Y_n^K = 0)(z-1) \right] \\ &= \frac{1}{z^{(K+1)-K}} \Big[e^{(K+1)\lambda_{ATM}(z-1)} Y_n(z) \\ &\quad + (z-1) \sum_{i=K}^{(K+1)-1} T_n^i(0) e^{(K+1-i)\lambda_{ATM}(z-1)} z^{i-K} \Big] \end{aligned} \tag{6}$$

Thus (5) holds for $h = K+1$. We next assume that (5) holds for $\acute{h}$, $K+1 \le \acute{h} \le N-1$. Then $Y_n^{\acute{h}+1}(z)$ is shown from (1) as

$$\begin{aligned} Y_n^{\acute{h}+1}(z) &= \frac{e^{\lambda_{ATM}(z-1)}}{z}\left[Y_n^{\acute{h}}(z) + P(Y_n^{\acute{h}}=0)(z-1)\right] \\ &= \frac{e^{\lambda_{ATM}(z-1)}}{z}\left[\frac{1}{z^{\acute{h}-K}}\left[e^{\acute{h}\lambda_{ATM}(z-1)}Y_n(z)\right.\right. \\ &+ \left.(z-1)\sum_{i=K}^{\acute{h}-1} T_n^i(0)e^{(\acute{h}-i)\lambda_{ATM}(z-1)}z^{i-K}\right] \\ &+ \left.P(Y_n^{\acute{h}}=0)(z-1)\right] \\ &= \frac{1}{z^{(\acute{h}+1)-K}}\left[e^{(\acute{h}+1)\lambda_{ATM}(z-1)}Y_n(z)\right. \\ &+ \left.(z-1)\sum_{i=K}^{(\acute{h}+1)-1} T_n^i(0)e^{((\acute{h}+1)-i)\lambda_{ATM}(z-1)}z^{i-K}\right]. \end{aligned} \tag{7}$$

So (5) holds for $\acute{h}+1$. Thus, by induction, (5) holds for $K+1 \le h \le N$. □

Using Theorem 1 and (2), the PGF for Y_{n+1} is given by

$$\begin{aligned} Y_{n+1}(z) &= \frac{1}{z^{N-K}}\left[e^{N\lambda_{\mathrm{ATM}}(z-1)}Y_n(z)\right. \\ &+ \left.(z-1)\sum_{i=K}^{N-1} T_n^i(0)e^{(N-i)\lambda_{\mathrm{ATM}}(z-1)}z^{i-K}\right]. \end{aligned} \tag{8}$$

We now assume that the system enters the steady state as $n \to \infty$. The PGF for the number of ATM packets in the buffer at the beginning of a cycle in the steady state is expressed by

$$Y(z) = \frac{(z-1)\sum_{i=K}^{N-1} T^i(0)e^{(N-i)\lambda_{\mathrm{ATM}}(z-1)}z^{i-K}}{z^{N-K} - e^{N\lambda_{\mathrm{ATM}}(z-1)}}, \tag{9}$$

where $T^i(0) \equiv \lim_{n\to\infty} T_n^i(0)$ is the probability that there is no ATM packet in the buffer at the end of the ith slot.

The numerator of $Y(z)$ in (9) includes the term $\sum_{i=M}^{N-1} T^i(0)e^{(N-i)\lambda_{\mathrm{ATM}}(z-1)}$ z^{i-K} whose coefficients are unknown so far. We can determine them by solving a set of $N-K-1$ linear equations for $T^i(0)$; $K \le i \le N-1$, together with the normalizing condition:

$$\lim_{z\to 1} Y(z) = \frac{\sum_{i=K}^{N-1} T^i(0)}{N-K-N\lambda_{\mathrm{ATM}}} = 1. \tag{10}$$

Now let us consider the zeros of the denominator of $Y(z)$ in (9), or the roots of the equation:

$$z^{N-K} - e^{N\lambda_{\mathrm{ATM}}(z-1)} = 0. \tag{11}$$

We apply Rouché's theorem to find the number of roots of (11), and then apply Lagrange's theorem to compute the roots explicitly.

Rouché's theorem [17, p.20]: If $f(z)$ and $g(z)$ are analytic functions of z inside and on a closed contour C on the complex z-plane, and if also $|g(z)| < |f(z)|$ on C, then $f(z)$ and $f(z) + g(z)$ have the same number of zeros inside C.

Theorem 2. The number of roots of (11) is $N - K$ if

$$N\lambda_{\mathrm{ATM}} < N - K. \tag{12}$$

Proof. In the present case, $f(z)$ and $g(z)$ in Rouché's theorem are given by

$$f(z) = z^{N-K} \quad ; \quad g(z) = -e^{N\lambda_{ATM}(z-1)}$$

On a circle $|z| = 1 + \epsilon$ for a small $\epsilon > 0$, we have

$$|f(z)| = (1+\epsilon)^{N-K} = 1 + (N-K)\epsilon + o(\epsilon)$$
$$|g(z)| \le 1 + N\lambda_{ATM}\epsilon + o(\epsilon)$$

Therefore, $|g(z)| < |f(z)|$ holds if

$$N\lambda_{ATM} < N - K.$$

Clearly, $f(z)$ has $N-K$ zeros inside $|z| = 1+\epsilon$. Therefore, equation (11) has $N-K$ roots inside $|z| = 1 + \epsilon$. □

It is obvious that one of the zeros of the denominator of $Y(z)$ is 1 due to the characteristic of the PGF. Then let such roots except 1 be $z_1, \ldots, z_{N-K-1}$. If z_k; $1 \le k \le N-K-1$, is substituted into $Y(z)$ in (9), then the numerator of $Y(z_k)$ must be 0. Thus we have

$$\sum_{i=K}^{N-1} T^i(0)e^{(N-i)\lambda_{\mathrm{ATM}}(z_k-1)} z_k^{i-K} = 0; \qquad 1 \le k \le N-K-1. \tag{13}$$

The $N-K-1$ roots are obtained explicitly by Lagrange's theorem.

Lagrange's theorem [17, p.20]: Let $f(z)$ and $g(z)$ be analytic on and inside a closed contour C surrounding a point a, and let ω be such that the inequality

$$|\omega g(z)| < |z - a|$$

is satisfied at all points z on C. Then the equation

$$z = a + \omega g(z)$$

in z has exactly one root inside C, and further, any function $f(z)$ which is analytic on and inside C can be expanded as a power series in ω by the formula

$$f(z) = f(a) + \sum_{n=1}^{\infty} \frac{\omega^n}{n!} \frac{d^{n-1}}{dz^{n-1}} \left(\frac{df(z)}{dz} \cdot g(z)^n \right) \Bigg|_{z=a}.$$

In our case, let

$$a = 0, \qquad \omega = e^{2\pi ki/(N-K)}; \quad 1 \le k \le N - K - 1$$

$$f(z) = z, \qquad g(z) = (e^{N\lambda_{\text{ATM}}(z-1)})^{1/(N-K)},$$

where $i := \sqrt{-1}$. According to Lagrange's theorem, the $N - K - 1$ zeros of the denominator of $Y(z)$ in (9) inside the unit circle are given by

$$z_k = \sum_{n=1}^{\infty} \frac{e^{\frac{2\pi kni}{N-K}}}{n!} \frac{d^{n-1}}{dz^{n-1}} \left(e^{N\lambda_{\text{ATM}}(z-1)} \right)^{\frac{n}{N-K}} \Bigg|_{z=0}; \qquad 1 \le k \le N - K - 1.$$

Thus, (10) and (13) determine the unknowns $T^i(0)$; $K \le i \le N - 1$, uniquely.

3.2.2 Average ATM Packet Waiting Time and Stability Condition
We are now in a position to calculate the average waiting time of an arbitrary ATM packet in the steady state and mention the stability condition.

Let $\overline{Y}$ be the average number of ATM packets in the buffer at the beginning of a cycle in the steady state. Differentiating $Y(z)$ in (9) with respect to z and then letting $z = 1$, it is given by

$$\begin{aligned} \overline{Y} = & \frac{\sum_{i=K}^{N-1} T^i(0) \left[i - K + (N - i)\lambda_{\text{ATM}} \right]}{2(N - K - N\lambda_{\text{ATM}})} \\ & - \frac{\sum_{i=K}^{N-1} T^i(0) \left[(N-K)(N-K-1) - N^2\lambda_{\text{ATM}}^2 \right]}{2(N - K - N\lambda_{\text{ATM}})^2}. \end{aligned} \tag{14}$$

Also, let $\overline{Y^h}$; $1 \le h \le N - 1$, be the average number of ATM packets in the buffer at the end of the hth slot in the steady state. From (1), these are given by

$$\overline{Y^h} = \begin{cases} \lambda_{ATM} + \overline{Y^{h-1}}; & 1 \le h \le K \\ \lambda_{ATM} + \overline{Y^{h-1}} + T^{h-1}(0) - 1; & K + 1 \le h \le N - 1, \end{cases} \tag{15}$$

Hence, $\overline{Y^h}$ is obtained as

$$\overline{Y^h} = \begin{cases} h\lambda_{\text{ATM}} + \overline{Y}; & 1 \le h \le K \\ h\lambda_{\text{ATM}} + \overline{Y} + \sum_{i=K}^{h-1} T^i(0) - (h - K); & K + 1 \le h \le N - 1. \end{cases} \tag{16}$$

It follows from Little's theorem that the average waiting time of an ATM packet in the steady state is given by

$$\begin{aligned}\overline{W} &= \frac{\overline{Y} + \sum_{h=1}^{N-1} \overline{Y^h}}{N\lambda_{\mathrm{ATM}}} \\ &= \frac{1}{N\lambda_{\mathrm{ATM}}}\left[N\overline{Y} + \lambda_{ATM}\frac{N(N-1)}{2} + \sum_{i=K}^{N-2}\sum_{j=K}^{i} T^j(0) \right. \\ &\quad \left. - \frac{(N-K-1)(N-K)}{2}\right] \\ &= \frac{N-1}{2} + \frac{\overline{Y}}{\lambda_{ATM}} + \frac{1}{N\lambda_{ATM}}\left[\sum_{j=K}^{N-2}(N-2-j)T^j(0) \right. \\ &\quad \left. - \frac{(N-K-1)(N-K)}{2}\right].\end{aligned} \tag{17}$$

The stability condition is given by (12), which means that the average number of ATM packets that arrive in a cycle is smaller than the number of slots available for transmitting them in a cycle.

4 Numerical and Simulation Results

This section presents the numerical and simulation results for the average waiting time of an ATM packet in the steady state in our traffic model.

Suppose that ITM traffic is generated by some real-time application such as streaming video, online meeting, and so on. Usually, the mean interarrival and holding times for the ITM traffic generated by such applications are much longer than those for the ATM traffic. As a result, the number of slots used to transmit the ITM traffic in a cycle remains constant over a long period of time consisting of many cycles while the queue of ATM packets reaches a steady state quickly and operates in the steady state for most of that period. Therefore, our static ITM traffic model can be used when the ITM traffic is generated by real-time applications.

The following conditions are assumed:

- Transmission speed : $C = 400$ Mbps.
- Speed of signal : $V = 2.0 \times 10^{-8}$ m/sec.
- Duration of a slot (unit of time) : $\tau = 25.0 \times 10^{-6}$ sec.
- Number of slots in a cycle : $N = 5$.
- Number of slots for the ITM in a cycle : $K = 1, 2, 3, 4$.
- Maximum distance between nodes : $d = 500$ m.
- Duration of an SG : $D_{\mathrm{SG}} = 2.5 \times 10^{-6}$ sec.
- Duration of an ARG : $D_{\mathrm{ARG}} = 5 \times 10^{-6}$ sec.
- Duration of an IG : Ignored.
- Size of an ATM packet : $L_{\mathrm{ATM}} = 1000$ bytes.
- Amount of the ITM traffic in a slot : $L_{\mathrm{ITM}} = 1000$ bytes.
- Size of an acknowledgment packet : $L_{\mathrm{ACK}} = 1$ byte.

We assume the following relations.

$$\tau = \max(D_{\text{SG}}, D_{\text{ARG}}) + \frac{(L_{\text{ATM}} + L_{\text{ACK}}) \times 8}{C} + \frac{2d}{V} = \frac{125 \times 10^{-6}}{N} \text{ sec} \tag{18}$$

$$B_{\text{ITM}} \approx \frac{C \cdot K}{N} \quad \text{bps} \tag{19}$$

where B_{ITM} indicates the bandwidth to transmit ITM traffic. Table 1 shows the number of MPEG-2 streams with 6 Mbps of peak rate carried in the ITM depending on the number of slots for the ITM.

Table 1. Number of MPEG-2 streams carried in the ITM.

K	B_{ITM}(bps)	Number of MPEG-2 streams
1	80×10^6	13
2	160×10^6	26
3	240×10^6	40
4	320×10^6	53

We have also simulated the system under the following conditions:

- The simulation is executed in discrete time.
- The number of nodes is 10.
- The arrival rate of packets at every node is the same.
- The arbitration method is taken into consideration.

A difference between the conditions for the numerical results and the simulation is the last one.

Figs. 3 and 4 plot the average waiting time of the ATM packet in the presence of the static ITM traffic. In these figures, ITM(A):K in the legend means the numerical result with K slot used for the ITM traffic, and ITM(S):K means the simulation result similarly. They show that the numerical results are in good agreement with the simulation results, which validates our analysis. We observe that as the arrival rate of ATM packets approaches the stability condition, the average waiting time diverges infinitely.

5 Concluding Remarks

As the demand for delay-sensitive applications such as an online conference system and MPEG video stream increases, so does the demand for networks with an advantage of dealing with such applications. One of the candidates for such a network is the IEEE 1394 network. Therefore, it seems important to evaluate its performance at this moment. In this paper, we have given a certain performance measure of that network by a simple queueing model under some assumptions. The numerical results based on our analysis have showed good agreement with simulation.

However, our discussion does not take into account some factors of the real system such as

- Arbitration scheme,

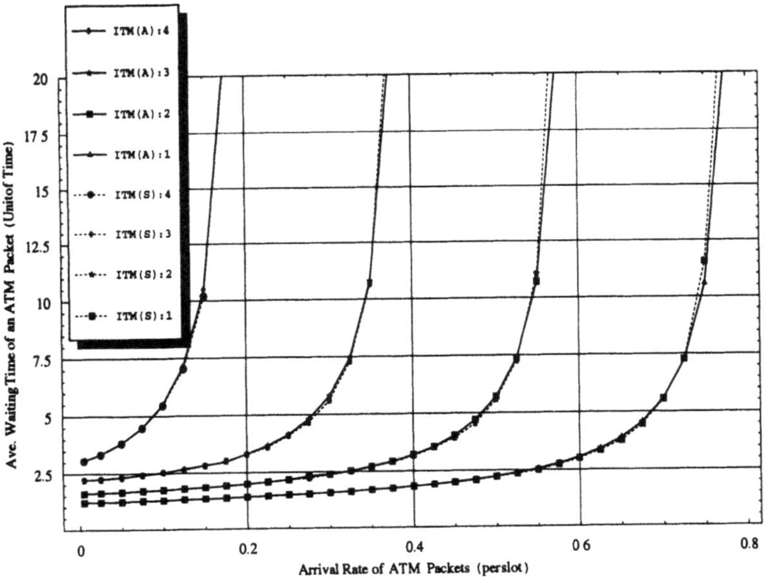

Fig. 3. Average waiting time of an ATM packet in the static ITM traffic model (simulation results with 95% confidence interval).

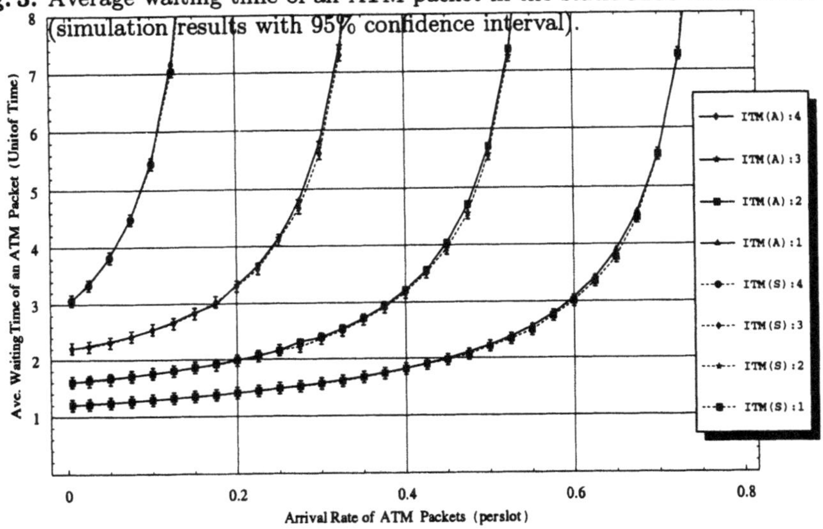

Fig. 4. The enlargement of Fig. 3.

- Variation in size of ATM packets,

and

- network topology that affects the duration of IG, SG and ARG.

In the future work, we should revise the present model so as to take these factors into consideration. After that, it would be needed to derive some performance measures analytically such as the waiting time and delay variation of an asynchronous packet.

Acknowledgement

We are thankful to Dr. Kotikalapudi Sriram of Lucent Technologies for sending us a copy of his dissertation [7] upon our request.

References

[1] Draft: P1394 Standard for a High Performance Serial Bus, The Institute of Electrical and Electronics Engineers, Inc. (IEEE), 1995.

[2] Motohiko Inada, *An introduction to the IEEE 1394*. Gijyutsu Hyouron Sha, Tokyo, 1998 (in Japanese).

[3] Nikkei Business Publications, Inc., *IEEE1394 the high speed interface in the digitization era*. Nikkei Business Publications, Tokyo, 1998 (in Japanese).

[4] Nikkei Business Publications, Inc., *Nikkei Communication*, No.277, pp.110–119, Nikkei Business Publications, Tokyo, September 1998 (in Japanese).

[5] Jong-Wook Jang, Sera Choi, and E. K. Park, "The design of resource assignment algorithm using multiple queues FIFO over residential broadband network" in the *Proceedings of SPIE All-Optical Networking 1999*, Vol.3843, pp.162–172, Boston, Massachusetts, September 7, 1999.

[6] Kotikalapudi Sriram, Pramod K. Varshney, and J. George Shanthikumar, "Discrete-time analysis of integrated voice/data multiplexers with and without speech activity detectors" *IEEE Journal on Selected Areas in Communications*, Vol.SAC-1, No.6, pp.1124–1132, December 1983.

[7] Kotkalapudi Sriram, "A Study of Multiplexing Schemes for Digitized Voice and Data," Ph.D. dissertation, Syracuse University, Syracuse, New York, August 1983.

[8] M. J. Ross and O. A. Mowafi, "Performance analysis of hybrid switching concepts for integrated voice/data communications" *IEEE Transactions on Communications*, Vol.COM-30, pp.1073–1087, May 1982.

[9] T.Bially, A. J. McLaughlin, and C. J. Weinstein, "Voice communication in integrated digital voice and data networks" *IEEE Transactions on Communications*, Vol.COM-28, pp.1478–1490, September 1980.

[10] M. J. Fischer and T. C. Harris, "A model for evaluating the performance of an integrated circuit-and packet-switched multiplex structure" *IEEE Transactions on Communications*, Vol.COM-24, pp.195–202, February 1976

[11] A. Leon-Garcia, R. H. Kwong, and G. F. Williams, "Performance evaluation methods for an integrated voice/data link" *IEEE Transactions on Communications*, Vol.COM-30, pp.1848–1858, August 1982

[12] C. J. Weinstein, M. L. Malpass, and M. J. Fischer, "Data traffic performance of an integrated circuit-and packet switched multiplex structure" *IEEE Transactions on Communications*, Vol.COM-22, pp.873–878, June 1980

[13] D. P. Gaver and J. P. Lehoczky, "Channels that cooperatively service a data stream and voice messages" *IEEE Transactions on Communications*, Vol.COM-30, pp.1153–1162, May 1982

[14] I. Gitman, W. N. Hsieh, and B. J. Occhiogrosso, "Analysis and design of hybrid switching networks" *IEEE Transactions on Communications*, Vol.COM-29, pp.1290–1300, September 1981

[15] M. J. Fischer and T. C. Harris, "An analysis of an integrated circuit and packet-switched telecommunications system" *IEEE Transactions on Communications*, Vol.COM-24, pp.195–202, February 1976

[16] Izzet Sahin, U. Narayan Bhat, "A stochastic system with scheduled secondary inputs" *Operations Research Letters*, pp.229–236, August 1989.
[17] Hideaki Takagi, *Queueing Analysis: A Foundation of Performance Evaluation, Volume1: Vacation and Priority Systems, Part1.* Elsevier, Amsterdam, 1991.

Author Index

Index

MIX
Papier aus verantwortungsvollen Quellen
Paper from responsible sources
FSC® C105338

If you have any concerns about our products,
you can contact us on
ProductSafety@springernature.com

In case Publisher is established outside the EU,
the EU authorized representative is:
Springer Nature Customer Service Center GmbH
Europaplatz 3, 69115 Heidelberg, Germany

Printed by Libri Plureos GmbH
in Hamburg, Germany